Combinatorics of Nonnegative Matrices

Translations of

MATHEMATICAL MONOGRAPHS

Volume 213

Combinatorics of Nonnegative Matrices

V. N. Sachkov
V. E. Tarakanov

Translated by
Valentin F. Kolchin

American Mathematical Society
Providence, Rhode Island

В. Н. Сачков, В. Е. Тараканов

КОМБИНАТОРИКА НЕОТРИЦАТЕЛЬНЫХ МАТРИЦ

Научное Издательство ТВП, Москва, 2000

Translated from the Russian by Dr. Valentin F. Kolchin

2000 *Mathematics Subject Classification.* Primary 05–02;
Secondary 05C50, 15–02, 15A48, 93–02.

Library of Congress Cataloging-in-Publication Data

Sachkov, Vladimir Nikolaevich.
[Kombinatorika neotritsatel′nykh matrits. English]
Combinatorics of nonnegative matrices / V. N. Sachkov, V. E. Tarakanov ; translated by Valentin F. Kolchin.
p. cm. — (Translations of mathematical monographs, ISSN 0065-9282 ; v. 213)
Includes bibliographical references and index.
ISBN 0-8218-2788-X (acid-free paper)
1. Non-negative matrices. 2. Combinatorial analysis. I. Tarakanov, V. E. (Valerii Evgen′evich) II. Title. III. Series.

QA188 .S1913 2002
512.9′434—dc21 2002074392

∞ The paper used in this book is acid-free and falls within the guidelines
established to ensure permanence and durability.
Visit the AMS home page at `http://www.ams.org/`

10 9 8 7 6 5 4 3 2 1 07 06 05 04 03 02

Contents

Preface

The subject of this book is nonnegative matrices. The variety of combinatorial properties of such matrices is widely discussed in mathematical literature, and there are lots of papers on this topic. However, there are only a few monographs devoted specially to the combinatorial properties of nonnegative matrices. The authors of the present book have tried to concentrate not on traditional algebraic and, in particular, spectral properties of nonnegative matrices, but rather on their relations to various mathematical structures studied in combinatorics. In addition to applications in graph theory, Markov chains, tournaments, and abstract automata, we consider relations between nonnegative matrices and such structures as coverings and minimal coverings of sets by families of subsets. Along with the study of combinatorial notions which can be interpreted using nonnegative matrices, considerable attention is given to the study of various properties of the matrices themselves and also of the classes formed by the matrices having a given structure. Asymptotic properties of nonnegative matrices as some of the parameters tend to infinity are also investigated.

In the book, both enumerative and extremal combinatorial problems are presented. In the study of enumerative problems, along with combinatorial methods we use the probabilistic approach. Among the extremal problems, we consider both problems directly related to matrices and problems where nonnegative matrices provide suitable tools of investigation.

In connection with our combinatorial approach, an essential role is played by the binary structure of a nonnegative matrix, that is, the arrangement of positive elements and zeros. It should be noted that this structure is related to the properties of matrices which, at first sight, are not of combinatorial character. One such example is the application of nonnegative matrices in the study of ergodicity of Markov chains in Chapter 6. The binary character of nonnegative matrices can be seen in the fact that many essential properties of such matrices are determined by the properties of their supports, that is, by the $(0,1)$-matrices which are obtained by replacing the positive elements of a matrix by ones. Therefore, we shall devote a lot of attention to $(0,1)$-matrices.

In analyzing $(0,1)$-matrices, we distinguish between local and global properties. Global properties characterize a matrix as a whole, whereas local properties are concerned with relations between parts of the matrix. Examples of global properties are the distribution of positive and zero elements in rows and columns, the term rank, and the cover rank. Local properties include, for example, various conditions on the inner products of rows and columns (viewed as vectors of the corresponding space over real numbers) and the presence (or absence) of submatrices of a given structure.

One more feature of the results on nonnegative matrices presented in this book consists of the fact that the consideration of seemingly pure combinatorial problems often leads to interesting results in some special areas of mathematics. For example, the nonstandard discrete limit distributions obtained in Chapter 4 in the problem on minimal coverings deserves, in our opinion, the attention of specialists in probability theory.

Although this book is mainly theoretical, the authors constantly kept in mind applications of nonnegative matrices. Among these numerous applications the most significant and promising ones are in the theory of Markov chains, in linear programming for constructing and analyzing economic models, and in information theory for designing reliable information devices. The authors hope that the methods and results presented here will be useful to specialists in these areas of research and engineering.

Trying to make the book self-contained, independent, whenever possible, of additional sources and understandable to a wide range of readers, the authors included some theorems which already are viewed as classical. The book also contains a significant number of results which have not been published in monographs, including a series of results obtained by the authors in the last few years.

In presenting the material, the authors usually tried to move from simple to more complicated problems, and placed more special results at the ends of the sections. Therefore, the beginning parts of some sections can be used as a textbook for students specializing in applied discrete mathematics and as the material for special courses and seminars in discrete mathematics.

Chapters 4–6 were written by V. Sachkov, and Chapters 1–3, by V. Tarakanov. However the authors would like to emphasize the unified approach to the material and to the ideas on which the book is based.

V. Sachkov, V. Tarakanov

List of Notation

$\mathbf{R}$ is the set of real numbers.
$\mathbf{R}^+$ is the set of all nonnegative real numbers.
$\mathbf{R}^n$ is the n-dimensional vector space over $\mathbf{R}$.
$\mathbf{Z}$ is the set of all integers.
N_n is the set $\{1, 2, \ldots, n\}$.
Z_n is the set $\{0, 1, \ldots, n-1\}$.
$|X|$ is the cardinality of a set X; a finite set X (subset, sample) is sometimes called an $|X|$-set ($|X|$-subset, $|X|$-sample).
$\mathcal{R}^{m,n}$ is the set of all nonnegative $m \times n$ matrices (matrices of size $m \times n$).
$\mathcal{B}^{m,n}$ is the set of all $(0,1)$-matrices of size $m \times n$.
$\subseteq$ is the symbol of inclusion for sets.
$\cap$ denotes the operation of intersection of sets.
$\cup$ denotes the operation of union of sets.
$\in$ is the symbol of inclusion of an element into a set.
$\setminus$ is the symbol of the operation of taking the difference of two sets.
$\emptyset$ is the empty set.
$\cong$ denotes isomorphism of groups.
$\mathfrak{S}_m$ is the group of all permutations of degree m.
$\{a\}$ is the cyclic group generated by the element a.
(a_{ij}) is the matrix with elements a_{ij}, $i = 1, 2, \ldots, m$, $j = 1, 2, \ldots, n$.
A^T is the transpose of a matrix A.
$\det A$ or $\Delta(A)$ is the determinant of a matrix A.
$\operatorname{per} A$ is the permanent of a matrix A.
$\mathbf{J}_{mn}$ or simply $\mathbf{J}$ is the $m \times n$ matrix with all elements equal to one.
$\mathbf{0}_{mn}$ or simply $\mathbf{0}$ is the $m \times n$ matrix with all elements equal to zero.
$\operatorname{diag}[d_1, d_2, \ldots, d_n]$ is the square matrix of order n with elements $d_1, d_2, \ldots, d_n$ on the principal diagonal and zeros in the remaining positions.
$\mathbf{I}_n = \operatorname{diag}[1, 1, \ldots, 1]$ is the identity matrix of order n.
$A[I, J]$, where $I \subseteq N_m$, $J \subseteq N_n$, is the submatrix of an $m \times n$ matrix A with rows numbers in I and column numbers in J.
(x, y) is the inner product of vectors $x = (x_1, \ldots, x_n)$, $y = (y_1, \ldots, y_n)$ equal to $x_1 y_1 + \ldots + x_n y_n$.
$\lfloor x \rfloor$ is the maximal integer which does not exceed a number $x \in \mathbf{R}$, in other words, the integer part of x.
$\lceil x \rceil$ is the minimal integer which is not less than $x \in \mathbf{R}$.
$(m_1, m_2, \ldots, m_r)$ is the greatest common divisor of integers $m_1, m_2, \ldots, m_r$.

CHAPTER 1

Matrices and Configurations

Introduction

This chapter serves as a kind of introduction to the book. In it we briefly outline the main topics discussed in the subsequent chapters.

In Section 1.1, we introduce the notions of a configuration with weights and its incidence matrix. In fact, any nonnegative matrix may be considered as the incidence matrix of some configuration. Similar ideas have appeared earlier. In this connection it is appropriate to recall the partial graph theory (see [**14**]). The notion of a configuration with weights allows us to relate the study of combinatorial properties of nonnegative matrices to various known results on combinatorial configurations, and to use methods of linear algebra and analysis in a wider context. Latter in this section we introduce the group of permutational automorphisms of a matrix and give some general properties of this structure (see [**108**]).

The principal global combinatorial characteristics of a nonnegative matrix, the term rank and the arrangement of positive elements in rows and columns, are studied in Section 1.2. We present the classical results (as applied to nonnegative matrices), the Frobenius–König, P. Hall, König, and Birkhoff theorems on the term rank and transversals for systems of subsets of a finite set. In this presentation, we follow the books [**87, 48, 75**]. Further, in this section we consider the classes of $(0,1)$-matrices (the Ryser classes) consisting of all matrices with given row and column sums. We give a criterion for the nonemptiness of a Ryser class based only on the parameters of the class. We consider applications of this criterion, in particular, an application to the extension problem for Latin rectangles. We trace the connection of Latin rectangles with square $(0,1)$-matrices that have the same numbers of ones in rows and columns, and with decompositions of such matrices into a sum of permutation matrices. The presentation of this section is based on the books [**48, 87, 12**], and the paper [**19**].

Combinatorial problems related to cyclic matrices (circulants) are considered in Section 1.3. We outline elementary properties of g-circulants, which are a natural generalization of cyclic matrices, and we consider the eigenvalues and the canonical form of a circulant. Here we follow [**2**] and [**120**]. Then we consider the groups of permutational automorphisms for circulants; in particular, we prove a generalization of a result of [**106**], which gives an easily verified sufficient condition for the coincidence of the group of permutational automorphisms of a circulant with a cyclic group and a dihedral group. In connection with cyclic block designs generated by difference sets, we show that a $(0,1)$-circulant with symmetric generating vector cannot be the incidence matrix of a cyclic block design. Turning to matrix equations for g-circulants, we give a criterion for a certain degree of g-circulant to be a matrix with equal elements. This most general criterion is obtained in [**70**]. In our presentation we use also the results of [**58, 64**]. To conclude the section, we show how the results on permutational automorphisms of circulants can be used in finding the number of $(0,1)$-matrices of an arbitrary order with two ones in each row and each column [**29**].

1.1. Definitions and examples

1.1.1. Nonnegative matrices. A nonnegative matrix is a matrix with nonnegative elements. We say that A is an $m \times n$ matrix, or a matrix of size $m \times n$, if it has m rows and n columns. The set of all $m \times n$ nonnegative matrices is denoted by $\mathcal{R}^{m,n}$. In the case $m = n$, we use the notation $\mathcal{R}_n$. In this book, we consider combinatorial properties of nonnegative matrices and, specifically, the relative position of the elements of such matrices. In the study of these properties, an important role is played by the binary structure of a matrix, that is, the positions of zero and nonzero elements. Let $A = (a_{ij})$ be a matrix in $\mathcal{R}^{m,n}$. The support ${}^sA = (\bar{a}_{ij})$ of the matrix A is the matrix consisting of zeros and ones, that is, the $(0,1)$-matrix with elements

$$\bar{a}_{ij} = \begin{cases} 1 & \text{if } a_{ij} > 0, \\ 0 & \text{if } a_{ij} = 0. \end{cases}$$

The set of all $(0,1)$-matrices of size $m \times n$ is denoted by $\mathcal{B}^{m,n}$, or by $\mathcal{B}_n$ if $m = n$. Many combinatorial properties of nonnegative matrices can be expressed in terms of the corresponding properties of the supports of these matrices. In this connection combinatorial properties of $(0,1)$-matrices, which are of interest by themselves, take an additional importance. Therefore we will give much attention to the combinatorics of $(0,1)$-matrices.

1.1.2. Configurations. Let S and $\mathcal{E}$ be sets, with S nonempty. Suppose that a real-valued function φ is defined on the product $\mathcal{E} \times S$ and for any $E \in \mathcal{E}$ there exists $s \in S$ such that $\varphi(E,s) > 0$. The triple $\mathcal{C} = (S, \mathcal{E}, \varphi)$ is called a configuration. The function φ is called the weight function of the configuration, and the value $\varphi(E,s)$ is the weight of s in E. If $\varphi(E,s) > 0$, then the elements E and s are called incident. It is usually supposed that the sets $\mathcal{E}$ and S are finite. Let $S = \{s_1, \dots, s_n\}$ and $\mathcal{E} = \{E_1, \dots, E_m\}$. Then the total weight of s in the configuration $\mathcal{C}$ is defined as

$$\varphi(\mathcal{C}, s) = \sum_{E \in \mathcal{E}} \varphi(E, s).$$

The condition on the weight function in the definition of a configuration can be written in the form

$$\sum_{j=1}^{n} \varphi(E, s_j) > 0 \tag{1.1.1}$$

for each $E \in \mathcal{E}$.

An example of a configuration is an arbitrary family $\mathcal{E}$ of nonempty subsets of a finite set S if we take the weight function equal to the characteristic function

$$\chi(E,s) = \begin{cases} 1 & \text{if } s \in E, \\ 0 & \text{if } s \notin E, \end{cases} \tag{1.1.2}$$

where $s \in S$, $E \in \mathcal{E}$. This configuration will be referred to as the configuration of subsets.

Another source of examples of configurations is provided by projective (or affine) planes $\mathcal{P}$, where S is the set of all points of $\mathcal{P}$ and $\mathcal{E}$ is the set of all lines, and $\varphi(E,s) = 1$ if E and s are incident and $\varphi(E,s) = 0$ otherwise. Identifying each line of the finite plane $\mathcal{P}$ with the set of points incident to the line, we obtain the configuration of subsets on the finite set of points of the projective plane.

Below we give other examples of configurations.

1.1.3. The incidence matrix of a configuration. Let us introduce one of the basic combinatorial notions. Let a configuration $\mathcal{C} = (S, \mathcal{E}, \varphi)$ with a weight function φ and nonempty sets $S = \{s_1, \ldots, s_n\}$ and $\mathcal{E} = \{E_1, \ldots, E_m\}$ be given. For any $s_j \in S$ and $E \in \mathcal{E}$, we set $a_{ij} = \varphi(E_i, s_j)$. The nonnegative $m \times n$ matrix $A(\mathcal{C}) = A = (a_{ij})$ is called the incidence matrix of the configuration $\mathcal{C}$.

To each element E_i, $i = 1, \ldots, m$, we associate the subset of S consisting of the elements s such that $\varphi(E_i, s) > 0$. By (1.1.1) each of these sets is nonempty. It is also clear that any $m \times n$ nonnegative matrix $A = (a_{ij})$ such that $\sum_{j=1}^{n} a_{ij} > 0$ for all $i = 1, \ldots, m$ generates a configuration $\mathcal{C} = (S, \mathcal{E}, \varphi)$ if we take n arbitrary elements $s_1, \ldots, s_n$ as S and m arbitrary elements $E_1, \ldots, E_m$ as $\mathcal{E}$, and define the weight function by setting $\varphi(E_i, s_j) = a_{ij}$, $i = 1, \ldots, m$, $j = 1, \ldots, n$.

The subsets of S which correspond to the elements $E_1, \ldots, E_m \in \mathcal{E}$ for a given weight function φ will be referred to as the support subsets of the configuration and, whenever it does not lead to confusion, will be denoted by the same letters $E_1, \ldots, E_m$. If, in addition to (1.1.1),

$$\sum_{i=1}^{m} a_{ij} > 0 \tag{1.1.3}$$

for all $j = 1, \ldots, n$, then, evidently,

$$\bigcup_{i=1}^{m} E_i = S.$$

The characteristic function (see (1.1.2)) takes only the values 0 and 1; therefore the incidence matrix of a configuration of subsets is a $(0,1)$-matrix. Note that sometimes, instead of the matrix A, the support ${}^{s}A$ is considered as the incidence matrix of a configuration $\mathcal{C}$ (see [**15**]).

1.1.4. Equivalence of configurations. An incidence matrix provides information about essential properties of the corresponding configuration in convenient and compact form. Note however that an incidence matrix is not determined by its configuration uniquely. Indeed, if a configuration $\mathcal{C} = (S, \mathcal{E}, \varphi)$, where $S = \{s_1, \ldots, s_n\}$ and $\mathcal{E} = \{E_1, \ldots, E_m\}$, has the incidence matrix A, then any renumbering of the elements of the sets S and $\mathcal{E}$ leads, without changing $\mathcal{C}$, to an incidence matrix which, in general, differs from A by permutations of rows and columns. The inverse assertion is also true. If by the procedure described above we construct a configuration $\mathcal{C} = (S, \mathcal{E}, \varphi)$ corresponding to a given nonnegative matrix A, then any permutation ρ of columns and any permutation π of rows of the matrix A leads to a configuration $\mathcal{C}' = (S, \mathcal{E}, \varphi')$ with the same sets S and $\mathcal{E}$, and the coincidence of the configurations becomes evident if we note that

$$\varphi'(E_{\pi i}, s_{\rho j}) = \varphi(E_i, s_j), \qquad i = 1, \ldots, m, \quad j = 1, \ldots, n.$$

Let $P = (p_{ij})$ be the permutation matrix of order m corresponding to a permutation π on the set N_m, that is,

$$p_{ij} = \begin{cases} 1 & \text{if } j = \pi i, \\ 0 & \text{otherwise.} \end{cases} \tag{1.1.4}$$

Let Q be the permutation matrix of order n corresponding to a permutation ρ on the set N_n. If $\mathcal{C} = (S, \mathcal{E}, \varphi)$ with $|S| = n$ and $|\mathcal{E}| = m$, then, as is easily seen, the renumbering of the elements of S in accordance with the permutation ρ and the elements of $\mathcal{E}$ in accordance with the permutation π replaces A with the incidence matrix

$$A' = P^{-1}AQ. \tag{1.1.5}$$

Matrices A and A' connected by relation (1.1.5), where P and Q are permutation matrices, are called permutationally equivalent matrices. Configurations $\mathcal{C}$ and $\mathcal{D}$ are called equivalent if the incidence matrices of these configurations are permutationally equivalent.

Below we consider a series of the most common configurations and their incidence matrices.

1.1.5. Stochastic and doubly stochastic configurations. Classes of configurations are distinguished by conditions imposed on the weight functions. Let the weight function φ of a configuration $\mathcal{C} = (S, \mathcal{E}, \varphi)$ be such that the incidence matrix $A = (a_{ij})$, where $a_{ij} = \varphi(E_i, s_j)$, $s_j \in S$, $E_i \in \mathcal{E}$, satisfies the conditions

$$\sum_{j=1}^{n} a_{ij} = 1, \qquad i = 1, \dots, m. \tag{1.1.6}$$

Such a configuration is called stochastic. This term is explained by the fact that for $m = n$ conditions (1.1.6) define a stochastic matrix.

If $m = n$ and, in addition to (1.1.6),

$$\sum_{i=1}^{n} a_{ij} = 1, \tag{1.1.7}$$

then the corresponding configuration is called doubly stochastic (since the incidence matrix of this configuration is a doubly stochastic matrix). Stochastic and doubly stochastic matrices play a significant role in several areas of mathematics, especially in probability theory. They have a number of interesting combinatorial properties, which are investigated in Chapters 3 and 5.

1.1.6. Latin configurations. Let a configuration $\mathcal{C} = (S, \mathcal{E}, \varphi)$ be such that $S = \{s_1, \dots, s_n\}$, $\mathcal{E} = \{E_1, \dots, E_n\}$, and the function φ takes only the values 0 and 1. Suppose also that

$$\sum_{i=1}^{n} a_{ij} = \sum_{j=1}^{n} a_{ij} = k, \tag{1.1.8}$$

where $a_{ij} = \varphi(E_i, s_j)$, $i, j = 1, \dots, n$, and $k \in N_n$. Evidently, the incidence matrix $A(\mathcal{C})$ has exactly k ones in each column and each row. Such a configuration is called a Latin configuration. Let us show that these configurations are closely related to such combinatorial objects as Latin rectangles. Recall that a Latin rectangle of size $k \times l$ on the set N_n is a rectangular table L with k rows and l columns which is filled by the numbers $1, \ldots, n$ in such a way that all numbers in each row are distinct and all numbers in each column are also distinct. If $k = l = n$, then a Latin rectangle is called a Latin square. Let a Latin rectangle L of size $k \times n$ on the set N_n be given. For this rectangle, we construct a Latin configuration $(S, \mathcal{E}, \varphi)$. For S we take the set N_n, for E_i we take the subset consisting of the numbers from the ith column, $i = 1, \ldots, n$, and for φ we take the characteristic function of the family of subsets $\{E_1, \ldots, E_n\}$ of the set N_n. Each element of N_n can appear in the columns of the rectangle L at most k times. The total number of elements in L is nk; therefore each element appears exactly k times. Thus, conditions (1.1.8) on the incidence matrix of a Latin configuration hold for the incidence matrix of the constructed configuration, and we obtain a Latin configuration. In Section 1.2 (see Theorem 1.2.11) we will show that the converse assertion is also true; namely, using a Latin configuration, we can construct a set of Latin rectangles.

Note that a Latin configuration is in fact a doubly stochastic configuration of a special form. Dividing each element of the incidence matrix by k yields a doubly stochastic matrix whose nonzero elements are all equal to $1/k$.

1.1.7. Cyclic configurations. A right (left) circulant of order n is a matrix of order n such that each row of the matrix is obtained by the cyclic shift of the previous row one step to the right (to the left). In what follows, a circulant means a right circulant. Sometimes the term cyclic matrix is used for circulant.

A configuration $\mathcal{C} = (S, \mathcal{E}, \varphi)$ with $m = |\mathcal{E}| = |S| = n$ is called cyclic if the incidence matrix of $\mathcal{C}$ is permutationally equivalent to a circulant.

Let $A = (\alpha_{ij})$ be a circulant of order n with the first row $(a_1, \ldots, a_n)$. Then

$$a_{ij} = \alpha_{j-i+1}, \qquad i, j = 1, \ldots, n,$$

where the indices are considered modulo n. If $\mathcal{C}$ is a cyclic configuration of subsets and the first row of the $(0,1)$-matrix $A(\mathcal{C})$ contains k ones and $n-k$ zeros, then each row and each column of $A(\mathcal{C})$ has the same property. Thus, any cyclic configuration of subsets is a Latin configuration.

The combinatorics of circulants is discussed in Section 1.3.

1.1.8. Balanced designs. We again consider configurations $\mathcal{C} = (S, \mathcal{E}, \varphi)$ such that the weight function φ takes only the values 0 and 1 and the incidence matrix $A = (a_{ij})$ has exactly k ones in each row. In addition, we require that each pair of columns of the matrix A satisfies the condition

$$\sum_{i=1}^{n} a_{is} a_{it} = \lambda, \tag{1.1.9}$$

where $s \neq t$ and λ is a positive integer independent of the choice of the columns. If $\mathcal{C}$ is a configuration of subsets (in this case $\varphi = \chi$), then condition (1.1.9) means that any pair of different elements appears in exactly λ subsets forming the configuration. In this case, each subset consists of k elements.

Let some element $a \in S$ appear in each subset of $\mathcal{E}$ exactly r times. Then the number of appearances of pairs of elements of the form $\{a, x\}$, $x \neq a$, in these subsets can be calculated in two different ways. Taking into account the appearances of a in all subsets, we obtain that this number is $r(k-1)$, and calculating this number by taking into account all appearances of a in pairs with x, we obtain $\lambda(n-1)$ (see (1.1.9)). From the equality of these numbers it follows that r does not depend on the choice of a, and is thus a parameter of the configuration.

A balanced incomplete block design is a configuration of subsets $\mathcal{C} = (S, \mathcal{E}, \varphi)$ such that the incidence matrix $A(\mathcal{C})$ contains exactly k ones in each row and condition (1.1.9) is fulfilled with some λ, $0 < \lambda \leq r - 1$. Balanced incomplete block designs are commonly called BIB designs. Block designs first appeared in the theory of experiment design, and initially were used to design statistical experiments for investigating crop yield. In these applications, the term block for the sets of $\mathcal{C}$ and the notation v for the cardinality of S and b for the cardinality of $\mathcal{E}$ are used. BIB designs are sometimes called (b, v, r, k, λ)-configurations. The main relations for the parameters of the BIB designs are

$$\lambda(v-1) = r(k-1) \tag{1.1.10}$$

and the obvious equality (see Problem 3 below)

$$bk = vr, \tag{1.1.11}$$

where $0 < \lambda \leq r - 1$ (the last conditions exclude from consideration the degenerate configurations with $v = k$ or $k = 1$).

The theory of BIB configurations and related configurations is well developed and contains many interesting results.

1.1.9. Graph configurations. Let a configuration $\mathcal{C} = (V, \mathcal{E}, \varphi)$, where $V = \{v_1, \dots, v_n\}$, $\mathcal{E} = \{E_1, \dots, E_m\}$, have the following property: the support ${}^sA$ of the incidence matrix A of $\mathcal{C}$ contains at most two nonzero elements in each row. Such a configuration is called a graph configuration. If in such a case $\varphi = \chi$ and, consequently, ${}^sA = A$, then the graph configuration is called a pseudograph and the matrix $A = (a_{ij})$ is the incidence matrix of the pseudograph. If in addition

$$\sum_{j=1}^{n} a_{ij} = 2 \tag{1.1.12}$$

and

$$\sum_{j=1}^{n} a_{ij}a_{kj} \leq 1 \tag{1.1.13}$$

for any $i \neq k$, $1 \leq i, k \leq m$, then the configuration $\mathcal{C}$ is called an undirected graph, or simply a graph. The elements of V are called the vertices and the elements of $\mathcal{E}$ are called the edges of the graph. The subsets $E_1, \dots, E_m$ do not necessarily cover the set of vertices V. The vertices corresponding to the zero column in the matrix A are called isolated. If a graph configuration with $\varphi = \chi$ satisfies only the equalities (1.1.12), then it is called a multigraph.

Graph theory is a wide area of combinatorics, where the methods of matrix theory play significant but not dominant role. Other relations between graphs and matrices are discussed in subsequent chapters.

1.1.10. Hypergraphs. The existence of well-developed terminology and methods of graph theory led to attempts extend them to arbitrary configurations. An important role in this direction was played by the French mathematician C. Berge and his colleagues (see [**13**]). They developed the language of hypergraphs, which can be used to formulate notions and results about various configurations and their incidence matrices. By a hypergraph one means any family of subsets of a given set. Of course, this language is especially useful in the cases where some results of graph theory are extended to configurations of a more general form.

1.1.11. Permutational automorphisms. Earlier we have introduced the notion of permutational equivalence of matrices and the notion of equivalence of configurations. Below we present some simple facts about permutational equivalence.

THEOREM 1.1.1. *Let A be an $m \times n$ matrix. The number of matrices in the equivalence class which contains A is equal to $m!n!/i_A$, where i_A is the number of pairs (P, Q) of permutation matrices P of order m and Q of order n such that $P^{-1}AQ = A$.*

PROOF. If for two pairs (P, Q) and (P', Q') we have

$$P^{-1}AQ = (P')^{-1}AQ',$$

then

$$((P(P')^{-1})^{-1}A(Q(Q')^{-1}) = A.$$

Here $P(P')^{-1}$ and $Q(Q')^{-1}$ are permutation matrices of order m and n respectively. On the other hand, if (P_j, Q_j), $j = 1, \dots, i_A$, are all pairs (P, Q) possessing this property, then for any matrix $A' = (P')^{-1}AQ'$ equivalent to A there are i_A new pairs of the form (P_jP', Q_jQ') that transfer A to the same A'. Therefore, in the rectangular table composed from $m!n!$ matrices $P^{-1}AQ$, where P runs over all permutation matrices of order m and Q runs over all permutation matrices of order n, each matrix has exactly i_A matrices coinciding with it. The theorem is proved. □

A pair of permutation matrices (P, Q), where P is of order m and Q is of order n, is called a permutational automorphism of an $m \times n$ matrix A if $P^{-1}AQ = A$. The following assertion can be verified directly.

THEOREM 1.1.2. *The family $\mathcal{G}(A)$ of all permutational automorphisms of a matrix A forms a group. The family of the first (respectively, the second) components of the elements of $\mathcal{G}(A)$ also forms a group $\mathcal{G}^{(1)}(A)$ (respectively, $\mathcal{G}^{(2)}(A)$).*

Thus, according to Theorems 1.1.1 and 1.1.2 the number of elements in the class of permutational equivalence that contains A is determined by the order of the group $\mathcal{G}(A)$. The identity element of $\mathcal{G}$ is a pair $(\mathbf{I}_m, \mathbf{I}_n)$, where $\mathbf{I}_k$ is the identity matrix of order k. The group $\mathcal{G}^{(1)}(A)$ (the group $\mathcal{G}^{(2)}(A)$) is called the group of admissible permutations of rows (columns) of the matrix A. Indeed, the elements of $\mathcal{G}^{(1)}(A)$ are the permutations P of rows such that for each of them there exists a permutation Q of columns which moves the columns to appropriate places of the matrix A. Below, the isomorphism of groups is denoted by the symbol $\simeq$.

THEOREM 1.1.3. *If the columns of a matrix A are pairwise distinct, then $\mathcal{G}(A)$ is isomorphic to $\mathcal{G}^{(1)}(A)$. If all rows of A are pairwise distinct, then $\mathcal{G}(A)$ is isomorphic $\mathcal{G}^{(2)}(A)$.*

PROOF. Consider the mapping $f \colon \mathcal{G}(A) \to \mathcal{G}^{(1)}(A)$ such that $f(P, Q) = P$ for $(P, Q) \in \mathcal{G}(A) = \mathcal{G}$. It is easy to see that this mapping is a group homomorphism. The kernel $\mathcal{H}^{(1)}$ of this homomorphism (that is, the set of all elements which are mapped to the identity element) consists of all automorphisms of the form $(\mathbf{I}_m, Q)$, and $\mathcal{G}/\mathcal{H}^{(1)} \simeq \mathcal{G}^{(1)}(A)$. Similarly, $\mathcal{G}/\mathcal{H}^{(2)} \simeq \mathcal{G}^{(2)}(A)$, where $\mathcal{H}^{(2)}$ consists of all automorphisms of the form $(P, \mathbf{I}_n)$, $P \in \mathcal{G}^{(1)}$. If all columns (rows) of A are pairwise distinct, then any admissible permutation P of rows uniquely determines an admissible permutation of columns (any admissible permutation Q of columns uniquely determines an admissible permutation of rows). Thus, if all columns are pairwise distinct, then $\mathcal{H}^{(1)} = (\mathbf{I}_m, \mathbf{I}_n)$ and $\mathcal{G}(A) \simeq \mathcal{G}^{(1)}(A)$. Similarly, if all rows in the matrix are pairwise distinct, then $\mathcal{G}(A) \simeq \mathcal{G}^{(2)}(A)$. The theorem is proved. □

1.1.12. Semigroups of nonnegative matrices and $(0,1)$-matrices. In analyzing combinatorial properties of matrices, algebraic considerations often appear to be useful. It is well known that the set $\mathcal{R}^{n,n}$ of square nonnegative matrices of order n is a semigroup under the usual multiplication. The semigroup $\mathcal{R}_n = \mathcal{R}^{n,n}$ is a semigroup with identity element (monoid), where the identity element is the matrix $\mathbf{I}_n$.

It is clear that the set $\mathcal{B}^{n,n} = \mathcal{B}_n$ of all square $(0,1)$-matrices of order n does not constitute a semigroup under the multiplication over $\mathbf{R}^+$. However, the set $\{0,1\}$ can be considered with the structure of the finite field $\mathcal{F}_2$ or the binary Boolean algebra B_2. Recall that in the first case the addition $\oplus$ on the set $\{0,1\}$ is defined as

$$a \oplus b \equiv \overline{a+b} \pmod 2, \tag{1.1.14}$$

where $a, b \in \{0,1\}$, $\overline{a+b}$ is the least nonnegative residue modulo 2, and the multiplication $\odot$ is defined as

$$a \odot b = a \cdot b. \tag{1.1.15}$$

In these formulas, $+$ and $\cdot$ are the usual operations of addition and multiplication.

In the case of the Boolean algebra B_2, the addition $\boxplus$ on the set $\{0,1\}$ is defined as

$$a \boxplus b = \begin{cases} 1 & \text{if } a+b \neq 0, \\ 0 & \text{if } a+b = 0, \end{cases} \tag{1.1.16}$$

and the multiplication $\boxdot$ is defined as in (1.1.15), that is,

$$a \boxdot b = a \cdot b, \qquad a, b \in \{0,1\}. \tag{1.1.17}$$

We denote the set of all matrices over $\mathcal{F}_2$ by $\mathcal{B}(\circ)$ and the set of all matrices over B_2 by $\mathcal{B}(\square)$. The set of all $(0,1)$-matrices of size $m \times n$ considered as matrices over $\mathcal{F}_2$ is denoted by $\mathcal{B}^{m,n}(\circ)$, and the same set where the $(0,1)$-matrices are considered as matrices over B_2 is denoted by $\mathcal{B}^{m,n}(\square)$. The result of multiplication of an $m \times r$ matrix $A = (a_{ij})$ by an $r \times n$ matrix $B = (b_{jk})$ is defined as the $m \times n$ matrix $C = (c_{ik})$ with elements

$$c_{ik} = \sum_{j=1}^{r}{}^{\oplus} a_{ij} \odot b_{jk}, \tag{1.1.18}$$

and the multiplication of the same matrices in $\mathcal{B}(\square)$ leads to the matrix $D = (d_{ik})$ such that

$$d_{ik} = \sum_{j=1}^{r}{}^{\boxplus} a_{ij} \boxdot b_{jk}. \tag{1.1.19}$$

Note that for both $\mathcal{F}_2$ and B_2 the distributivity holds. It is not difficult to show that the set $\mathcal{B}_n(\circ) = \mathcal{B}^{n,n}(\circ)$ (the set $\mathcal{B}_n(\square) = \mathcal{B}^{n,n}(\square)$) is a semigroup under the multiplication defined by (1.1.18) (respectively, by (1.1.19)).

The algebraic operations introduced in this subsection are useful in the study of combinatorial properties of nonnegative and $(0,1)$-matrices. Depending on the questions under consideration, a nonnegative matrix is considered as a matrix over some field, or ring, or Boolean algebra, and the corresponding semigroup properties can be exploited.

Problems.

1. Let S be a finite set consisting of n elements, and $\mathcal{E}$ the set of all n-dimensional vectors over a finite field with q elements. Construct a weight function φ such that the incidence matrix $A = A(\mathcal{C})$ of the configuration $(S, \mathcal{E}, \varphi)$ consists of q^n different rows. Describe the structure of the support of A.
2. Let the Latin configuration $\mathcal{C}_1$, $\mathcal{C}_2$, and $\mathcal{C}_3$ be determined, respectively, by the incidence matrices

$$A(\mathcal{C}_1) = \begin{pmatrix} 1 & 1 \\ 1 & 1 \end{pmatrix}, \quad A(\mathcal{C}_2) = \begin{pmatrix} 0 & 1 & 1 & 0 & 0 \\ 0 & 0 & 1 & 1 & 0 \\ 0 & 0 & 0 & 1 & 1 \\ 1 & 0 & 0 & 0 & 1 \\ 1 & 1 & 0 & 0 & 0 \end{pmatrix}, \quad A(\mathcal{C}_3) = \begin{pmatrix} 1 & 1 & 1 & 0 \\ 0 & 1 & 1 & 1 \\ 1 & 1 & 0 & 1 \\ 1 & 0 & 1 & 1 \end{pmatrix}.$$

   For each of the matrices $\mathcal{C}_i$, find a Latin rectangle L_i, $i = 1, 2, 3$, using the construction described in Section 1.1.6.
3. Prove equality (1.1.11).
4. Let

$$A = \begin{pmatrix} 1 & 1 & 0 & 0 \\ 0 & 1 & 1 & 0 \\ 0 & 0 & 1 & 1 \\ 1 & 0 & 0 & 1 \end{pmatrix}.$$

   What is the number of different matrices in the equivalence class containing A?
5. Find the group $\mathcal{G}$ of permutational automorphisms for the following $(0,1)$-matrices:
   (a) $A = \mathbf{J}_{mn}$, where $\mathbf{J}_{mn}$ is the $m \times n$ matrix with all elements equal to one,
   (b) the matrix A such that the column sums are pairwise different and there are no identical rows,
   (c) the matrix

$$A = \begin{pmatrix} 0 & 0 & 1 & 1 & 0 \\ 0 & 1 & 0 & 1 & 1 \\ 0 & 1 & 1 & 0 & 1 \\ 0 & 0 & 1 & 1 & 0 \\ 1 & 0 & 0 & 0 & 1 \end{pmatrix}.$$

6. Prove that the set of all doubly stochastic matrices of order n is a semigroup under multiplication.

1.2. Term rank. Arrangement of positive elements

The special features of the structure of nonnegative matrices essentially depend on some general characteristics which can be attributed to one of two classes, the classes of local and global characteristics. Global characteristics describe the matrix itself, and local characteristics are connected with relations between separate parts of the matrix. Among the global characteristic, two stand out, and they will be considered in this section. These characteristics are the term rank and the row and column sums (of the matrix or of the support of the matrix).

1.2.1. Term rank. In linear algebra, an important global characteristic of a matrix is the rank, that is, the maximum number of linearly independent (as vectors) rows or columns. This function plays an essential role in combinatorial considerations as well. However there is a more specific notion which characterizes the combinatorial nature of nonnegative matrices and is essential in studying the properties of such matrices. Let $A = (a_{ij})$ be an $m \times n$ nonnegative matrix. The term rank of a matrix is the maximum number of positive elements of the matrix such that no two of them are in the same row or the same column. The term rank of the matrix A is denoted by $\rho(A)$. If $m \leq n$, then it is clear that $0 \leq \rho(A) \leq m$. Two elements of a matrix are called collinear if they lie on the same line, that is, belong to the same row or the same column. A diagonal of an $m \times n$ nonnegative matrix is an arbitrary set of $\min\{m,n\}$ noncollinear elements of the matrix. A diagonal is called positive if all elements of it are positive. First of all, we are interested in positive diagonals; therefore we will sometimes use the term diagonal instead of positive diagonal. In the same sense we use the term partial diagonal in a nonnegative matrix of size $m \times n$ for a set of k positive pairwise noncollinear elements, $k \leq \min\{m,n\}$. It is clear that an $m \times n$ matrix A with $m \leq n$ has a positive diagonal if and only if $\rho(A) = m$. For example, if A is a permutation matrix of order n, then it has a unique positive diagonal and $\rho(A) = n$.

1.2.2. Transversals. Let $\mathcal{C} = (S, \mathcal{E}, \chi)$ be a configuration of subsets, that is, let $\mathcal{E} = \{E_1, \dots, E_m\}$ be a family of subsets of a set $S = \{s_1, \dots, s_n\}$ (χ is the characteristic function of sets from $\mathcal{E}$). The family $\{E_1, \dots, E_m\}$ is said to have a transversal, or a system of distinct representatives, if there exists a function $\sigma\colon N_m \to S$ such that $\sigma(i) \in E_i$ and $\sigma(i) \neq \sigma(j)$ for $i \neq j$, $i, j = 1, \dots, m$. Note that although the definition has been given for the case of finite sets, the transversals can be defined for infinite sets S and $\mathcal{E}$ as well. It is clear that transversals exist only if $m \leq n$.

THEOREM 1.2.1. *A family of subsets $\mathcal{E} = \{E_1, \dots, E_m\}$ of an n-set S, $m \leq n$, has a transversal if and only if the incidence matrix A of the configuration $(S, \mathcal{E}, \chi)$ has a positive diagonal.*

PROOF. Let a matrix $A = (a_{ij})$ have a positive diagonal. An element a_{ij} is positive if and only if $s_j \in E_i$. The condition of positivity of a diagonal means that there exist pairwise distinct $j_1, \dots, j_m \in N_n$ such that the elements $a_{1j_1}, \dots, a_{mj_m}$ are positive. But then the set of elements $s_{j_1}, \dots, s_{j_m}$ is a transversal of the family $\{E_1, \dots, E_m\}$.

The converse assertion is obvious. □

Thus, the equality $\rho(A) = m$ for the incidence matrix A of a configuration of subsets $\mathcal{C} = (S, \mathcal{E}, \chi)$, where $\mathcal{E} = \{E_1, \dots, E_m\}$, $m \leq n$, is equivalent to the existence of a transversal of $\mathcal{E}$.

1.2.3. Numerical functions on matrices. Permanent. Let $A = (a_{ij})$ be an $m \times n$ matrix, $m \leq n$. Consider the set S_n^m of all ordered m-samples

$$\sigma = (\sigma(1), \dots, \sigma(m))$$

without repetitions from the set N_n. In the study of matrices, both in combinatorics and in linear algebra, an important role is played by the following functions. For

A and a function $f(\sigma)$ on S_n^m taking complex values, let

$$F(A) = \sum_{\sigma \in S_n^m} f(\sigma) a_{1\sigma(1)} \dots a_{m\sigma(m)}, \tag{1.2.1}$$

where $\sigma = (\sigma(1), \dots, \sigma(m))$. The functions on matrices of the form (1.2.1) are called Schur functions. The most common Schur functions are the following two functions.

1. Let $m = n$. In this case, σ is a permutation on the set N_n. Define the function f on the set of permutations of degree n, setting

$$f(\sigma) = \begin{cases} 1 & \text{if } \sigma \text{ is an even permutation,} \\ -1 & \text{if } \sigma \text{ is an odd permutation.} \end{cases}$$

   Then $F(A)$ defined by (1.2.1) is the determinant $\Delta(A)$ of the matrix A.
2. Let $f(\sigma) \equiv 1$. Then (1.2.1) defines the permanent $\operatorname{per} A$ of the matrix A:

$$\operatorname{per} A = \sum_{\sigma \in S_n^m} \prod_{i=1}^{m} a_{i\sigma(i)}. \tag{1.2.2}$$

It is clear that the permanent of a nonnegative matrix differs from zero if and only if A has a positive diagonal. By Theorem 1.2.1 the family $\{E_1, \dots, E_m\}$ of subsets of the set S of n elements has a transversal if and only if the permanent of the incidence matrix A of the family is not equal to zero. In its turn, the last property is equivalent to the equality $\rho(A) = m$.

Let two subsets $\alpha \subseteq N_m$ and $\beta \subseteq N_n$ be given. For an $m \times n$ matrix A, we denote by $A[\alpha, \beta]$ the submatrix of A consisting of the elements of A from the rows with indices in α and the columns with indices in β, and by $A(\alpha, \beta)$ we denote the submatrix obtained from A by deleting the rows with indices in α and columns with indices in β. If $\alpha = \{i\}$, $\beta = \{j\}$, then instead of $A(\{i\}, \{j\})$ the notation $A(i, j)$ will be used. The two properties formulated below directly follow from the definition of permanent.

1. Let $A = (a_{ij})$ be an $m \times n$ matrix, $m \leq n$. Then

$$\operatorname{per} A = \sum_{j=1}^{n} a_{ij} \operatorname{per} A(i, j), \qquad i = 1, \dots, m. \tag{1.2.3}$$

   This formula, as well as the similar formula for determinants, is called the decomposition of a permanent with respect to the ith row. For $m = n$, the decomposition with respect to a column is defined similarly.
2. Let an $m \times n$ matrix A with $m \leq n$ be of the form

$$A = \begin{pmatrix} B & C \\ \mathbf{0} & D \end{pmatrix},$$

   where B and D are submatrices such that the numbers of rows do not exceed the numbers of columns, and $\mathbf{0}$ is a zero submatrix. Then

$$\operatorname{per} A = \operatorname{per} B \operatorname{per} D. \tag{1.2.4}$$

Let us list some other simple but useful properties of permanents.

3. If A is a square matrix of order n, and A^T is its transpose, then $\operatorname{per} A = \operatorname{per} A^T$.
4. Permanents are linear with respect to rows and columns.

5. Any permutation of rows and columns of a matrix does not change the permanent.

1.2.4. A condition for the existence of a positive diagonal. As has been shown, the existence of a positive diagonal in the nonnegative matrix A is equivalent to the inequality $\operatorname{per} A > 0$. Moreover, the following assertion is true.

THEOREM 1.2.2. *For a nonnegative $m \times n$ matrix A with $m \leq n$, the number of positive diagonals is equal to* $\operatorname{per} {}^sA$, *where ${}^sA$ is the support of A.*

The assertion follows directly from the definition of a permanent (see (1.2.2)).

The properties listed above allow us to relate the presence of a positive diagonal in a nonnegative matrix with the structure of the matrix.

Let d be a positive number and A a nonnegative matrix. We write $A \geq d$ or $A > d$ if all elements of A are, respectively, no less (or greater) than d (of course, these notations will be used only when they cannot be confused with the numerical inequalities). In particular, the inequality $A \geq 0$ means that the matrix A is nonnegative.

THEOREM 1.2.3 (Frobenius–König Theorem). *Let $A = (a_{ij})$ be a nonnegative $m \times n$ matrix, $m \leq n$. Each diagonal of A contains a zero element if and only if A contains a zero $s \times t$ submatrix such that $s + t = n + 1$.*

PROOF. We reduce the theorem to the case of square matrices. We complement the matrix A by adding $n - m$ rows consisting of ones and obtain the square matrix $\tilde{A}$. It is clear that A has a positive diagonal if and only if $\tilde{A}$ has. Moreover, the zero submatrices of A and $\tilde{A}$ are located in the same places.

Thus, let A be a square matrix of order n.

Let us prove the sufficiency. Suppose that A contains a zero $s \times t$ submatrix such that $s + t = n + 1$. A permanent does not depend on permutations of rows and columns (see property 5 in Section 1.2.3); therefore we can assume that this submatrix is located in the lower left corner of the matrix, that is, the matrix is of the form

$$A = \begin{pmatrix} B & C \\ \mathbf{0} & D \end{pmatrix},$$

where B is an $(n - s) \times t$ submatrix and $\mathbf{0}$ stands for the zero $s \times t$ submatrix. Any diagonal of a square matrix contains one element from each of the first t columns. Therefore, if we are looking for a positive diagonal, then the elements of these t columns have to belong to different rows with numbers $1, \ldots, n - s$. However, $n - s = t - 1 < t$, and this condition cannot be satisfied. Thus, there is no positive diagonal.

Let us prove the necessity. We use induction. Let $\operatorname{per} A = 0$. For $n = 1$, the assertion is obvious. Let the assertion be proved for matrices of all orders less than n. For the zero matrix of order n the assertion is obvious as well. Let A contain a positive element a_{ij}. Then, using the decomposition of $\operatorname{per} A$ with respect to the ith row, we obtain

$$\operatorname{per} A = \sum_{j=1}^{n} a_{ij} \operatorname{per} A(i, j) = 0.$$

Since $a_{ij} > 0$, we see that $\operatorname{per} A(i, j) = 0$. The order of $A(i, j)$ equals $n - 1$; therefore, by the induction hypothesis, the assertion is true, that is, $A(i, j)$ contains a zero $s_1 \times t_1$ submatrix such that $s_1 + t_1 = n_1 + 1 = n$.

We rearrange the rows and columns of A so that this zero submatrix is located in the lower left corner. We obtain a matrix B with $\operatorname{per} B = \operatorname{per} A$ such that

$$B = \begin{pmatrix} C & E \\ \mathbf{0} & D \end{pmatrix},$$

where $\mathbf{0}$ is the zero $s_1 \times t_1$ matrix such that $s_1 + t_1 = n$, C is an $(n - s_1) \times t_1$ matrix, and D is an $s_1 \times (n - t_1)$ matrix. Thus, C and D are square matrices of orders t_1 and s_1, respectively. Since $\operatorname{per} B = \operatorname{per} A = 0$, according to (1.2.4)

$$\operatorname{per} B = \operatorname{per} C \operatorname{per} D = 0,$$

that is, either $\operatorname{per} C = 0$, or $\operatorname{per} D = 0$. Without loss of generality, we assume that $\operatorname{per} C = 0$. Then, by the induction hypothesis, C contains a zero $u \times v$ submatrix such that $u + v = t_1 + 1$. Let it be located in the rows of the matrix B with numbers $i_1, \ldots, i_u$ and in the columns with numbers $j_1, \ldots, j_v$. Consider the submatrix

$$B[\{i_1, \ldots, i_u, t_1 + 1, t_1 + 2, \ldots, n\}, \{j_1, \ldots, j_v\}].$$

This is the zero $(u + n - t_1) \times v$ submatrix, where $u + n - t_1 + v = n + 1$. Thus, we find a zero $s \times t$ submatrix of B, where $s = u + n - t_1$, $t = v$, $s + t = n + 1$. The matrix A differs from B only by the location of rows and columns; therefore A contains a zero submatrix of the same size. □

1.2.5. The structure of doubly stochastic matrices. As an application of the results obtained above we consider the doubly stochastic configurations and describe the structure of the incidence matrices of such configurations.

THEOREM 1.2.4. *A doubly stochastic matrix contains a positive diagonal.*

PROOF. Let A be a doubly stochastic matrix of order n. Suppose that A contains no positive diagonals. Then by the Frobenius–König Theorem some permutations of rows and columns of A lead to a matrix A' of the form

$$A' = \begin{pmatrix} B & C \\ \mathbf{0} & D \end{pmatrix},$$

where $\mathbf{0}$ stands for the zero $q \times r$ submatrix with $q + r \geq n + 1$. The permutations of rows and columns do not change the row and column sums of the matrix. Therefore A' is a doubly stochastic matrix. Hence the sums of all elements of A', B, and D are equal to n, r, and q, respectively, and thus, $r + q \leq n$. This contradiction proves the theorem. □

Theorem 1.2.4 means that the term rank of a doubly stochastic matrix of order n equals n.

Now we prove the main theorem on the structure of doubly stochastic matrices.

THEOREM 1.2.5. *Let A be a doubly stochastic matrix of order n. Then there exist r permutation matrices $P_1, \ldots, P_r$ and r nonnegative numbers $\alpha_1, \ldots, \alpha_r$, $\alpha_1 + \cdots + \alpha_r = 1$, such that*

$$A = \sum_{k=1}^{r} \alpha_k P_k.$$

PROOF. Denote by $\sigma(A)$ the number of positive elements of A. The matrix A is doubly stochastic; therefore $\sigma(A) \geq n$. If $\sigma(A) = n$, then A is a permutation matrix and the assertion is true with $r = 1$. Let $\sigma(A) > n$. Suppose that the theorem is proved for all doubly stochastic matrices of order n with fewer than

$\sigma(A)$ positive elements. By Theorem 1.2.4 the matrix $A = (a_{ij})$ has a positive diagonal $(a_{1\tau(1)}, \ldots, a_{n\tau(n)})$ for some permutation τ on N_n. Since $\sigma(A) > n$, A is not a permutation matrix and not all elements of the diagonal are equal to one. Let $P = (p_{ij})$ be the permutation matrix corresponding to τ, that is, $p_{ij} = 1$ if and only if $j = \tau(i)$. Set $a = \min_i a_{i\tau(i)}$. It is clear that $0 < a < 1$ and $A - aP > 0$. Consider the matrix

$$B = (b_{ij}) = (A - aP)/(1 - a).$$

The matrix B is doubly stochastic, since

$$\begin{aligned}\sum_{j=1}^{n} b_{ij} &= \sum_{j=1}^{n} (a_{ij} - ap_{ij})/(1 - a) \\ &= \sum_{j=1}^{n} a_{ij}/(1 - a) - a \sum_{j=1}^{n} p_{ij}/(1 - a) = 1, \qquad i = 1, \ldots, n,\end{aligned}$$

and similarly,

$$\sum_{i=1}^{n} b_{ij} = 1, \qquad j = 1, \ldots, n.$$

In addition, it is clear that $\sigma(B) \leq \sigma(A) - 1$. By the induction hypothesis,

$$B = \sum_{j=1}^{r-1} \beta_j P_j$$

for permutation matrices $P_1, \ldots, P_{r-1}$ and nonnegative numbers $\beta_1, \ldots, \beta_{r-1}$ such that $\beta_1 + \cdots + \beta_{r-1} = 1$. Thus,

$$A = (1 - a)B + aP = \sum_{j=1}^{r-1} (1 - a)\beta_j P_j + aP = \sum_{j=1}^{r} \alpha_j P_j,$$

where $P_r = P$, $\alpha_j = (1 - a)\beta_j$, $j = 1, \ldots, r - 1$, and $\alpha_r = a$. It is obvious that the α_j, $j = 1, \ldots, r$, are nonnegative and

$$\sum_{j=1}^{r} \alpha_j = \sum_{j=1}^{r-1} (1 - a)\beta_j + a = (1 - a) \sum_{j=1}^{r-1} \beta_j + a = 1.$$

The theorem is proved. □

1.2.6. The König Theorem. One of the fundamental combinatorial results is the theorem which connects the value of the term rank (in the modern terminology) of a nonnegative matrix with the arrangement of the positive elements of the matrix. This theorem was proved by König.

A row or a column of a matrix will be referred to as a line. The cover rank of a matrix is the minimum number of lines on which all positive elements of the matrix are located.

THEOREM 1.2.6. *Let A be a nonnegative $m \times n$ matrix. Then the cover rank of the matrix A equals $\rho(A)$.*

PROOF. We denote by ρ' the cover rank of the matrix A. It is natural to set $\rho' = \rho(A) = 0$ for the zero matrix A. Assume now that A is not the zero matrix. By the definition of the term rank, there exists a submatrix of A containing a positive partial diagonal with $\rho = \rho(A)$ elements. No two of these elements lie on one line of A; therefore $\rho' \geq \rho$.

Let us prove that $\rho \geq \rho'$. Let there exist e rows and f columns of A containing all positive elements, $e + f = \rho'$. Rearrange these rows and columns, putting them in the beginning positions. Of course, the new matrix has the same values of ρ and ρ'. Thus, we obtain the matrix

$$\begin{pmatrix} A_1 & A_2 \\ A_3 & \mathbf{0} \end{pmatrix},$$

where A_1, A_2, A_3 are $e \times f$, $e \times (n-f)$, $(m-e) \times f$ matrices, respectively, and $\mathbf{0}$ stands for the zero $(m-e) \times (n-f)$ matrix. One of the matrices A_2 and A_3 can be empty (if $e = 0$ or $f = 0$). Since $e + f \leq \min\{m, n\}$, a permanent is defined for the matrices A_2 and A_3^T.

If A_2 and A_3 are nonempty, then $\operatorname{per} A_2$ and $\operatorname{per} A_3^T$ are nonzero. Indeed, suppose that $\operatorname{per} A_2 = 0$. Then by Theorem 1.2.3 there exists a zero $e_1 \times g$ submatrix of A_2 such that

$$e_1 + g = n - f + 1. \tag{1.2.5}$$

Without loss of generality we assume that this submatrix is located in the lower right corner of A_2, so that the total matrix is of the form

$$\begin{pmatrix} A_1 & \boxed{\begin{matrix} A_2' & A_2'' \\ A_2''' & \mathbf{0} \end{matrix}} \\ A_3 & \mathbf{0} \end{pmatrix}. \tag{1.2.6}$$

Now we take the first $e - e_1$ rows and the first $n - g$ columns of the matrix (1.2.6). It is clear that these lines contain all positive elements of the matrix. On the other hand, $e - e_1 + n - g < e + f$, since it follows from (1.2.5) that $n - g - e_1 < f$. This contradicts the minimality of $\rho' = e + f$. Thus, $\operatorname{per} A_2 > 0$ and $\rho(A_2) = e$.

The matrix A_3^T is considered similarly. Applying Theorem 1.2.3 to A_3^T, one can show that $\operatorname{per} A_3^T \neq 0$ and, consequently, $\rho(A_3) = f$. It is clear that

$$\rho(A) \geq \rho(A_2) + \rho(A_3) = e + f = \rho'.$$

Thus, $\rho = \rho'$, and the theorem is proved. □

König's Theorem plays an important role in combinatorial considerations, especially in graph theory. However, generally speaking, this theorem does not simplify the problem of finding the term rank of a matrix. Estimates of the term rank are considered in Chapter 2.

1.2.7. The Ph. Hall Theorem on transversals. For a given configuration of subsets on a finite set there exists a set-theoretic criterion for the existence of a transversal of the configuration. This criterion was found by Philip Hall, and is one of the best known combinatorial results. In fact, the Ph. Hall Theorem is a set-theoretic analogue of the König Theorem, and these two theorems are equivalent.

THEOREM 1.2.7 (Ph. Hall Theorem). *Let $\mathcal{C} = (S, \mathcal{E}, \chi)$ be a configuration of subsets of a finite set S, $\mathcal{E} = \{E_1, \dots, E_m\}$. The family $\mathcal{E}$ of subsets has a transversal if and only if for any k-sample $i_1, \dots, i_k$ without repetitions from the set N_m*

the inequalities

$$\left|\bigcup_{s=1}^{k} E_{i_s}\right| \geq k, \qquad k = 1, \dots, m, \tag{1.2.7}$$

are fulfilled.

PROOF. In the proof, we use the König Theorem.

First we prove the sufficiency. Suppose that a transversal exists, but conditions (1.2.7) are not fulfilled, that is, for some k there are subsets $E_{i_1}, \dots, E_{i_k} \in \mathcal{E}$ such that

$$\left|\bigcup_{s=1}^{k} E_{i_s}\right| = k' < k.$$

Consider the incidence matrix A of $\{E_1, \dots, E_m\}$ and denote by $\rho' = \rho'(A)$ the cover rank of A. Rearrange the rows and columns of A so that the rows corresponding to $E_{i_1}, \dots, E_{i_k}$ and the columns corresponding to the elements contained in these subsets are located in the beginning positions. Then we obtain a matrix of the form

$$\begin{pmatrix} B_1 & \mathbf{0} \\ B_2 & B_3 \end{pmatrix},$$

where B_1 is a $k \times k'$ matrix. Thus, all ones of A can be covered by $k' + m - k < m$ lines, that is, $\rho' < m$, while the existence of a transversal (see Theorem 1.2.1) implies that $\rho = \rho' = m$. We obtain a contradiction to Theorem 1.2.6.

Let us prove the necessity. Suppose that conditions (1.2.7) are fulfilled, but there is no transversal and, consequently, by Theorem 1.2.1 $\rho(A) < m$. By the König Theorem, $\rho' < m$. Therefore, there exist e rows and f columns such that they cover all ones of the incidence matrix A and $\rho' = e + f < m$. Thus, the matrix A is permutationally equivalent to a matrix of the form

$$\begin{pmatrix} A_1 & A_2 \\ A_3 & \mathbf{0} \end{pmatrix},$$

where A_3 is an $(m - e) \times f$ matrix. Thus, the subsets corresponding to all $m - e$ lower rows contain in total at most f different elements. Since $f < m - e$, we obtain a contradiction to conditions (1.2.7). The theorem is proved. □

We deduced the Ph. Hall Theorem from the König Theorem, but in fact, these assertions are equivalent, and it is possible to deduce Theorem 1.2.6 from Theorem 1.2.7 (see Problem 2).

1.2.8. Ryser classes. For a nonnegative $m \times n$ matrix $A = (a_{ij})$, let $R = (r_1, \dots, r_m)$ be the vector of row sums

$$r_i = \sum_{j=1}^{n} a_{ij}, \qquad i = 1, \dots, m,$$

and $S = (s_1, \dots, s_n)$ the vector of column sums

$$s_j = \sum_{i=1}^{m} a_{ij}, \qquad j = 1, \dots, n.$$

Let $A \in \mathcal{B}^{m,n}$ be a $(0,1)$-matrix, considered as a matrix over integers, and let $R = (r_1, \dots, r_m)$, $S = (s_1, \dots, s_n)$ be two vectors with nonnegative integer coordinates. An $m \times n$ matrix A belongs to the Ryser class $\mathfrak{A}(R, S)$ if the vectors of row and column sums of A coincide with R and S, respectively. If the size $m \times n$ of a matrix is fixed, then instead of $\mathfrak{A}^{m,n}(R,S)$ we will use the notation $\mathfrak{A}(R,S)$. Sometimes it is useful to consider the classes of nonnegative matrices with given vectors of row and column sums of the supports of the matrices. Therefore it is reasonable to consider the class of nonnegative matrices such that the supports of the matrices from the class belongs to some Ryser class.

It is not difficult to see that any $m \times n$ matrix is permutationally equivalent to a matrix with vectors of the row sums R and the column sums S forming two nonincreasing sequences

$$R = (r_1, \dots, r_m), \qquad r_1 \geq \cdots \geq r_m,$$

and

$$S = (s_1, \dots, s_n), \qquad s_1 \geq \cdots \geq s_n.$$

In what follows we will often use this fact. A vector $X = (x_1, \dots, x_k)$ such that $x_1 \geq \cdots \geq x_k$ will be called a monotone vector. The Ryser class $\mathfrak{A}(R,S)$ with monotone vectors R and S is called a monotone class. If $R = S = (k, \dots, k)$ and $m = n$, then $\mathfrak{A}(R,S)$ is denoted by $\mathfrak{A}(k)$.

Let $R = (r_1, \dots, r_m)$ and $S = (s_1, \dots, s_n)$ be two vectors with positive integer coordinates. The natural question arises whether a $(0,1)$-matrix with the vectors of row sums R and column sums S exists. In other words, under what conditions is the Ryser class $\mathfrak{A}(R,S)$ nonempty? Below we give a criterion for nonemptiness of the Ryser class $\mathfrak{A}(R,S)$ for given vectors R and S. For this purpose we introduce two auxiliary matrices which are constructed using the given vectors R and S with nonnegative integer coordinates. These matrices will play an essential role in our further considerations.

1.2.9. Maximal and structure matrices. Let

$$R = (r_1, \dots, r_m), \qquad S = (s_1, \dots, s_n)$$

be two vectors with nonnegative integer coordinates and

$$r_i \leq n, \quad s_j \leq m, \quad i = 1, \dots, m, \quad j = 1, \dots, n, \quad r_1 + \cdots + r_m = s_1 + \cdots + s_n.$$

We consider the $(0,1)$-matrix of size $m \times n$ with rows constructed in the following way: the ith row begins with r_i ones and ends with $n - r_i$ zeros, $i = 1, \dots, m$. This matrix is called maximal and denoted by $A(R; n)$.

We introduce a partial order on the set of monotone vectors with nonnegative integer coordinates as follows: a vector $S = (s_1, \dots, s_n)$ does not exceed a vector $S^* = (s_1^*, \dots, s_p^*)$ if $n = p$ and

$$s_1 + \cdots + s_n = s_1^* + \cdots + s_n^*, \tag{1.2.8}$$

$$s_1 + \cdots + s_i \leq s_1^* + \cdots + s_i^*, \qquad i = 1, \dots, n-1. \tag{1.2.9}$$

It is clear that under condition (1.2.8), condition (1.2.9) is equivalent to the inequalities

$$s_{i+1} + \cdots + s_n \geq s_{i+1}^* + \cdots + s_n^*, \qquad i = 1, \dots, n-1. \tag{1.2.10}$$

The relation introduced above will be denoted by the symbol $\prec$. The relation $\prec$ is reflexive, antisymmetric, and transitive, so that $\prec$ is indeed a partial order. The relation $\prec$ is extended to arbitrary nonnegative vectors of equal sizes with integer coordinates. Let $\tilde{S}$, $\tilde{S}^*$ be two such vectors, and S, S^* the vectors obtained from them by arranging the coordinates in nonincreasing order. Then we set $\tilde{S} \prec \tilde{S}^*$ if $S \prec S^*$.

If the class $\mathfrak{A}^{m,n}(R,S)$ is nonempty and $A(R;n)$ is the maximal matrix with the vector of column sums $\bar{S} = (\bar{s}_1, \dots, \bar{s}_n)$, then the vectors $\bar{S}$ and the vector $S = (s_1, \dots, s_n)$ of column sums of the class $\mathfrak{A}(R,S)$ satisfy the relation $S \prec \bar{S}$. Note also that for the maximal matrix $A(R;n)$

$$\bar{s}_j = |\{i \colon r_i \geq j,\ i = 1, \dots, m\}|, \qquad j = 1, \dots, n, \tag{1.2.11}$$

that is, $(r_1, \dots, r_m)$ and $(\bar{s}_1, \dots, \bar{s}_n)$ are conjugate partitions of the number $r_1 + \dots + r_m$.

Let $I \subseteq N_m$ and $J \subseteq N_n$ be two subsets consisting of k and l elements, respectively, $0 \leq k \leq m$, $0 \leq l \leq n$ (I and J can be empty). For integer-valued vectors $R \geq 0$ and $S \geq 0$ of dimensions m and n, respectively, consider the number

$$t_{R,S}(I,J) = |I||J| + \sum_{i \notin I} r_i - \sum_{j \in J} s_j. \tag{1.2.12}$$

If R and S are fixed, then instead of $t_{R,S}(I,J)$ we write $t(I,J)$. If R and S are monotone, then it is clear that $t(N_k, N_l)$ is minimal among all numbers $t(I,J)$ with $|I| = k$, $|J| = l$. In this case, we denote this number by t_{kl}. The $(m+1) \times (n+1)$ matrix $T = (t_{kl})$ is called the structure matrix for the monotone vectors R and S.

1.2.10. A criterion for nonemptiness of Ryser classes. The criterion is given by the following theorem.

THEOREM 1.2.8. *Let $R = (r_1, \dots, r_m)$, $S = (s_1, \dots, s_N)$ be two nonnegative integer-valued vectors such that $r_1 + \dots + r_m = s_1 + \dots + s_n$. Then the following assertions are equivalent:*

(1) *The class $\mathfrak{A}^{n,n}(R,S)$ is nonempty.*
(2) *$t(I,J) \geq 0$ for all $I \subseteq N_m$, $J \subseteq N_n$.*
(3) *$S \prec \bar{S}$ and $r_k \leq n$, $k = 1, \dots, m$, where $\bar{S} = (\bar{s}_1, \dots, \bar{s}_n)$ is the vector of column sums of the matrix $A(R;n)$.*

PROOF. Since any matrix can be obtained from some matrix of a monotone class by permutation of rows and columns, it is sufficient to prove the theorem for the case where R and S are monotone.

Let us prove that $1 \Rightarrow 2$. Let the class $\mathfrak{A}(R,S)$ be nonempty. Consider a matrix $A \in \mathfrak{A}(R,S)$ and choose arbitrarily two subsets $I \subseteq N_m$ and $J \subseteq N_n$ of cardinalities k and l, respectively. The submatrix $A[I,J]$ is denoted by X, $A(I,J)$ by Y, and $A(\bar{I},J)$ by Z, where $\bar{I} = N_m \setminus I$. We denote by $\sigma_0(U)$ and $\sigma_1(U)$ the number of zeros and ones, respectively, in a matrix U. Then

$$\sigma_1(Y) = \sum_{i \notin I} r_i - \sigma_1(Z).$$

In turn,

$$\sigma_1(Z) = \sum_{j \in J} s_j - \sigma_1(X).$$

Hence,

$$\sigma_1(Y) = \sum_{i \notin I} r_i - \sum_{j \in J} s_j + \sigma_1(X). \tag{1.2.13}$$

Adding $\sigma_0(X)$ to both sides of (1.2.13) and taking into account the equality $\sigma_o(X)+\sigma_1(X) = kl = |I||J|$, we obtain

$$\sigma_0(X) + \sigma_1(Y) + t(I, J) \geq 0.$$

Let us prove that $2 \Rightarrow 3$. Let k be an arbitrary integer such that $1 \leq k \leq n$. Let $J = N_k$, and let I be the set of indices i such that $r_i > k$. Then it follows from assertion 2 that

$$\sum_{j=1}^{k} s_j \leq |I||J| + \sum_{i \notin I} r_i = k|I| + \sum_{i \notin I} r_i = \sum_{i=1}^{m} \min\{r_i, k\} = \sum_{j=1}^{k} \bar{s}_j.$$

Hence $S \prec \bar{S}$.

Take an arbitrary $k \in N_m$ and set $I = \{k\}$ and $J = N_n$. Then it follows from assertion 2 that

$$0 \leq n + \sum_{i \neq k} r_i - \sum_{j=1}^{n} s_j = n - r_k,$$

and assertion 3 is proved.

Let us prove that $3 \Rightarrow 1$. Let $S \prec \bar{S}$ and $r_k \leq n$, $k = 1, \ldots, m$. We give a method to construct a matrix A from the class $\mathfrak{A}(R, S)$ beginning with $A(R; n)$. The construction is illustrated by the example from Section 1.2.11 below. We use induction and suppose that the assertion is proved for all matrices with fewer than n columns. It follows from (1.2.10) and the monotonicity of S that $s_n \geq \bar{s}_n$ and $\bar{s}_1 \geq s_1 \geq s_n$. Therefore, in $A(R; n)$, we can take s_n rows with the greatest row sums and transfer the extreme right ones of these rows to the nth column. If we have to choose between several rows with the same row sum, let us agree to take the ones from the lowest rows. As a result, we arrive at a matrix of the form $(\bar{A}_{n-1} \vdots A_1)$, where A_1 is the column-vector with s_n ones and A_{n-1} is the maximal $m \times (n-1)$ matrix with monotone (because of our choice) vector of column sums. We denote this vector by $E = (e_1, \ldots, e_{n-1})$.

Let us show that $S_{n-1} = (s_1, \ldots, s_{n-1}) \prec E$. For this purpose, we consider the connection between the components of S and E. The ones in $A(R; n)$ are located as a connected set, and if we draw the boundary of this set, then in view of the monotonicity of $\bar{S}$ we obtain a line increasing from left to right with breaks at the points where neighbouring components of the vector $\bar{S}$ are not equal to one another. Therefore the matrix $A(R; n)$ is of the form

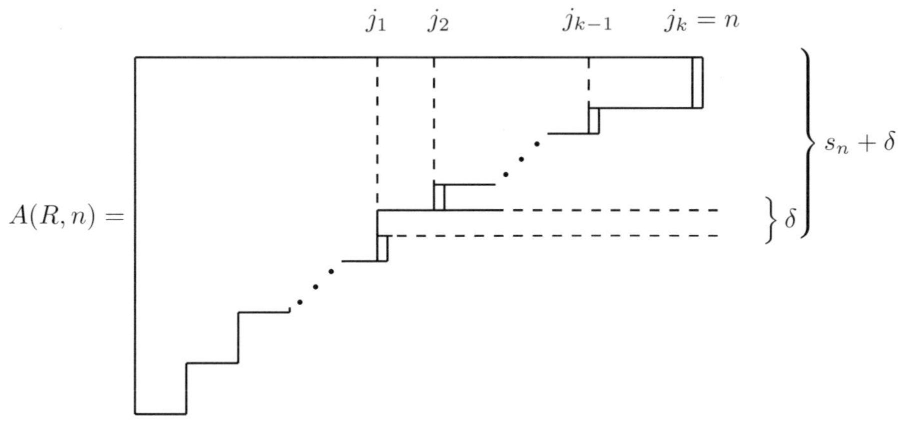

In this picture, the double vertical lines show the locations of the ones which have to be transferred to the nth column in order to form the column A_1. The number of the extreme left column from which an element equal to one is taken to the column A_1 is denoted by j_1. The locations of the breaks, beginning with j_1, are denoted by $j_1, j_2, \ldots, j_k = n$. Thus,

$$\bar{s}_{j_t+1} = \bar{s}_{j_t+2} = \cdots = \bar{s}_{j_{t+1}}, \qquad t = 1, \ldots, k-1. \tag{1.2.14}$$

We denote $\bar{s}_{j_1} - s_n$ by δ, that is,

$$s_n = \bar{s}_{j_1} - \delta, \tag{1.2.15}$$

where $0 \le \delta < \bar{s}_{j_1} - \bar{s}_{j_1+1}$. Further,

$$e_j = \bar{s}_j, \qquad j \ne j_1, j_2, \ldots, j_{k-1}. \tag{1.2.16}$$

Finally,

$$e_{j_t} = \begin{cases} \bar{s}_{j_{t+1}}, & t = 2, 3, \ldots, k-1, \\ \bar{s}_{j_{t+1}} + \delta, & t = 1. \end{cases} \tag{1.2.17}$$

The matrix $(\bar{A}_{n-1} \vdots A_1)$ contains the same number of ones as the matrix $A(R; n)$; therefore

$$e_1 + \cdots + e_{n-1} + s_n = \bar{s}_1 + \cdots + \bar{s}_n = s_1 + \cdots + s_n.$$

Hence, $e_1 + \cdots + e_{n-1} = s_1 + \cdots + s_{n-1}$, that is, condition (1.2.8) is fulfilled for E and S_{n-1}.

In order to prove the relation $S_{n-1} \prec E$, it suffices to show that (1.2.10) is fulfilled as well, that is,

$$e_j + e_{j+1} + \cdots + e_{n-1} \le s_j + s_{j+1} + \cdots + s_{n-1}, \qquad j = 2, 3, \ldots, n-1.$$

Consider two cases.

In the first case, $j_1 < j \le n-1$. Then there exists $t \ge 2$ such that $j_{t-1} < j \le j_t$. Since, according to (1.2.17), $e_{j_t} = \bar{s}_{j_{t+1}}$ for $t > 1$ and, according to (1.2.16), $e_j = \bar{s}_j$ for the remaining indices, we obtain

$$e_j + e_{j+1} + \cdots + e_{n-1} = \bar{s}_j + \bar{s}_{j+1} + \cdots + \bar{s}_{j_t-1} + \bar{s}_{j_t+1} + \cdots + \bar{s}_n.$$

On a segment between two neighbouring breaks, the components of $\bar{S}$ have the same value (see (1.2.14)); therefore

$$\bar{s}_j + \cdots + \bar{s}_{j_t-1} = \bar{s}_{j+1} + \cdots + \bar{s}_{j_t},$$

that is,

$$e_j + e_{j+1} + \cdots + e_{n-1} = \bar{s}_{j+1} + \bar{s}_{j+2} + \cdots + \bar{s}_n. \tag{1.2.18}$$

By the condition, $S \prec \bar{S}$; that is, according to (1.2.10),

$$\bar{s}_{j+1} + \cdots + \bar{s}_n \le s_{j+1} + \cdots + s_n.$$

By the monotonicity of S,

$$s_{j+1} + \cdots + s_n \le s_j + \cdots + s_{n-1},$$

and hence, in view of (1.2.18),

$$e_j + \cdots + e_{n-1} \le s_j + \cdots + s_{m-1};$$

that is, we obtain the required inequality (1.2.10).

In the second case, $1 \le j \le j_1$. Again by (1.2.16) and (1.2.17) we obtain

$$e_j + \cdots + e_{n-1} = \bar{s}_j + \cdots + \bar{s}_{j_1-1} + \delta + \bar{s}_{j_1+1} + \cdots + \bar{s}_n \tag{1.2.19}$$

Since $\delta = \bar{s}_{j_1} - s_n$ (see (1.2.15)) and $S \prec \bar{S}$, it follows from (1.2.19) that

$$\begin{aligned} e_j + \cdots + e_{n-1} &= \bar{s}_j + \cdots + \bar{s}_{j_1-1} + \bar{s}_{j_1} + \bar{s}_{j_1+1} + \ldots \bar{s}_n - s_n \\ &\le s_j + s_{j+1} + \cdots + s_n - s_n \\ &= s_j + s_{j+1} + \cdots + s_{n-1}, \end{aligned} \tag{1.2.20}$$

that is, again we obtain (1.2.10).

Thus, we have shown that $S_{n-1} \prec E$. The row sums of $\bar{A}_{n-1}$ do not exceed $n-1$. Therefore, by the induction hypothesis, beginning with the monotone maximal matrix $\bar{A}_{n-1}$, we can construct a $(0,1)$-matrix A_{n-1} of size $m \times (n-1)$ with the column sums $s_1, \ldots, s_{n-1}$ and the row sums equal to the corresponding row sums of $\bar{A}_{n-1}$. The matrix $\bar{A} = (A_{n-1} \vdots A_1)$ has the vector of column sums $S = (s_1, \ldots, s_n)$ and the vector of row sums $R = (r_1, \ldots, r_m)$. Thus, $A \in \mathfrak{A}^{m,n}(R,S)$, and the theorem is proved. □

1.2.11. An application of the nonemptiness criterion. The construction described in the proof of Theorem 1.2.8 shows the significance of the maximal matrix. This proof can be considered as an algorithm for constructing some canonical matrix $\tilde{A}$ of the class $\mathfrak{A}^{m,n}(R,S)$. Nonemptiness of this class is guaranteed by the condition $S \prec \bar{S}$, which can be checked directly. The number of steps of the algorithm equals $n-1$. At each step, one column of the matrix $\tilde{A}$ is constructed, beginning with the nth and ending with the second column. We illustrate the construction by the example where

$$m = n = 5, \quad R = (4,3,3,2,2), \quad S = (4,4,2,2,2).$$

In this case,

$$A(R;5) = \begin{pmatrix} 1 & 1 & 1 & 1 & 0 \\ 1 & 1 & 1 & 0 & 0 \\ 1 & 1 & 1 & 0 & 0 \\ 1 & 1 & 0 & 0 & 0 \\ 1 & 1 & 0 & 0 & 0 \end{pmatrix}$$

and $\bar{S} = (5,5,3,1,0)$. It is clear that $S \prec \bar{S}$.

The process of constructing the matrix $\tilde{A} \in \mathfrak{A}(R,S)$ starting from $A(R;5)$ is realized according to the algorithm as shown below:

$$\begin{pmatrix} 1 & 1 & 1 & 1 & 0 \\ 1 & 1 & 1 & 0 & 0 \\ 1 & 1 & 1 & 0 & 0 \\ 1 & 1 & 0 & 0 & 0 \end{pmatrix} \to \begin{pmatrix} 1 & 1 & 1 & 0 & \vdots & 1 \\ 1 & 1 & 1 & 0 & \vdots & 0 \\ 1 & 1 & 0 & 0 & \vdots & 1 \\ 1 & 1 & 0 & 0 & \vdots & 0 \\ 1 & 1 & 0 & 0 & \vdots & 0 \end{pmatrix} \to \begin{pmatrix} 1 & 1 & 0 & \vdots & 1 & 1 \\ 1 & 1 & 0 & \vdots & 1 & 0 \\ 1 & 1 & 0 & \vdots & 0 & 1 \\ 1 & 1 & 0 & \vdots & 0 & 0 \\ 1 & 1 & 0 & \vdots & 0 & 0 \end{pmatrix}$$

$$\to \begin{pmatrix} 1 & 1 & \vdots & 0 & 0 & 1 \\ 1 & 1 & \vdots & 0 & 1 & 0 \\ 1 & 1 & \vdots & 0 & 0 & 1 \\ 1 & 0 & \vdots & 1 & 0 & 0 \\ 1 & 0 & \vdots & 1 & 0 & 0 \end{pmatrix} \to \begin{pmatrix} 1 & \vdots & 1 & 0 & 1 & 1 \\ 1 & \vdots & 1 & 0 & 1 & 0 \\ 1 & \vdots & 1 & 0 & 0 & 1 \\ 1 & \vdots & 0 & 1 & 0 & 0 \\ 0 & \vdots & 1 & 1 & 0 & 0 \end{pmatrix}.$$

Consider another example. Let $m = 5$, $n = 4$, $R = (4,4,2,1,1)$, $S = (4,4,3,1)$. The matrix $A(R;4)$ is of the form

$$\begin{pmatrix} 1 & 1 & 1 & 1 \\ 1 & 1 & 1 & 1 \\ 1 & 1 & 0 & 0 \\ 1 & 0 & 0 & 0 \\ 1 & 0 & 0 & 0 \end{pmatrix},$$

that is, $\bar{S} = (5,3,2,2)$. The vector $\bar{S}$ does not dominate S, and therefore according to Theorem 1.2.8 the class $\mathfrak{A}(R,S)$ is empty.

As the next example we consider the class $\mathfrak{A}^{n,n}(R,S)$ of square matrices of order n with $R = (k,\dots,k)$ and $S = (k_1,\dots,k_n)$, where $k_1 + \dots + k_n = kn$, $1 \le k \le n$. In this case, the first k columns of $A(R;n)$ consist of ones and the remaining $n-k$ columns consist of zeros, that is, $\bar{S} = (n,\dots,n,0,\dots,0)$. It is clear that $S \prec \bar{S}$, that is, the class $\mathfrak{A}^{n,n}(R,S)$ is nonempty for any $k, k_1, \dots, k_n$. In particular, the class $\mathfrak{A}^n(k)$ with $k_1 = \dots = k_n$ is nonempty.

1.2.12. The Ryser class $\mathfrak{A}^n(k)$ and Latin configurations. In conclusion we consider the application of the results obtained in this section to Latin rectangles and matrices of the classes $\mathfrak{A}^n(k)$. As noted above, the class $\mathfrak{A}^n(k)$ is nonempty for any $k \in N_n$. In Section 1.1, we have defined a Latin configuration and have shown that the incidence matrix of such a configuration is a matrix of the class $\mathfrak{A}^n(k)$, where $k \times n$ is the size of the corresponding Latin rectangle constructed on the numbers of N_n. The Latin configurations are similar to doubly stochastic ones, and so they possess many properties similar to the properties of the doubly stochastic configurations.

THEOREM 1.2.9. *Any matrix of the class $\mathfrak{A}^n(k)$ contains a positive diagonal.*

The proof is similar to that of Theorem 1.2.4, and we leave it to the reader (see Problem 3). The following analogue of Theorem 1.2.5 is also true.

THEOREM 1.2.10. *Let $A = (a_{ij})$ be an arbitrary matrix of the class $\mathfrak{A}^n(r)$, $1 \leq r \leq n$. Then there exist permutation matrices $P_1, \ldots, P_r$ such that*

$$A = P_1 + \cdots + P_r. \tag{1.2.21}$$

PROOF. We use induction on r. For $r = 1$, the matrix A itself is a permutation matrix. We suppose that the theorem is proved for all $s < r$, and prove it for a matrix A in $\mathfrak{A}^n(r)$. By Theorem 1.2.9, such a matrix has a positive diagonal $(a_{1\pi(1)}, \ldots, a_{n\pi(n)})$, where π is a permutation on N_n. Let P be the permutation matrix corresponding to π. Then the matrix $A - P$ belongs to $\mathfrak{A}^n(r-1)$, and, consequently,

$$A - P = \sum_{i=1}^{r-1} P_i$$

for some permutation matrices $P_1, \ldots, P_{r-1}$. Hence

$$A = P + \sum_{i=1}^{r-1} P_i,$$

which proves the theorem. □

Theorem 1.2.10 gives a closer relation between matrices in $\mathfrak{A}^n(k)$ and Latin rectangles. Two Latin rectangles L and L' on the numbers of N_n are called equivalent if L' can be obtained from L by permutations of rows and columns and by renumbering the elements of the rectangle according to some permutation σ on N_n (that is, when each number $i \in N_n$ located in any place is replaced by $\sigma(i)$).

THEOREM 1.2.11. *To each $k \times n$ Latin rectangle L on the numbers of N_n, one can associate a Latin configuration with the incidence matrix $A(L)$ of the class $\mathfrak{A}^n(k)$. Conversely, using any Latin configuration on a set of n elements consisting of k-subsets, one can construct a $k \times n$ Latin rectangle on the numbers of N_n. In this case, equivalent configurations correspond to equivalent Latin rectangles.*

PROOF. Let a $k \times n$ Latin rectangle L on the numbers of N_n be given. In Section 1.1, it was shown that, taking the columns of L, we obtain a Latin configuration with incidence matrix $A \in \mathfrak{A}^n(k)$. A permutation of rows of L leads to L' with the same incidence matrix A, a permutation of columns leads to the corresponding permutation of rows of A, and a renumbering of the elements leads to a permutation of columns.

Conversely, let a Latin configuration $\mathcal{C} = (S, \mathcal{E}, \chi)$ consisting of k-subsets of the n-set S be given. As was shown in Section 1.1, the incidence matrix $A(\mathcal{C}) = A$ of the configuration $\mathcal{C}$ belongs to $\mathfrak{A}^n(k)$; that is, A contains k ones in each row and each column. By Theorem 1.2.10 there exists a representation

$$A = P_1 + \cdots + P_k \tag{1.2.22}$$

of the matrix A as a sum of permutation matrices. Consider a rectangular $k \times n$ table. In the ith row, $i = 1, \ldots, k$, we write sequentially the numbers of the columns, where the ones of the first, second, ..., nth row stand in the matrix P_i. It is clear that as a result we obtain a $k \times n$ Latin rectangle on the elements of N_n, since each row of the table is a permutation of the numbers $1, \ldots, n$ and each column of the table is a sample without repetitions from the set N_n. A change of the order of terms in (1.2.22) leads to a permutation of the rows of the rectangle. When we rearrange the rows of A according to some permutation π, this operation is equivalent to the multiplication of A from the left by P^{-1}, where P is the permutation matrix corresponding to π. Multiplying both sides of (1.2.22) by P^{-1} from the left, we obtain

$$P^{-1}A = P^{-1}P_1 + \cdots + P^{-1}P_k. \tag{1.2.23}$$

We denote $P^{-1}A$ by A' and $P^{-1}P_i$ by P_i', $i = 1, \dots, k$. If, starting with A', we construct a Latin rectangle with the use of (1.2.23) by the method described above, then the numbers in each row of L' are the numbers in the corresponding row of L rearranged by the permutation π, the same for each row. Thus, L' differs from L only by the permutation π of the columns. In the same way we can show that the rearrangement of the columns of A according to some permutation σ on N_n (that is, passing from A to AQ, where Q is the permutation matrix of σ) leads to a rectangle L'' which differs from L only by the rearrangement of the numbers contained in L according to the permutation σ. □

As is seen from the proof of Theorem 1.2.11, an $r \times n$ Latin rectangle on N_n can be obtained from the matrix $A \in \mathfrak{A}^n(r)$ directly, without considering the configuration $\mathcal{C}$, by writing out the permutation matrices $P_1, \dots, P_r$ generated by the rows of L. The inverse correspondence is not unique, due to nonuniqueness of the representation of a matrix as a sum of permutation matrices. Therefore the number of matrices in the class $\mathfrak{A}^n(r)$ is less than the number of Latin rectangles. This fact becomes clearly visible in the case of Latin squares. Indeed, each Latin square of order n can be obtained by some decomposition of the form (1.2.21) of only one matrix $\mathbf{J}_{nn}$ into a sum of n permutation matrices. The question of how many decompositions of the form (1.2.21) of the matrices $A \in \mathfrak{A}^n(r)$ are possible is of great interest in connection with enumerative problems for Latin rectangles.

1.2.13. Extension of Latin rectangles. Let an $r \times s$ Latin rectangle on the elements of N_n, $r, s \le n$, be given. The question is, when can it be extended to a Latin rectangle of greater size? In particular, this question arises in enumerating Latin rectangles. The answer can be found by applying the results obtained above.

THEOREM 1.2.12. *Any $r \times n$ Latin rectangle can be extended to a Latin square of order n.*

PROOF. Let $r < n$. We consider the rows of a Latin rectangle L as the second rows of the permutations $\pi_1, \dots, \pi_r$ on the set N_n. Let $P_1, \dots, P_r$ be the corresponding permutation matrices. The matrix $A = P_1 + \cdots + P_r$ belongs to the class $\mathfrak{A}^n(r)$. Therefore, $\mathbf{J} - A \in \mathfrak{A}^n(n-r)$. By Theorem 1.2.10 there exist permutation matrices $P_{r+1}, \dots, P_n$ such that $\mathbf{J} - A = P_{r+1} + \cdots + P_n$, or

$$\mathbf{J} = (P_1 + \cdots + P_r) + (P_{r+1} + \cdots + P_n). \tag{1.2.24}$$

Using the decomposition (1.2.24), we construct a Latin square by the method of Theorem 1.2.11. The first r rows form the rectangle L. The theorem is thus proved. □

The problem of extending an $r \times s$ rectangle is slightly more complicated in the case $s < n$. First we prove an auxiliary result, which generalizes Theorem 1.2.10. Take an n-set X and an integer $m \le n$. A one-to-one mapping $\sigma\colon N_m \to X$ (that is, an m-sample without repetitions from X) is called an m-permutation of X. If $m = n$, then an m-permutation is a usual permutation on the elements of X. To each m-permutation of an n-set $X = \{x_1, \dots, x_n\}$ we can put in correspondence an

m-permutation matrix, the $(0,1)$-matrix $P = (p_{ij})$ of size $m \times n$ such that $p_{ij} = 1$ if and only if $\sigma(i) = x_j$. Conversely, using an m-permutation matrix of size $m \times n$, we can determine an m-permutation on the set N_n in the same way as it is done for permutation matrices and permutations. Note that any m rows of a permutation matrix of order n form an m-permutation matrix.

THEOREM 1.2.13. *Let A be a $(0,1)$-matrix of size $m \times n$, $m < n$. If the vector of row sums of A is $R = (r, \dots, r)$ and the components of the vector of column sums $S = (s_1, \dots, s_n)$ satisfy the condition*

$$0 \le r - s_j \le n - m, \qquad j = 1, \dots, n, \tag{1.2.25}$$

then there exist m-permutation matrices $P_1, \dots, P_r$ of size $m \times n$ such that $A = P_1 + \cdots + P_r$.

PROOF. We consider the $(n-m)$-dimensional vector $R' = (r, \dots, r)$ and the n-dimensional vector $S' = (r - s_1, \dots, r - s_n)$. The vector $\bar{S}'$ of the column sums of the maximal matrix $A(R'; n)$ is $(n-m, \dots, n-m, 0, \dots, 0)$ with r nonzero components. From the condition (1.2.25) and the equality $mr = s_1 + \cdots + s_n$ it follows that $S' \prec \bar{S}'$. Therefore, by Theorem 1.2.8, the class $\mathfrak{A} = \mathfrak{A}(R', S')$ is nonempty. We take an arbitrary matrix $A' \in \mathfrak{A}$ and write it under the matrix A. It is clear that

$$A_1 = \begin{pmatrix} A \\ A' \end{pmatrix} \in \mathfrak{A}^n(r).$$

By Theorem 1.2.10 $A = Q_1 + \cdots + Q_r$, where the Q_i, $i = 1, \dots, r$, are permutation matrices. Taking the first m rows from each of these matrices, we obtain the required decomposition of A into a sum of m-permutation matrices. □

THEOREM 1.2.14. *Let L be an $r \times s$ Latin rectangle on the numbers $1, \dots, n$, and let $N(i)$ be the number of appearances of i in L, $i = 1, \dots, n$. The Latin rectangle L can be extended to a Latin square of order n if and only if*

$$N(i) \ge r + s - n. \tag{1.2.26}$$

PROOF. Let us prove the necessity. We denote by T_k the set of all elements of N_n which do not appear in the kth row of L, $k = 1, \dots, r$. Each T_k consists of $n - s$ elements. We denote by $M(i)$ the number of appearances of i in the subsets $T_1, \dots, T_r$. If L can be extended to a Latin square, then, obviously, $M(i) \le n - s$. It is clear that $M(i) + N(i) = r$. Therefore, $N(i) = r - M(i) \ge r + s - n$, that is, (1.2.26) is fulfilled.

Let us prove the sufficiency. Let (1.2.26) be fulfilled. Consider the incidence matrix A of the configuration $\{T_1, \dots, T_r\}$ on N_n. The matrix A is a $(0,1)$-matrix of size $r \times n$. The vector of row sums of A is $(n - s, \dots, n - s)$ and the vector of column sums is $(M(1), \dots, M(n))$. Note that the ones of the kth row of A are located only in the columns with numbers which are not contained in the kth row of the rectangle L, $k = 1, \dots, r$. Since $M(i) = r - N(i)$, by (1.2.26) we have $0 \le n - s - M(i)$, $i = 1, \dots, n$, and since L is a Latin rectangle, $N(i) = r - M(i) \le s$, or $s + M(i) \ge r$. Hence

$$0 \le n - s - M(i) \le n - r,$$

that is, the conditions of Theorem 1.2.13 are fulfilled for the matrix A. Thus, there exist r-permutation matrices $P_1, \dots, P_{n-s}$ such that $A = P_1 + \cdots + P_{n-s}$.

We denote by $\sigma_1, \dots, \sigma_{n-s}$ the r-permutations on N_n corresponding to the matrices $P_1, \dots, P_{n-s}$. Let σ be one of these r-permutations and $P = (p_{kj})$ the matrix corresponding to this r-permutation, that is, for $k = 1, \dots, r$,

$$p_{kj} = \begin{cases} 1, & j = \sigma(k), \\ 0, & j \neq \sigma(k). \end{cases}$$

Then $X = \{\sigma(1), \dots, \sigma(r)\}$ is an r-subset of N_n, and according to the property of the matrix A mentioned above, $\sigma(k)$ is the number which does not appear in the kth row of the rectangle L. We construct such r-subsets $X_j = \{\sigma_j(1), \dots, \sigma_j(r)\}$ for each P_j, $j = 1, \dots, n-s$. Adding to L the $n-s$ columns that are vector-columns with elements of the sets $X_j, j = 1, \dots, r$, we obtain the following $r \times n$ Latin rectangle on the elements of N_n, containing L:

$$\begin{pmatrix} & \sigma_1(1) & \sigma_2(1) & \dots & \sigma_{n-s}(1) \\ L & & & \dots & \\ & \sigma_1(r) & \sigma_2(r) & \dots & \sigma_{n-s}(r) \end{pmatrix}. \tag{1.2.27}$$

Indeed, in (1.2.27) the elements in each column are distinct and $(\sigma_1(k), \dots, \sigma_{n-s}(k))$ is a row consisting of pairwise distinct elements which are not contained in the kth row of L, $k = 1, \dots, r$. By Theorem 1.2.12 any Latin rectangle of size $r \times n$ can be extended to a Latin square of order n. □

In conclusion, we describe the connection between the global characteristics of the matrices considered in this section. For the family of matrices which constitute a nonempty monotone Ryser class $\mathfrak{A} = \mathfrak{A}(R, S)$ with $R = (r_1, \dots, r_m)$, $S = (s_1, \dots, s_n)$, we define the maximal term rank $\bar{\rho} = \bar{\rho}(\mathfrak{A})$ as the maximum $\rho(A)$ over all $A \in \mathfrak{A}$. Then

$$\bar{\rho} = \min_{i,j}(t_{ij} + i + j), \quad i = 0, 1, \dots, m, \quad j = 0, 1, \dots, n,$$

where $t_{ij} = t(N_i, N_j)$ (see (1.2.12)). The characteristic $\bar{\rho}$ is easily calculated by comparing $(m+1)(n+1)$ numbers t_{ij}. This result will be proved in Chapter 2.

Problems.

1. Consider two partitions of an n-set S into m pairwise distinct parts
$$S = \bigcup_{i=1}^{m} X_i = \bigcup_{i=1}^{m} Y_i, \quad |X_i \cap X_j| = |Y_i \cap Y_j| = 0, \quad i \neq j, \quad i, j = 1, \dots, m.$$
Prove that if all these parts consist of the same number of elements, then the families $\{X_1, \dots, X_m\}$ and $\{Y_1, \dots, Y_m\}$ have a common transversal.
2. Derive the König Theorem (Theorem 1.2.6) from the Ph. Hall Theorem (Theorem 1.2.7).
3. Prove Theorem 1.2.9.
4. Consider the following pairs of vectors with integer components:
   (a) $R = (4, 4, 2, 1)$, $S = (4, 2, 2, 2, 1)$,
   (b) $R = (3, 1, 5, 4)$, $S = (4, 1, 4, 3, 1)$,
   (c) $R = (2, 4, 1, 2, 2)$, $S = (3, 1, 5, 2)$,
   (d) $R = (4, 3, 3, 2)$, $S = (3, 3, 3, 3)$.
   Find the cases in which the class $\mathfrak{A}(R, S)$ is nonempty. For each nonempty class write out a $(0,1)$-matrix belonging to this class.

5. Let the $(0,1)$-matrix

$$A = \begin{pmatrix} 1 & 1 & 0 & 1 & 0 \\ 1 & 0 & 0 & 1 & 1 \\ 0 & 0 & 1 & 1 & 1 \\ 1 & 1 & 1 & 0 & 0 \\ 0 & 1 & 1 & 0 & 1 \end{pmatrix}$$

be given. Find a decomposition of A into a sum of permutation matrices of order 5. For this decomposition write out all equivalent Latin rectangles with elements of the first row arranged in the natural order (see Theorem 1.2.11).

6. Let L be a Latin rectangle of size $r \times n$ on the elements of N_n, and let S_i be the r-set consisting of the elements of the ith column of L, $i - 1, \dots, r$. What is the number of all extensions of L to a Latin rectangle of size $(r+1) \times n$?

7. Let the 3×4 Latin rectangle L on N_6 be given partially, that is,

$$L = \begin{pmatrix} 1 & 2 & 5 & 6 \\ 5 & 3 & * & 2 \\ * & 5 & 6 & 3 \end{pmatrix},$$

where the elements $*$ are not determined. Find the number of ways to allocate the numbers from N_6 so that the rectangle L can be extended to a Latin square of order 6. For one of these ways, write out the Latin square obtained by extending L.

1.3. Combinatorial theory of cyclic matrices

Circulant matrices, or circulants, which are also called cyclic matrices, have been defined in Section 1.1. They have many interesting properties. Circulants are widely used not only in combinatorial investigations, but in various applications including information theory (coding theory), and in studies of dynamic systems and crystal structures in physics. In this section, we investigate circulants from the point of view of combinatorial theory. Many results of this section concern not only nonnegative and $(0,1)$-matrices but, more generally, matrices over arbitrary fields or binary Boolean algebras.

1.3.1. g-circulants. We generalize the definition of a circulant given in Section 1.1. An $n \times n$ matrix is called a g-circulant matrix, or a g-circulant, if each row of the matrix is obtained by g cyclic shifts of the preceding row to the right. Here g may be an arbitrary integer, for example, zero (then all rows coincide), an integer greater than n (the number of shifts is taken modulo n), and even negative (then g shifts to the right means $|g|$ shifts to the left). If $g = 1$, we obtain the definition of a circulant given in Section 1.1; that is, a circulant is a 1-circulant according to the new definition.

It is clear that a circulant is determined by the first row. We denote by $C_g(a)$ the g-circulant of order n with the first row $a = (\alpha_1, \dots, \alpha_n)$. The matrix $C_g(a)$ is called the g-circulant generated by the vector a, which in turn is called the generating vector corresponding to the g-circulant. Often it is convenient to number the coordinates of generating vectors of order n not by the numbers $1, \dots, n$, but by $0, 1, \dots, n-1$. It is obvious that a characteristic property of a g-circulant matrix $A = (a_{ij})$ is the validity of the equalities

$$a_{ij} = a_{i-1,j-g}, \tag{1.3.1}$$

where i, j run over all numbers of rows and columns modulo n.

In the study of g-circulants, an important role is played by the $n \times n$ permutation matrix

$$P_n = \begin{pmatrix} 0 & 1 & 0 & \dots & 0 \\ 0 & 0 & 1 & \dots & 0 \\ & & & \dots & \\ 0 & 0 & 0 & \dots & 1 \\ 1 & 0 & 0 & \dots & 0 \end{pmatrix}. \tag{1.3.2}$$

Note that the lower index of P_n can be omitted if we consider matrices of a fixed order.

Let us introduce two operations on the set of square matrices of order n. For such a matrix A we will consider:

• the matrix obtained by the cyclic upward shift of all the rows of A (the first row becomes the last one), and

• the matrix obtained by the cyclic right shift of all columns of A (the last column becomes the first one).

It is easy to see that the first operation can be realized by the multiplication of the matrix P_n by A from the left and the second by the multiplication of P_n by A from the right.

THEOREM 1.3.1. *A matrix $A = (a_{ij})$ of order n is a g-circulant if and only if*

$$P_n A = A P_n^g, \tag{1.3.3}$$

where P_n is defined by (1.3.2).

PROOF. Indeed, the row with number $i-1$ in $P_n A$ is the row with number i in A. On the other hand, it is obvious that each row of AP_n^g is obtained from the corresponding row of A by g rightward cyclic shifts. Therefore, if (1.3.3) is true, then, comparing the elements in the entry $(i-1, j)$ of both sides of the equality, we obtain $a_{ij} = a_{i-1,j-g}$ for all $i, j \in N_n$; that is, (1.3.1) is true. The converse assertion is also obvious. □

1.3.2. Elementary properties. First we formulate the following evident but useful property of a 1-circulant A of order n: if $a = (\alpha_0, \alpha_1, \dots, \alpha_{n-1})$ is the generating vector of A, then

$$A = \sum_{k=0}^{n-1} \alpha_k P_n^k, \tag{1.3.4}$$

where P_n is defined by (1.3.2).

The following property concerns the transposed matrix A^T.

THEOREM 1.3.2. *If A is a g-circulant of order n and $(g, n) = 1$, then A^T is a g'-circulant, where g' is the smallest positive solution of the congruence $gx \equiv 1 \pmod{n}$.*

PROOF. Let A be a g-circulant and $(g, n) = 1$. Then $P_n A = A P_n$ by Theorem 1.3.1, and by transposing we obtain $A^T P_n^{-1} = P_n^{-g} A^T$. Hence,

$$A^T P_n^{-g'} = P_n^{-gg'} A^T,$$

and, finally,

$$P_n A^T = A^T P_n^{g'},$$

that is, A^T is a g'-circulant. □

COROLLARY 1.3.3. *If A is a circulant, then A^T is a circulant as well, and the generating vector of A^T is the generating vector of A written in the reverse order beginning with the first coordinate.*

Note that for a g-circulant A with $(g,n)>1$ the matrix A^T, in general, is not a circulant. For example, consider the 2-circulant

$$A=\begin{pmatrix} a & b & c & d\\ c & d & a & b\\ a & b & c & d\\ c & d & a & b \end{pmatrix}.$$

It is clear that if, for example, $a\neq b,d$, then A^T is not a g-circulant for any g.

THEOREM 1.3.4. *If A is a g-circulant of order n and B is an h-circulant of order n, then AB is a gh-circulant.*

PROOF. It follows from the conditions of the theorem and Theorem 1.3.1 that

$$P_nA=AP_n^g, \qquad P_nB=BP_n^h.$$

The second equality means that

$$P_n^gB=BP_n^{gh}.$$

Therefore,

$$P_nAB=AP_n^gB=ABP^{gh};$$

that is, AB is a gh-circulant by Theorem 1.3.1. □

Let $C=(c_{ij})$ be a circulant of order n with generating vector $c=(c_{11},\dots,c_{1n})$. The period of C is the minimal positive integer m such that after m cyclic shifts to the right of the coordinates of c we obtain the same vector c. In other words, the period of a circulant C of order n is the minimal positive integer m such that

$$cP_n^m=c, \tag{1.3.5}$$

where c is the generating vector of C. It is clear that (1.3.5) is equivalent to the condition

$$CP_n^m=C \tag{1.3.6}$$

for the circulant C.

Obviously, the period divides the order n of the circulant and either coincides with n or is at most $n/2$. The circulant with period equal to n is called primitive.

Let $A=(a_{ij})$ be a matrix of order q and $B=(b_{ij})$ a matrix of order m. We denote by $A\times B$ the Kronecker product of A by B, that is, the matrix

$$\begin{pmatrix} a_{11}B & a_{12}B & \dots & q_{1q}B\\ a_{21}B & a_{22}B & \dots & 2_{2q}B\\ & & \dots & \\ a_{q1}B & a_{q2}B & \dots & a_{qq}B \end{pmatrix};$$

the order of this matrix is qm.

THEOREM 1.3.5 ([**120**]). *Let the period of a circulant $C=(c_{ij})$ of order n be equal to m. Then $C=\mathbf{J}\times D$, where $q=n/m$ and D is a primitive circulant of order m.*

PROOF. The first row of the circulant C of period m is, obviously, the n/m times repeated sequence $(c_{11}, \dots, c_{1m})$, since (1.3.5) implies that $c_{1j} = c_{1,j-m}$. By the definition of a circulant, any other row of C has the same property. Further, from (1.3.6) and Corollary 1.3.3 it follows that

$$C^T = (CP^m)^T = P^{-m}C^T = C^T P^{-m}.$$

Thus, we have the equality

$$C^T P^m = C^T,$$

which means that each column vector of C is the n/m times repeated sequence of the first m coordinates of this vector. We denote by D the matrix $(c_{ij})_{i,j=1}^m$. It is clear that D is a circulant of order m, and C can be written in the form $\mathbf{J} \times D$, where $q = n/m$. The circulant D is primitive, since otherwise m is not the minimal positive integer which satisfies the equality $cP^m = c$. The theorem is thus proved. □

Thus, we see that the study of circulants is reduced in great part to the study of primitive circulants.

1.3.3. Eigenvalues and eigenvectors. For a cyclic matrix, we can easily solve the problem of finding the canonical form over the field of complex numbers, including finding the eigenvalues and eigenvectors.

THEOREM 1.3.6. *Let $A = C(a)$ be a circulant of order n with generating vector $a = (\alpha_0, \alpha_1, \dots, \alpha_{n-1})$, and let ζ be a primitive root of degree n of 1. Then the numbers*

$$\sum_{i=o}^{n-1} \alpha_i \zeta^{ih}, \qquad h = 0, 1, \dots, n-1, \tag{1.3.7}$$

form the set of eigenvalues of the matrix A.

PROOF. Consider the vectors $c_0, c_1, \dots, c_{n-1}$, where

$$c_h = (1, \zeta^h, \dots, \zeta^{h(n-1)}), \qquad h = 0, 1, \dots, n-1. \tag{1.3.8}$$

First we find the eigenvalues of the matrix P_n. It is clear that

$$P_n c_h^T = \zeta^h c_h^T, \qquad h = 0, 1, \dots, n-1; \tag{1.3.9}$$

that is, ζ^h, $h = 0, 1, \dots, n-1$, are the eigenvalues of P_n. Since ζ is a primitive root of degree n of 1, all degrees ζ^h, $h = 0, 1, \dots, n-1$, are distinct and the vectors $c_0, c_1, \dots, c_{n-1}$ are linearly independent. Thus, they form the set of all eigenvectors of P_n. Further, from (1.3.9) it follows that

$$P_n^k c_h^T = \zeta^{kh} c_h^T, \qquad k = 2, 3, \dots.$$

Using the representation (1.3.4), we obtain

$$Ac_h^T = \left(\sum_{k=0}^{n-1} \alpha_k P_n^k \right) c_h^T = \left(\sum_{k=0}^{n-1} \alpha_k \zeta^{kh} \right) c_h^T. \tag{1.3.10}$$

The vectors $c_0, c_1, \dots, c_{n-1}$ are linearly independent; therefore by (1.3.10) they form the set of all eigenvectors of A, and consequently, the numbers

$$\sum_{k=0}^{n-1} \alpha_k \zeta^{kh}, \qquad h = 0, 1, \dots, n-1,$$

form the set of eigenvalues of A. The theorem is proved. □

Consider the matrix $U = (c_0^T, c_1^T, \dots, c_{n-1}^T)$, where the c_h are defined by (1.3.8). The columns of U are the eigenvectors of the circulant A of order n. The linear independence of these vectors implies the nonsingularity of U. Let $(\alpha_0, \alpha_1, \dots, \alpha_{n-1})$ be the generating vector of A, and let

$$\lambda_h = \sum_{k=0}^{n-1} \alpha_k \zeta^{kh}, \qquad h = 0, 1, \dots, n-1.$$

Taking into account (1.3.10) and the action of A on the columns $c_0^T, c_1^T, \dots, c_{n-1}^T$ of the matrix U, we obtain $AU = UD$, where $D = \operatorname{diag}[\lambda_0, \lambda_1, \dots, \lambda_{n-1}]$, or

$$U^{-1}AU = D. \tag{1.3.11}$$

Thus, the canonical form (1.3.11) of the circulant A is determined directly by the generating vector.

1.3.4. Automorphism groups. We denote by $\operatorname{Aut}_p A$ the group of permutational automorphisms of the matrix A (see Section 1.1, Theorem 1.1.2). It follows from the definition of the circulant C that $\operatorname{Aut}_p C$ contains the cyclic subgroup generated by the permutation matrix $P = P_n$, where n is the order of C, with the generating element $\pi = (P, P)$. Indeed, by the characteristic property of the circulant C (see (1.3.3) with $g = 1$) we have

$$\begin{aligned}\pi^k(C) &= P^{-k}CP^k = P^{-(k-1)}(P^{-1}CP)P^{k-1}\\ &= P^{-(k-1)}CP^{k-1} = \cdots = C, \qquad k = 0, 1, \dots, n-1.\end{aligned}$$

Thus, the description of the group $\operatorname{Aut}_p(C)$ for the circulant C should begin with the question of whether C has automorphisms different from powers of π. Below we introduce a class of circulants which have such automorphisms.

An n-dimensional vector $c = (\gamma_1, \dots, \gamma_n)$ is called symmetric if c written in the reverse order coincides with c up to a cyclic shift, that is, there exists $k \in N_n$ such that

$$(\gamma_n, \gamma_{n-1}, \dots, \gamma_1) = (\gamma_k, \gamma_{k+1}, \dots, \gamma_{k-1}). \tag{1.3.12}$$

It is clear from the definition that if c is symmetric, then any vector cP^h, $h = 1, 2, \dots, n$, is symmetric as well. Thus, a vector $c = (\gamma_1, \dots, \gamma_n)$ is symmetric if and only if

$$c^\tau = cP^k \tag{1.3.13}$$

for some $k \in N_n$, where

$$c^\tau = (\gamma_1, \gamma_n, \gamma_{n-1}, \dots, \gamma_2). \tag{1.3.14}$$

Therefore, we obviously see that

$$C^\tau = CP^k \tag{1.3.15}$$

for the circulant C with the symmetric generating vector c, where C^τ denotes the circulant with the generating vector c^τ defined by (1.3.14). Note that, according to Corollary 1.3.3, $C^\tau = C^T$.

Let us show that circulants with symmetric generating vectors are somewhat similat to symmetric matrices.

THEOREM 1.3.7. *If the generating vector c of a circulant C satisfies* (1.3.15) *with even k, then the matrix $CP^{k/2}$ is symmetric.*

PROOF. Since C^T is the circulant with generating vector c^τ, it follows from (1.3.13) that

$$C^T = CP^{2(k/2)},$$

that is,

$$C^T P^{-k/2} = CP^{k/2}$$

and

$$(CP^{k/2})^T = CP^{k/2}.$$

The theorem is thus proved. □

COROLLARY 1.3.8. *If C is a circulant of odd order n with symmetric generating vector, then C is permutationally equivalent to a symmetric circulant matrix.*

Indeed, for an odd n the congruence $2x \equiv k \pmod{n}$ has a unique solution in N_n for any $k \in N_n$, and the assertion follows directly from the theorem, since any cyclic shift of the columns of a circulant gives a circulant.

A dihedral group is a finite group G of order $2n$ with generators a of order n and b of order 2 which satisfy the relation

$$bab = a^{-1}.$$

The cyclic subgroup $\{a\}$ is a normal subgroup of G. A dihedral group of order $2n$ exists for any $n \geq 2$.

THEOREM 1.3.9. *Let the period of the generating vector $c = (\gamma_1, \dots, \gamma_n)$ of a circulant C of order n be equal to n. Then $\operatorname{Aut}_p C$ contains a subgroup isomorphic to the dihedral group of order $2n$ with the normal subgroup generated by the automorphism $\pi = (P, P)$ if and only if c is a symmetric vector.*

PROOF. Since C is a primitive circulant, all rows of C are pairwise distinct, and so are the columns. Therefore according to Theorem 1.1.3 its group of permutational automorphisms is isomorphic to the group of admissible permutations of columns. In particular, the automorphism π can be represented as the operation of multiplication of C by P.

Suppose that the generating vector c of the circulant C is symmetric. Then by (1.3.15) we have

$$C^T = CP^h \tag{1.3.16}$$

for some $h \in N_n$; that is, the operation of transposition yields the matrix that can be obtained from C by an admissible permutation of columns. It follows from (1.3.16) that in the case under consideration the transposition is a permutational automorphism of the circulant C. The same is true for any circulant CP^k, $k \in N_n$. If we denote this automorphism by τ and take into account that $\pi(C) = CP$, then

$$\tau\pi\tau(C) = \tau\pi(C^T) = \tau(C^T P) = P^{-1}C = CP^{-1} = \pi^{-1}(C),$$

that is, $\tau\pi\tau = \pi^{-1}$. Since C is primitive, the order of π equals n. Thus, the group $\operatorname{Aut}_p C$ contains the dihedral group of C of order $2n$ with generators π, τ.

Conversely, let $\operatorname{Aut}_p C$ contain a subgroup isomorphic to the dihedral group with cyclic normal subgroup $\{\pi\}$; that is, let there exist an element $\eta \in \operatorname{Aut}_p C$ such that $\eta^2 = e$ (e is the identity element of the group) and $\eta\pi\eta = \pi^{-1}$. We will consider the permutational automorphisms of the circulant C as permutations of indices of the columns of C; in particular, let $\pi = (12\dots n)$, $\pi^{-1} = (1n(n-1)\dots 2)$. Now if $\eta(1) = k \neq 1$, then $\pi^{-k}\eta(1) = 1$. If we denote $\pi^{-k}\eta$ by η', then we see that $(\eta')^2 = e$ and $\eta'\pi\eta' = \pi^{-1}$. Thus, we can take such an element η' as η, and therefore from the beginning we can assume that $\eta(1) = 1$. Further, it is easy to see that a permutation η on the set N_n such that $\eta^2 = e$, $\eta(1) = 1$, and $\eta\pi\eta = (1n\dots 2)$, is determined uniquely. Indeed, for any $i \in N_n$, $i \neq 1$, we have

$$\eta(\eta(i) + 1) = i - 1,$$

that is,

$$\eta(i) - \eta(i-1) \equiv -1 \pmod{n}.$$

Hence, representing the permutation by transpositions, we obtain

$$\eta = (2\,n)(3\,(n-1))\dots(k\,(n+2-k))\dots.$$

Since η should be an admissible permutation of columns, C contains a row of the form $c^\tau = (\gamma_1, \gamma_n, \dots, \gamma_2)$. If this row has the index t, then

$$(\gamma_{n-t+2}, \gamma_{n-t+3}, \dots, \gamma_{n-t+1}) = (\gamma_1, \gamma_n, \dots, \gamma_2).$$

Thus, the generating vector c of the circulant C is symmetric (see (1.3.12) with $n - t + 3 = k$). □

1.3.5. Cases of the coincidence of $\operatorname{Aut}_p C$ **with a cyclic group and a dihedral group.** In the case of $(0,1)$-circulants, it is possible to single out subclasses of cyclic matrices whose groups of permutational automorphisms are isomorphic to cyclic groups or dihedral groups. We introduce some additional notation.

We denote by $\Gamma = (\gamma_{ij})$ the matrix of the inner products $\gamma_{ij} = (a_i, a_j)$, $i, j = 1, 2, \dots, m$, where $a_1, \dots, a_m$ are row vectors of a matrix A of size $m \times n$. Obviously, Γ is a symmetric matrix of order m. If η is a function defined on the elements of the matrix $A = (\alpha_{ij})$, then along with the notation $\eta(\alpha_{ij})$ we use the notation α_{ij}^η, the notation A^η for the matrix $((\eta(\alpha_{ij}))$, and the notation a_i^η for the ith row of the matrix A^η. For example, if $a_i = (\alpha_{i1}, \dots, \alpha_{in})$ and η is a permutation of rows of the matrix $A = (\alpha_{ij})$, then $a_i^\eta = (\alpha_{\eta^{-1}i,1}, \dots, \alpha_{\eta^{-1}i,n})$.

LEMMA 1.3.10. *If η is an admissible permutation of rows of the matrix A with row vectors $a_1, \dots, a_m$, then the elements of the matrix Γ do not change under the action of η; that is, $\gamma_{ij}^\eta = \gamma_{ij}$, where γ_{ij}^η is the inner product of the ith and jth rows of the matrix A^η.*

PROOF. If η is an admissible permutation of rows of the matrix A, then there exists a permutation of the columns of the matrix A^η which transfers A^η to A. But a permutation of columns cannot change the inner products of rows. Thus,

$$\gamma_{ij}^\eta = (a_i^\eta, a_j^\eta) = (a_i, a_j) = \gamma_{ij}$$

for any admissible permutation η of rows. □

THEOREM 1.3.11. *Let C be a primitive $(0,1)$-circulant of order $n > 2$ with row vectors $c_1, \dots, c_n$, $\gamma_{1i} = (c_1, c_i)$, $i = 1, \dots, n$, and let there exist an index $t \neq 1$ such that $\gamma_{1t} \neq \gamma_{1k}$ for $k \neq t$, $k \neq n+2-t$ and $(t-1, n) = 1$. If the generating vector c_1 of the circulant C is not symmetric, then $\operatorname{Aut}_p C$ is isomorphic to the cyclic group of order n generated by the automorphism $\pi = (P, P)$. If c_1 is symmetric, then $\operatorname{Aut}_p C$ is isomorphic to the dihedral group of order $2n$ introduced in Theorem 1.3.9.*

PROOF. Since C is primitive, all rows and columns of C are pairwise distinct and $\mathrm{Aut}_p C$ is isomorphic to the group of admissible permutations of rows. In turn, the elements of this group can be considered as permutations on the set of the numbers of rows. Let η' be an admissible permutation of rows and $\eta'(1) = r \neq 1$. Since $\pi \in \mathrm{Aut}_p C$ and π can be represented as $\pi\colon C \to P_n C$, we have $\pi(1) = r - 1$, $\pi^r(r) = 1$, and $\pi^r \eta'(1) = 1$. We set $\pi^r \eta' = \eta$. Since $\eta(1) = 1$, either $\eta(t) = t$ or $\eta(t) = n + 2 - t$. Indeed, by Lemma 1.3.10

$$(c_1, c_t) = \gamma_{1t} = \gamma_{1t}^{\eta} = (c_1^{\eta}, c_t^{\eta}) = (c_{\eta^{-1}(1)}, c_{\eta^{-1}(t)}) = (c_1, c_{\eta^{-1}(t)}) = \gamma_{1k},$$

where $k = \eta^{-1}(t)$, and, similarly,

$$(c_1, c_{n+2-t}) = (c_1, c_{\eta^{-1}(n+2-t)}) = \gamma_{1l},$$

where $l = \eta^{-1}(n + 2 - t)$. Thus, according to the conditions of the theorem there exist only two possibilities:

(a) $k = t$, or
(b) $k = n + 2 - t$, $l = t$.

Note that for the elements of the matrix $\Gamma = (\gamma_{ik})$, which are the inner products of rows of the circulant, the equalities

$$\gamma_{r,r+t-1} = \gamma_{1t}, \qquad r = 1, 2, \ldots, n,$$

hold. Indeed, under the same shifts of two rows their inner product does not change, that is,

$$(c_r, c_{r+t-1}) = (c_1, c_t), \qquad r = 1, 2, \ldots, n.$$

Hence

$$\eta(r + t - 1) - \eta(r) \equiv \pm(t - 1) \pmod{n}, \qquad r = 1, 2, \ldots, n, \tag{1.3.17}$$

since by Lemma 1.3.10

$$\gamma_{1t} = \gamma_{r,r+t-1} = (c_r, c_{r+t-1}) = (c_{\eta(r)}, c_{\eta(r+t-1)}) = (c_1, c_s) = \gamma_{1s},$$

where

$$s \equiv \eta(r + t - 1) - (\eta(r) - 1) \pmod{n},$$

and therefore $\eta(r+t-1) - (\eta(r) - 1)$ is congruent either to t or to $n+2-t$ modulo n. That is, (1.3.17) is fulfilled.

First we consider alternative (b). Let

$$\eta(t) = n + 2 - t, \qquad \eta(n + 2 - t) = t,$$

that is, η transposes the rows with numbers t and $n + 2 - t$. The equality (1.3.17) with $r = t$ implies that $\eta(2t - 1) \equiv n + 2 - (2t - 1) \pmod{n}$, since $\eta(2t - 1) \neq 1$ in view of $t \neq 1$. Suppose now that the congruence

$$\eta(1 + u(t - 1)) \equiv n + 2 - (u(t - 1) + 1) \pmod{n} \tag{1.3.18}$$

is true for $u = 3, 4, \ldots, d - 1$, and prove it for $u = d$. As has been proved above (see (1.3.17)),

$$\eta(1 + d(t - 1)) \equiv n + 2 - ((d - 1)(t - 1) + 1) \pm (t - 1) \pmod{n},$$

that is, either

$$\eta(1 + d(t - 1)) \equiv n + 2 - ((d - 2)(t - 1) + 1) + 1) \pmod{n}$$

or

$$\eta(1 + d(t - 1)) \equiv n + 2 - (d(t - 1) + 1) \pmod{n}.$$

Since $n > 2$, only the second alternative can be realized. Indeed, in the first case we obtain a contradiction, because, by the induction assumption,

$$,\eta(1 + d(t-1)) \equiv \eta(1 + (d-2)(t-1)) \pmod n$$

and hence $2(t-1) \equiv 0 \pmod n$ and $2 \equiv 0 \pmod n$, so that $(t-1, n) = 1$. Thus, (1.3.18) is proved. Any h, $1 < h \leq n$, can be represented as $1 + d(t-1)$ with an appropriate d which satisfies the congruence $1 + d(t-1) \equiv h \pmod n$. Therefore, for any $h \in N_n$, $h \neq 1$, the equality $\eta(h) = n + 2 - h$ is true. Thus,

$$\eta = (2n)(3(n-1))\dots(h(n+2-h))\dots;$$

that is, η is the permutation of the second order from Theorem 1.3.9. Therefore, $\eta' = \pi^{-r}\eta$, that is, under the condition (b) an arbitrary permutational automorphism belongs to a subgroup isomorphic to the dihedral group generated by the elements η and π. By Theorem 1.3.9 the generating vector of the circulant is symmetric. This means that for a nonsymmetric generating vector alternative (b) is impossible.

Consider alternative (a). In this case, $\eta(t) = t$. By similar but slightly simpler reasoning than in case (b) we obtain that $\eta' = \pi^{-r}$. Thus, for a nonsymmetric generating vector the corresponding automorphism belongs to the cyclic group $\{\pi\}$. The theorem is proved. □

Note that all conditions of the theorem are essential. For example, consider the circulant C of order 8 with the generating vector $(1, 1, 1, 1, 0, 0, 1, 0, 0)$. For this circulant, the sequence of inner products $(\gamma_{11}, \dots, \gamma_{18})$ is equal to $(4, 2, 1, 2, 2, 2, 1, 2)$. Taking $t = 3$, we obtain $(t-1, n) > 1$, and so one of the conditions of Theorem 1.3.9 is violated. It is easy to see that the permutation of rows $\theta = (24)(37)(68)$ is admissible. It is obvious that θ does not belong to the dihedral group generated by the elements $\eta = (28)(37)(46)$ and $\pi = (1\,2\,3\,4\,5\,6\,7\,8)$, since all elements of order 2 of the dihedral group are of the form $\pi^r\eta$ and, except the element η itself, contain the involution of the form $(1r)$, $r \neq 1$. On the other hand, for the circulant of order 10 with the generating symmetric vector $(1, 1, 1, 0, 0, 0, 1, 0, 0, 0)$ the dihedral group is the group of all permutational automorphisms of the circulant. In order to prove this, we can use Theorem 1.3.9 with $t = 4$. Indeed,

$$(\gamma_{11}, \dots, \gamma_{1,10}) = (4, 2, 1, 0, 2, 2, 2, 0, 1, 2)$$

and $\gamma_{14} = \gamma_{18} = 0$ differ from all remaining inner products.

COROLLARY 1.3.12. *The group of automorphisms of the circulant $C^{(n)}$ of order n with the generating vector consisting of $r \geq 2$ sequentially placed ones and of zeros in the remaining places is isomorphic to the dihedral group if $n - r \geq 2$, and also if $n = 3$ and $r = 2$.*

Indeed, the generating vector of $C^{(n)}$ is symmetric, has period n, and the conditions of Theorem 1.3.11 on the inner products for the values of r and n from the corollary are fulfilled for $t = 2$.

1.3.6. Circulants and difference sets. The circulants are related to interesting combinatorial configurations called symmetric block designs and difference sets.

Let C be a balanced incomplete block design with parameters v, b, r, k,λ (see Section 1.1.8). The inequality

$$v \leq b,$$

called the Fisher inequality, holds (see [**87**], [**108**] or [**48**]). If $v = b$, then such a block design is called symmetric. Symmetric block designs are studied intensively in combinatorial mathematics and have numerous applications, especially in coding theory and in the theory of experiment design.

Let v be a positive integer. A subset D of the set $Z_v = \{0, 1, \dots, v-1\}$ is called a difference set if any nonzero element $x \in Z_v$ can be represented exactly λ times in the form

$$x \equiv d - d' \pmod{v}, \qquad d, d' \in Z_v.$$

If D consists of k elements, then v, k, λ are called the parameters of the difference set. In order to exclude trivial cases, it is usually assumed that $k - \lambda \geq 2$.

Let $D = \{d_1, \dots, d_k\}$ be a difference set for Z_v. Take $v-1$ subsets of the form $D+i = \{d_1+i, \dots, d_k+i\}$, adding the number i to each element of D and reducing the results modulo v, $i = 1, 2, \dots, v-1$. It is easy to see that if we take the sets $D, D+1, \dots, D+v-1$ as blocks, then we obtain a symmetric block design with parameters v, k, λ. Indeed, if $a, b \in Z_v$, $a \neq b$, then the inclusion $\{a, b\} \in D+i$ for some i is equivalent to the congruences $a \equiv d_{k_1} + i \pmod{v}$, $b \equiv d_{k_2} + i \pmod{v}$, $d_{k_1}, d_{k_2} \in D$, or

$$a - b \equiv d_{k_1} - d_{k_2} \pmod{v}. \tag{1.3.19}$$

On the other hand, if (1.3.19) is fulfilled for given $a, b \in Z_v$ and some $d_{k_1}, d_{k_2} \in D$, then the block $D+i$ which contains a, b is determined uniquely by the condition $i \equiv a - d_{k_1} \pmod{v}$ (or, what is equivalent, $i \equiv b - d_{k_2} \pmod{v}$). Thus, any pair of distinct elements $\{a, b\}$ belongs to exactly λ subsets of the form $D+i$.

Thus, each difference set D in Z_v is associated with some symmetric block design. Such block designs differ from the other symmetric block designs by the structure of their incidence matrices. If the elements of $Z_v = \{0, 1, \dots, v-1\}$ and the blocks of the design $D, D+1, \dots, D+v-1$ are arranged as they were written above, then the incidence matrix C of such configuration is a circulant of order v: the inclusion $a \in D+i$ implies that $a+1 \in D+i+1$. Thus, a difference set gives a cyclic configuration, that is, a configuration whose incidence matrix is permutationally equivalent to a cyclic matrix.

Consider some examples of difference sets. The set $D = \{0, 1, 3\}$ in Z_7 is a difference set with parameters $v = 7$, $k = 3$, $\lambda = 1$. It gives a finite projective plane of order 2. In the same way the sets $D_1 = \{0, 1, 3, 9\}$ in Z_{13} and $D_2 = \{0, 1, 4, 14, 16\}$ in Z_{21} are difference sets with $\lambda = 1$ and give projective planes of order 3 and 4 respectively. The set $D_3 = \{0, 1, 2, 4, 5, 8, 10\}$ gives an example of a difference set in Z_{15} with $\lambda = 3$.

Like any cyclic configuration, a block design corresponding to a difference set in Z_v has a cyclic automorphism group of order v. It is easy to verify that the incidence matrix A of a block design satisfies the fundamental matrix equation

$$AA^T = (k - \lambda)\mathbf{I}_v + \lambda \mathbf{J}_{vv}, \tag{1.3.20}$$

where v, k, λ are the parameters of the difference set. It is clear that any $(0,1)$-circulant satysfying (1.3.20) can be considered as the incidence matrix of a block design in Z_v with cyclic automorphism group. Indeed, we can label the rows and the columns of C by the numbers $0, 1, \dots, v-1$ and take the set of numbers of the columns containing 1 in the first row as the basic block.

In particular, the circulant solution of (1.3.20) with $\lambda = 1$ gives the incidence matrix of a finite projective plane with v points (see [**87**] or [**48**]). This finite projective plane is called cyclic, since it has a cyclic automorphism group of order v.

Note that along with classical difference sets based on the additive cyclic group Z_v, difference sets based on an arbitrary group can be defined in perfect analogy. In contrast to the latter sets, the difference sets introduced above are called cyclic.

1.3.7. Multipliers of a difference set. Difference sets D and D' are called equivalent if there is an $h \in Z_v$ such that the corresponding cyclic incidence matrices C and C' satisfy the equality $C' = CP^h$, where P is the permutation matrix defined in (1.3.2). If D is a difference set in Z_v and C is the corresponding $(0,1)$-circulant, then $D+l$ for any $l \in Z_v$ is the difference set equivalent to D: if C_l is the circulant corresponding to $D+l$, then

$$C_l = CP^l.$$

Let $D = \{d_1, \dots, d_k\}$ be a difference set in Z_v, and let $t \in Z_v$, $(t, v) = 1$. Then, as is easily seen, the set tD is also a difference set with the same parameters. If for $t \in Z_v$ there exists $s \in Z_v$ such that $tD \equiv D + s \pmod v$, then t is called a multiplier of the difference set D. It is clear that for the corresponding incidence matrices $C^{(t)}$ and C the equality $C^{(t)} = CP^s$ holds. The multiplier t generates an automorphism of the symmetric block design corresponding to the difference set D. The multipliers play a significant role in constructing difference sets. All known difference sets have a nontrivial multiplier $t \equiv 1 \pmod v$.

THEOREM 1.3.13. *A $(0,1)$-circulant C with symmetric generating vector c cannot be the incidence matrix of a cyclic block design.*

PROOF. Let C be a primitive $(0,1)$-circulant of order v with symmetric generating vector c. We number the rows and the columns of C by $0, 1, \dots, v-1$. Let $c = (\gamma_0, \gamma_1, \dots, \gamma_{v-1})$. Then, for some $h \in Z_v$,

$$c^{\tau} = cP^h, \tag{1.3.21}$$

where $c^{\tau} = (\gamma_0, \gamma_{v-1}, \gamma_{v-2}, \dots, \gamma_1)$ (see (1.3.3) and (1.3.14)). The columns are numbered by $0, 1, \dots, v-1$; thus, if there exists a difference set $D = \{d_1, \dots, d_k\}$, where d_i is the number of the column containing 1 of the vector c, $i = 1, \dots, k$, then there exists also a difference set $-D$ corresponding similarly to c^{τ}. Then it follows from (1.3.21) and the equivalent equality $C^{\tau} = CP^h$ that -1 is a factor of D. We obtain a contradiction, since, as it is shown in [**18**], there is no difference set with factor -1.

Since the incidence matrix of a nontrivial (that is, with $k > \lambda$) symmetric block design cannot contain two equal rows, whereas a nonprimitive circulant contains two equal rows, the assertion of the theorem is proved. □

COROLLARY 1.3.14. *If a cyclic $(0,1)$-matrix C is such that CP^h is a symmetric matrix for some $h \in Z_v$, then C cannot be the incidence matrix for any cyclic block design.*

For an extended discussion of difference sets, see [**48**] and [**11**].

1.3.8. Equations for circulants. As we mentioned in Section 1.3.6, the circulant solution of equation (1.3.20) with $\lambda = 1$ gives a finite cyclic projective plane. However, g-circulant solutions of matrix equations similar to (1.3.20) are of interest not only for the problems on existence and construction of block design, but also for the study of other combinatorial configurations, in particular, for graph theory.

Consider the equation

$$A^m = d\mathbf{I} + \lambda\mathbf{J} \tag{1.3.22}$$

with the right-hand side as in (1.3.20); here m is a positive integer, while d and λ are rational numbers. Our interest is with rational g-circulant solutions of this equation, and primarily with solutions in the form of cyclic $(0, 1)$-matrices. It follows from the properties of g-circulants that the parameters m, λ, d cannot take arbitrary values. For example, for $d = 0$, $g = 1$, comparing the spectra (the sets of the eigenvalues) of the left-hand side and the right-hand side of (1.3.22), we see that there is no λ such that a certain degree of a nonnegative integer-valued 1-circulant with generating vector containing two distinct numbers satisfies (1.3.22).

A valuable tool for the study of circulants is provided by the Hall polynomials, introduced by M. Hall for investiging difference sets. Here and below we enumerate the rows and columns of a matrix A by numbers $0, 1, \ldots, n-1$. If $A = (a_{ij})$ is a g-circulant of order n, then the polynomial

$$\theta_A(x) = \sum_{i=0}^{n-1} a_{0i} x^i \tag{1.3.23}$$

is called the Hall polynomial. It is clear that the Hall polynomial determines the first row and, consequently, all rows of a g-circulant. In the cases where it is clear which circulant A is determined by the Hall polynomial, the index A is omitted.

We know (see Theorem 1.3.4) that the product of a g-circulant and an h-circulant is a gh-circulant (the number gh is reduced modulo n, where n is equal to the common order of the circulants). The analogue of this property for Hall polynomials is described by the following theorem.

THEOREM 1.3.15. *If $A = (a_{ij})$ is a g-circulant and $B = (b_{ij})$ is an h-circulant, each of order n, and $\theta_A(x)$ and $\theta_B(x)$ are the Hall polynomials of A and B, then*

$$\theta_{AB}(x) \equiv \theta_A(x^h)\theta_B(x) \pmod{(x^2 - 1)} \tag{1.3.24}$$

for the Hall polynomial $\theta_{AB}(x)$ of the product of A and B.

PROOF. Since B is an h-circulant, the equalities

$$b_{jk} = b_{0,k-jh}, \qquad j, k = 0, 1, \ldots, n-1,$$

hold (see (1.3.1)). Therefore the elements of the first row of the gh-circulant $AB = C = (c_{ij})$ satisfy the equalities

$$c_{0k} = \sum_{j=0}^{n-1} a_{0j} b_{jk} = \sum_{j=0}^{n-1} a_{0j} b_{0,k-jh}.$$

It remains to observe that the coefficient of x^k in the polynomial

$$\left(\sum_{j=0}^{n-1} a_{0j} x^{jh}\right)\left(\sum_{j=0}^{n-1} b_{0j} x^j\right)$$

is equal to c_{0k}, $k = 0, 1, \ldots, n-1$ (the exponents of x are reduced modulo n). □

THEOREM 1.3.16. *If A is a g-circulant of order n with the Hall polynomial $\theta_A(x)$, then A^m is a g^m-circulant with the Hall polynomial*

$$\theta_{A^m}(x) \equiv \theta_A(x)\theta_A(x^g)\ldots\theta_A(x^{g^{m-1}}) \pmod{(x^n - 1)}, \qquad m = 1, 2, \ldots,$$

where the exponents of g are reduced modulo n.

This theorem is proved by induction with the use of Theorem 1.3.15.

THEOREM 1.3.17 ([**34**]). *Equation* (1.3.22) *has a solution equal to a g-circulant A of order n with the Hall polynomial $\theta(x)$ if and only if*

$$\theta(x)\theta(x^g)\dots\theta(x^{g^{m-1}}) \equiv d+\lambda(1+x+\dots+x^{n-1}) \pmod{(x^n-1)} \tag{1.3.25}$$

and for $d \neq 0$, in addition,

$$g^m \equiv 1 \pmod{n}. \tag{1.3.26}$$

PROOF. Let the g-circulant A of order n satisfy (1.3.22). Then A^m is a g^m-circulant (see Theorem 1.3.4) and the first rows of A^m and $d\mathbf{I}+\lambda\mathbf{J}$ coincide. The first row of the g^m-circulant A^m is given by the Hall polynomial, which by Theorem 1.3.16 coincides with the left-hand side of (1.3.25). Thus, (1.3.25) is true. If $d \neq 0$, then the matrix $d\mathbf{I}+\lambda\mathbf{J}$ is a 1-circulant and not an h-circulant for $1 < h < n$. Therefore the matrix A^m must be a 1-circulant, but this is possible only in the case where (1.3.26) is fulfilled.

Conversely, let (1.3.25) and (1.3.26) for $d \neq 0$ be fulfilled for the g-circulant A. By Theorem 1.3.4, A^m is a g-circulant. By (1.3.25) the first row of A^m coincides with the first row of the matrix $d\mathbf{I}+\lambda\mathbf{J}$, which is a 1-circulant for $d \neq 0$ and an h-circulant for any h for $d = 0$. If $d \neq 0$ and (1.3.26) is fulfilled, then A^m is a 1-circulant, which obviously must coincide with $d\mathbf{I}+\lambda\mathbf{J}$. Thus, in all cases, A satisfies equation (1.3.22). The theorem is proved. □

COROLLARY 1.3.18. *If an integer-valued g-circulant A of order n satisfies equation* (1.3.22)*, then*

$$c^m = d + \lambda n,$$

where c is the integer equal to the value $\theta(1)$ of the Hall polynomial $\theta(x)$ of A.

1.3.9. On consistency of (1.3.22) **for** $d = 0$**.** For $d = 0$, the results obtained above can be refined.

THEOREM 1.3.19 ([**117**]). *The power A^m of a g-circulant matrix*

$$A = (a_{ij}), \quad i, j = 0, 1, \dots, n-1, \quad \sum_{j=0}^{n-1} a_{0j} \neq 0,$$

is a matrix with equal elements if and only if each root of degree n of 1 not equal to 1 is a root of the polynomial

$$\theta(x)\theta(x^g)\dots\theta(x^{g^{m-1}}), \tag{1.3.27}$$

where $\theta(x)$ is the Hall polynomial of A.

PROOF. If the power A^m of a g-circulant A is a matrix with equal elements, then (1.3.22) with $d = 0$ is fulfilled for some λ. Then by Theorem 1.3.17 the congruence (1.3.25) must be true. In this case, $\lambda \neq 0$ if $\sum_{j=0}^{n-1} a_{0j} \neq 0$, since on setting $x = 1$ in (1.3.25) we obtain $\left(\sum_{j=0}^{n-1} a_{0j}\right)^m = \lambda n$. Such a λ exists if and only if the Hall polynomial of A^m, which by Theorem 1.3.16 is equal to $\theta(x)\theta(x^g)\dots\theta(x^{f^{m-1}})$, is divisible by $1+x+\dots+x^{n-1}$. The roots of this polynomial are all nth roots of unity except the root 1, and all them must be roots of $\theta_{A^m}(x)$.

Conversely, if for the Hall polynomial $\theta(x)$ of the g-circulant A all roots of degree n of 1, except the root 1, are the roots of the polynomial (1.3.27), then (1.3.25) is fulfilled and A^m is a matrix with equal elements. The theorem is proved. □

Let D be the set of all positive divisors of n, including 1 and n. In order to formulate a consistency criterion for equation (1.3.22), we introduce the following operation on D: if g is a positive integer, then for any $\delta \in D$ let $T(\delta) = \delta/(\delta, g)$. It is clear that $T(\delta) = \delta$ if $(\delta, g) = 1$. For a prime number p, we denote by $\nu_p(a)$ the maximum degree of p dividing the integer a, that is,

$$a = \pm \prod_{p|a} p^{\nu_p(a)}.$$

Then

$$\nu_p(T(\delta)) = \begin{cases} 0, & \nu_p(\delta) \leq \nu_p(g), \\ \nu_p(\delta) - \nu_p(g), & \nu_p(\delta) > \nu_p(g). \end{cases}$$

It is obvious that the result of the operation T can be obtained by testing sequentially all prime divisors of g and $T^k(\delta) = \delta/(\delta, g^k)$, where T^k is the kth iteration of T. We denote by $h_\delta(x)$ the δth cyclotomic polynomial, that is, the monic polynomial (with the leading term equal to one) whose roots are the primitive δth roots of unity, i.e., their orders in the group of roots of unity are equal to δ.

THEOREM 1.3.20. *Let $c_0, c_1, \dots, c_{n-1}$ be rational numbers, $\sum_{i=0}^{n-1} c_1 \neq 0$, and let C be a g-circulant of order n with the generating vector $(c_0, c_1, \dots, c_{n-1})$. Let D_0 be the set of all divisors δ of n such that $\delta \neq 1$ and $h_\delta(x)$ divides*

$$\theta(x) = \sum_{i=0}^{n-1} c_i x^i.$$

Then the matrix C^m consists of equal elements if and only if for each $\delta \in D$, $\delta \neq 1$, there exists $k \in Z_m$ such that $T^k(\delta) \in D_0$, where $T(\delta) = \delta/(\delta, g)$.

PROOF. By Theorem 1.3.19, C^m has equal elements if each nth root of unity ζ, $\zeta \neq 1$, is a root of the polynomial $\theta(x)\theta(x^g)\dots\theta(x^{g^{m-1}})$, or, what is the same, ζ^{g^k} is a root of $\theta(x)$ for some $k \in Z$. If the order of ζ in the group of roots of 1 is δ, then, as is easy to see, the order of ζ^{g^k} is $\delta/(\delta, g^k) = T^k(\delta)$. It is clear that $T^k(\delta) = t \neq 1$. Otherwise $\delta = (\delta, g^k) = 1$, and then $\zeta^{g^k} = 1$, $\theta(1) = 0$, which contradicts the condition $\sum_{i=0}^{n-1} c_i \neq 0$. Thus, a root of 1 of order t is a root of $\theta(x)$. The tth cyclotomic polynomial is irreducible over the field of rational numbers, and therefore the polynomial $\theta(x)$ with rational coefficients, which has a common root with $h_t(x)$, is divisible by $h_t(x)$. Consequently, $t \in D_0$.

Conversely, let $\zeta \neq 1$ be an nth root of unity, and δ the order of ζ in the group of roots of unity. Then $\delta \mid n$ and $\delta \neq 1$. By the condition of the theorem there exists $k \in Z_m$ such that $T^k(\delta) \in D_0$. As was noted above, the order of ζ^{g^k} is $t = T^k(\delta)$, and $t \neq 1$, since 1 does not belong to D_0. The roots of the tth cyclotomic polynomial $h_t(x)$ are all primitive roots of degree t of 1, including ζ^{g^k}. On the other hand, by the conditions of the theorem, $h_t(x)$ divides $\theta(x)$. Consequently, ζ^{g^k} is a root of $\theta(x)$ and ζ is a root of $\theta(x^{g^k})$. Thus, any root of degree n of 1, except 1, is a root of the polynomial (1.3.27). Then, by Theorem 1.3.19, C^m consists of equal elements. The theorem is proved. □

The criterion given in Theorem 1.3.20 is the most general known condition for the existence of a g-circulant solution of the equation $A^m = \lambda\mathbf{J}$.

In conclusion we mention the relation between the g-circulant $(0,1)$-matrices with an arbitrary g and configurations which generalize difference sets. A subset $A = \{a_1, \dots, a_k\}$ of Z_v is called a (v, k, λ, g)-additive set if any element $x \in Z_v$ is represented exactly λ times in the form

$$x \equiv a_i + ga_j \pmod v, \qquad a_i, a_j \in A.$$

It is clear that for $g = v-1$ we obtain a difference set. Additive sets were introduced and investigated by C. W. H. Lam, who showed in [**65**] that the existence of a g-circulant matrix C of order v satisfying the matrix equation $C^2 = \lambda\mathbf{J}$ is equivalent to the existence of a (v, k, λ, g)-additive set with $k^2 = \lambda v$.

1.3.10. Some other questions on circulants. A number of well known enumeration problems can be reduced to questions on $(0,1)$-circulants, among which are such classical problems as *problème des rencontres* and *problème des ménages*, whose solutions require the calculation of permanents of $(0,1)$-circulants of a special form. We illustrate the application of $(0,1)$-circulants to enumeration problems by finding the number of $(0,1)$-matrices in the Ryser class $\mathfrak{A}^n(2)$, which will be done with the use of the results obtained above.

First we obtain the canonical form of a $(0,1)$-matrix in $\mathfrak{A}^n(2)$. Let matrices $A_1, A_2, \dots, A_s$ of orders $n_1, n_2, \dots, n_s$, respectively, be given. We denote by $A_1 \oplus A_2 \oplus \cdots \oplus A_s$ the block-diagonal matrix

$$A = \begin{pmatrix} \boxed{A_1} & & & 0 \\ & \boxed{A_2} & & \\ & & \ddots & \\ 0 & & & \boxed{A_s} \end{pmatrix} \tag{1.3.28}$$

of order $n = n_1 + \cdots + n_s$. The matrix A is called the direct sum of the matrices $A_1, \dots, A_s$.

LEMMA 1.3.21. *A matrix A in $\mathfrak{A}^n(2)$ is permutationally equivalent to a matrix of the form $A_1 \oplus \cdots \oplus A_r$ for some r, where A_k is a matrix of order n_k, $A_k = \mathbf{I}_{n_k} + Q_k$, and Q_k is a permutation matrix corresponding to a cycle of length n_k on a set of n_k elements, $k = 1, \dots, r$, $n_1 \geq n_2 \geq \cdots \geq n_r$, $n_1 + \cdots + n_r = n$.*

PROOF. By Theorem 1.2.12, $A = R_1 + R_2$, where R_1, R_2 are permutation matrices of order n. Obviously, it is possible to rearrange the rows and the columns of A so that the ones of the matrix R_1 occupy the principal diagonal, that is, A is permutationally equivalent to a matrix A' of the form $A' = \mathbf{I} + P'$, where P' is a permutation matrix corresponding to some permutation π. Let $\pi = \xi_1 \dots \xi_r$ be the decomposition of π into the product of r independent cycles of lengths $n_1, \dots, n_r$, and let $n_1 \geq \cdots \geq n_r$. For each k, denote by I_k and J_k respectively the sets of indices of rows and columns containing elements 1 of the matrix P' corresponding to the cycle ξ_k, $k = 1, 2, \dots, r$. It is clear that $|I_k| = |J_k| = n_k$ and, moreover, $I_k = J_k$, $k = 1, 2, \dots, r$. Now we rearrange the rows of A' in such a way that first come the rows with numbers in the set I_1, next in I_2, and so on, and, finally, in I_r. Similarly we rearrange the columns of A': first come the columns with numbers in J_1, next in J_2, and so on, and, finally, in J_r. Since $I_k = J_k$, $k = 1, 2, \dots, r$, the rows and the columns can be rearranged with the use of the same permutation ρ

on N_n. If Q is the permutation matrix corresponding to the permutation ρ, then the matrix A' transfers to the matrix

$$A'' = Q^{-1}A'Q = I + Q^{-1}P'Q.$$

It is clear that for each $k = 1, 2, \ldots, r$ the submatrix

$$A''[\{m_{k-1}+1, \ldots, m_{k-1}+n_k\}, \{m_{k-1}+1, \ldots, m_{k-1}+n_k\}],$$

where $m_k = n_0 + n_1 + \cdots + n_k$ (we set $m_0 = n_0 = 0$), is a matrix of order n_k of the class $\mathfrak{A}^{n_k}(2)$. Since the cyclic structure of a permutation does not change under renumbering the elements, each of these matrices is of the form $\mathbf{I}_{n_k} + Q_k$, where Q_k is the matrix corresponding to a cycle with n_k elements. The lemma is proved. □

LEMMA 1.3.22. *Let A be an $m \times m$ matrix of the form $\mathbf{I} + R$, where R is a permutation matrix corresponding to a cycle of length m. Then there exists an $m \times m$ permutation matrix Q such that $Q^{-1}AQ = \mathbf{I} + P_m$, where P_m is the permutation matrix of the cycle $(1\,2\ldots m)$.*

PROOF. Let $\xi = (i_1 i_2 \ldots i_m)$. Consider the permutation

$$\rho = \begin{pmatrix} 1 & 2 & \ldots & m \\ i_1 & i_2 & \ldots & i_m \end{pmatrix}.$$

It is easy to see that $\rho^{-1}\xi\rho = (12\ldots m)$. A product of permutations corresponds to the product of their permutation matrices. If the permutation matrix Q represents the permutation ρ, then $Q^{-1}AQ = \mathbf{I} + P_m$, and the lemma is proved. □

THEOREM 1.3.23. *Any matrix in $\mathfrak{A}^n(2)$ is permutationally equivalent to a matrix of the form $A_1 \oplus \cdots \oplus A_r$ for some r, where $A_i \in \mathfrak{A}^{n_i}(2)$, $n_1 + \cdots + n_r = n$, $n_1 \geq \cdots \geq n_r$, $A_i = \mathbf{I}_{n_i} + P_{n_i}$, P_{n_i} is a matrix of the form (1.3.2), $i = 1, 2, \ldots, r$.*

The theorem follows directly from Lemmas 1.3.21 and 1.3.22.

Thus, each matrix in $\mathfrak{A}^n(2)$ is permutationally equivalent to a direct sum of circulants with generating vector of the form $(1, 1, 0, \ldots, 0)$. We denote by $C^{(n)}$, $n = 2, 3, \ldots$, the circulant with generating vector $(1, 1, 0, \ldots, 0)$ of length n.

We need another obvious assertion about the groups of permutation automorphisms of matrices.

THEOREM 1.3.24. *Let $\{A_1, \ldots, A_q\}$ be $(0,1)$-matrices of orders $n_1, \ldots, n_q$, respectively, and let a $(0,1)$-matrix A of order n be representable as the direct sum of s matrices, where each of the first r_1 blocks of equal sizes is A_1, each of next r_2 blocks of equal sizes is A_2, and so on, finally, each of the r_q blocks of equal sizes is A_q, $r_1 + r_2 + \cdots + r_q = s$, $n_1 r_1 + \cdots + n_q r_q = n$, and the blocks in different groups are pairwise nonequivalent. Then the group $\mathrm{Aut}_p A$ is isomorphic to the direct product of the groups*

$$\mathfrak{S}_{r_1} \times \cdots \times \mathfrak{S}_{r_q} \times (\mathrm{Aut}_p A_1)^{r_1} \times \cdots \times (\mathrm{Aut}_p A_q)^{r_q},$$

where $\mathfrak{S}_m$ is the symmetric group of degree m.

THEOREM 1.3.25 ([**77**]). *The number of matrices in the class $\mathfrak{A}^n(2)$ is equal to*

$$(n!)^2 \sum (2^{r_1+\cdots+r_q} r_1! \ldots r_q! \, n_1^{r_1} \ldots n_q^{r_q})^{-1},$$

where the summation is over all unordered partitions $\{n_1^{r_1}, n_2^{r_2}, \ldots, n_q^{r_q}\}$ with r_k parts of size $n_k \geq 2$ such that $r_1 n_1 + \cdots + r_q n_q = n$, $k = 1, 2, \ldots, q$.

PROOF. By Theorem 1.3.23, each matrix A in $\mathfrak{A}^n(2)$ is permutationally equivalent to the canonical form, that is, to the block-diagonal matrix with r_1 blocks $C^{(n_1)}$, r_2 blocks $C^{(n_2)}$, ..., r_q blocks $C^{(n_q)}$ sequentially placed on the diagonal of the matrix, where $n_1 > \cdots > n_q \geq 2$. The canonical forms are in one-to-one correspondence with unordered partitions of n into q summands such that $n = r_1 n_1 + \cdots + r_q n_q$, $n_i \geq 2$, $i = 1, \ldots, q$. Thus, the number of canonical forms coincides with the number of such partitions. By Theorem 1.3.24, the group of permutational automorphisms of each canonical matrix is the direct product of the automorphism groups of blocks, that is,

$$\mathrm{Aut}_p A \cong \mathfrak{S}_{r_1} \times \cdots \times \mathfrak{S}_{r_q} \times (\mathrm{Aut}_p C^{(n_1)})^{r_1} \times \cdots \times (\mathrm{Aut}_p C^{(n_q)})^{r_q}.$$

By Corollary 1.3.12, $\mathrm{Aut}_p C^{(m)}$ is isomorphic to the dihedral group of order $2m$ for $m \geq 3$. The group $\mathrm{Aut}_p C^{(2)}$ obviously has the order 4. Therefore the order of $\mathrm{Aut}_p A$ equals $r_1! \ldots r_q! (2n_1)^{r_1} \ldots (2n_q)^{r_q}$. Then by Theorem 1.1.1 the number of matrices in the equivalence class containing A is equal to

$$(n!)^2 (2^{r_1 + \cdots + r_q} r_1! \ldots r_q! \, n_1^{r_1} \ldots n_q^{r_q})^{-1}.$$

The sum over all partitions of n of the form indicated above gives the total number of matrices in $\mathfrak{A}^n(2)$. ☐

The connection with enumerative problems attracts considerable interest to the study of the permanents of $(0,1)$-circulants with particular attention to the permanents of circulants of the form $\sum_{i=0}^{k-1} P_n^i$, which are close to the Toeplitz matrices for small values of k (see [**72**]). Among the recent papers we mention [**76, 77, 99**], where advances in the study of permanents of such circulants are made.

The study of nonnegative circulants as matrices over the Boolean algebra is another interesting direction in this area of combinatorics. Such a view on nonnegative matrices is useful in problems of estimating some graph constants. The algebraic properties of circulants, for example, the problem of describing idempotents in semigroups of Boolean matrices, are also close to this direction.

Problems.

1. A matrix of order n is called a left g-circulant, where $0 < g < n$, if each row of the matrix is obtained from the preceding row by g shifts to the left. Show that any g-circulant of order n is also a left h-circulant for some $h \in N_n$. For left g-circulants, prove the analogues of formula (1.3.1) and Theorems 1.3.1 and 1.3.4.
2. Find all $(0,1)$-circulants of order 6 that can be reduced to a diagonal matrix with integer elements.
3. Find the group of permutational automorphisms for the following three $(0,1)$-matrices:

$$\begin{pmatrix} 1 & 1 & 0 & 0 & 1 \\ 0 & 0 & 1 & 1 & 1 \\ 1 & 0 & 0 & 1 & 1 \\ 1 & 1 & 1 & 0 & 0 \\ 0 & 1 & 1 & 1 & 0 \end{pmatrix},$$

   the circulant C of order 9 with the generating vector $c = (100011001)$, and the circulant C of order 6 with the generating vector $c = (010010)$.
4. Let C and $\bar{C}$ be two $(0,1)$-circulants with the generating vectors $(\alpha_1, \ldots, \alpha_n)$ and $(\bar{\alpha}_1, \ldots, \bar{\alpha}_n)$, respectively, where $\bar{\alpha}_i = 1 - \alpha_i$, $i = 1, \ldots, n$. Show that the groups of permutational automorphisms $\mathrm{Aut}_p C$ and $\mathrm{Aut}_p \bar{C}$ are isomorphic.

5. Describe the groups $\operatorname{Aut}_p C$ for all symmetric $(0,1)$-circulants C of order 7.
6. Let C be a g-circulant of order n with the generating vector $c = (\gamma_1, \dots, \gamma_n)$ and $(g, n) = 1$. Show that if for some number λ and some positive integer m the relation $C^m = \lambda \mathbf{J}$ is fulfilled, then $\gamma_1 = \cdots = \gamma_n$.

CHAPTER 2

Ryser Classes

Introduction

In this chapter, we study classes of nonnegative matrices of identical form and with identical numbers of positive elements in rows and columns. Many of the combinatorial properties of the matrices in these classes depend on the row and column sums of the supports of the matrices. Thus, the basic objects of the study are classes $\mathfrak{A}(R,S)$ consisting of $(0,1)$-matrices with given integer-valued vectors R,S as the vectors of row and column sums, respectively.

A constructive description of the classes $\mathfrak{A}(R,S)$ in Section 2.1 is based on the existence in any nonempty monotone class of some canonical matrix $\tilde{A}$ defined in Section 1.2. An important role is played by the interchange operation combining the matrices of a class by means of the interchange graph, which, according to the Ryser Theorem, is connected. Note that, along with the traditional use of the canonical matrix $\tilde{A}$ in the proofs of the nonemptiness of the class and the connectivity of the interchange graph, it can be applied to the enumeration of the matrices in the class. Using the graph representation of a nonempty class, the distance between matrices of the class is introduced as the distance between these matrices in the interchange graph. The distance between matrices A and B is expressed in terms of the number of positions of nonidentical elements in A and B and the packing circuit number of some digraph $O(A)-O(B)$. Then the problems related to connectivity of the interchange graph are investigated. We prove that, with few exceptions, the interchange graph of any nonempty class $\mathfrak{A}(R,S)$ is 3-connected. The presentation in Section 2.1, as well as the whole of Chapter 2, is much influenced by [**19**]. In a number of cases, we follow [**87**]. The proof of 3-connectivity of the interchange graph is based on [**92**].

In Section 2.2, the classes $\mathfrak{A}(R,S)$ are considered from another point of view. The positions in the matrices of the class are divided into two types, invariant positions, that is, the positions with the same elements in all matrices of the class, and noninvariant positions, for which this is not the case. A more general invariant set of positions is defined as the set with all row sums (or all column sums) identical for all matrices of the class. For the matrices which have invariant positions there exists a representation in some standard form reducing the study of the given class to the study of smaller classes. Here the main role is played by the structure matrix T of a nonempty class $\mathfrak{A}^{m,n}(R,S)$. This $(m+1)\times(n+1)$ nonnegative integer-valued matrix was defined in Section 1.2. We consider properties of the structure matrix such as the unimodality of the sequences of elements in rows and columns, the location of zeros, and its connection with the location of invariant positions, and calculate the minimal value of the elements in a row (in a column) and find its position. These features of the structure matrix T form a basis for the standard representation mentioned above. In conclusion we present a result on the rank of the structure matrix. The exposition in this section follows (with certain simplifications) [**19**] and [**87**]. In the discussion of the rank and the trace of the matrix T, we use the results of [**96**] and [**93**].

Estimates of the term rank of classes $\mathfrak{A}(R,S)$ are considered in Section 2.3. The term rank of any matrix of the class $\mathfrak{A}$ lies within the interval between the minimal value $\tilde{\rho} = \tilde{\rho}(\mathfrak{A})$ and the maximal value $\bar{\rho} = \bar{\rho}(\mathfrak{A})$ of the term ranks of the matrices in $\mathfrak{A}$, and can be equal to any integer in this interval. Among the matrices of rank $\bar{\rho}$ an important role is played by special matrices $A_{\bar{\rho}}$ with ones in the positions $(1,\bar{\rho}), (2,\bar{\rho}-1), \dots, (\bar{\rho},1)$. It is proved that any matrix A of rank ρ can be transformed by interchanges into a matrix B with ones on the positions $(1,\rho), (2,\rho-1), \dots, (\rho,1)$ in such a way that each of the matrices obtained during these interchanges has ρ independent ones. This assertion is a generalization of the well-known Ryser Theorem on matrices with maximal term rank. One of the most significant results of Chapter 2 is the formula for the maximal term rank, which expresses the rank in terms of elements of the structure matrix. Then we formulate conditions under which $\tilde{\rho}$ and $\bar{\rho}$ coincide for a given class of $(0,1)$-matrices. In particular, we prove that if a class does not contain a matrix with positive diagonal and invariant ones, then these two characteristics cannot coincide. A relation between $\tilde{\rho}$ and $\bar{\rho}$ is given. We prove that if this relation holds for some values, then there exist classes with these values of $\tilde{\rho}$ and $\bar{\rho}$. In conclusion, it is shown that for any two matrices whose term ranks are equal to the maximal term rank $\bar{\rho}$ of the common class of these matrices, there exists a path in the interchange graph of this class connecting these vertices and passing only through matrices with term rank $\bar{\rho}$. In Section 2.3 we mostly follow [**19**] and [**87**]. The proof of connectivity of the subgraph induced in the interchange graph by the matrices of maximal term rank is based on [**111**].

2.1. A constructive description of Ryser classes

In Chapter 2, we study the classes $\mathfrak{A}(R,S)$ of $(0,1)$-matrices with given row and column sums. The definition of these classes was given in Section 1.2. These classes were introduced in the late 1950s by Ryser, Fulkerson, Haber, and Gale, and now are being investigated intensively. Ryser contributed significantly to the study of the classes $\mathfrak{A}(R,S)$, and these classes will be referred to as Ryser classes. As mentioned in Chapter 1, the combinatorial problems related to Ryser classes are also significant for nonnegative matrices of general form; the connecting link is the notion of the support of a $(0,1)$-matrix.

2.1.1. The canonical matrix of a class. Let $R = (r_1, \dots, r_m)$ and $S = (s_1, \dots, s_n)$ be integer-valued nonnegative vectors. We assume (even if it is not specified) that these vectors are monotone, that is, $r_1 \geq \dots \geq r_m$ and $s_1 \geq \dots \geq s_n$. This assumption does not lead to a loss of generality, because the notions and quantities related to matrices which are considered in this section are invariant under permutations of rows and columns. For investigation of the class $\mathfrak{A} = \mathfrak{A}(R,S)$ of $(0,1)$-matrices with given vectors of row sums R and column sums S, we have introduced in Section 1.2 the maximal matrix $A(R;n)$ and the structure matrix. In the proof of Theorem 1.2.8 about the condition for nonemptiness of a Ryser class, we have found that, starting with the matrix $A(R;n)$, one can construct, by a certain special method, a matrix $\tilde{A} \in \mathfrak{A}(R,S)$ for an arbitrary nonempty class $\mathfrak{A}(R,S)$. This fact can serve as a starting point for a constructive description of the set of all $(0,1)$-matrices of the class $\mathfrak{A}(R,S)$. An algorithm for constructing the matrix $\tilde{A}$ was also described in the proof of Theorem 1.2.8. The matrix $\tilde{A} = \tilde{A}(R,S)$ will be called the canonical matrix of the nonempty class $\mathfrak{A}^{m,n}(R,S)$. Consider some simple properties of the canonical matrix.

THEOREM 2.1.1. *Let $\mathfrak{A}^{m.n}(R,S)$ be the Ryser class determined by monotone vectors $R = (r_1,\dots,r_m)$ and $S = (s_1,\dots,s_n)$, and let $\tilde{A} = \tilde{A}(R,S)$ be the canonical matrix of the class. Then $\tilde{A}[N_m, N_l]$ for any $l = 1,\dots,n-1$ is also the canonical matrix for some Ryser class $\mathfrak{A}^{m,l}(R',S')$, where $S' = (s_1,\dots,s_l)$.*

PROOF. Let $\mathfrak{A}^{m,n}(R,S)$ and $\tilde{A}(R,S)$ be given. For $n = 1$, the assertion of the theorem is obviously true. In the proof, we use induction on the number of columns. Assume that the assertion is true for all classes of matrices with less than n columns. Deleting the last column of $\tilde{A}$, we obtain the matrix $\tilde{A}[N_m, N_{n-1}]$. Apply the algorithm for constructing $\tilde{A}$ with the use of $A(R;n)$, described in Section 1.2. Suppose that the first step of the algorithm is correct; that is, starting from $A(R;n)$, we have constructed the matrix A' whose last column sum is s_n. Then in the first $n-1$ columns of A' we obtain the maximal matrix $A(R';n-1)$ for some vector $R' = (r'_1,\dots,r'_m)$. It is not difficult to write the coordinates of R'. If $r_{s_n} > r_{s_n+1}$, then

$$r'_i = \begin{cases} r_i - 1, & i = 1,\dots,s_n, \\ r_i, & i = s_n+1,\dots,m; \end{cases} \tag{2.1.1}$$

otherwise, if there exist integers $s \geq 0$ and $t > 0$ such that

$$r_{s_n-s-1} > r_{s_n-s} = r_{s_n-s+1} = \cdots = r_{s_n} = r_{s_n+1} = \cdots = r_{s_n+t} > r_{s_n+t+1}$$

(we assume that $r_0 = r_1 + 1, r_{m+1} = 0$), then

$$r'_i = \begin{cases} r_i - 1, & i = 1,\dots,s_n-s-1, \quad i = s_n+t-s,\dots,s_n+t, \\ r_i & \text{otherwise.} \end{cases} \tag{2.1.2}$$

Let $S' = (s_1,\dots,s_{n-1})$. In the proof of Theorem 1.2.8, it was shown that $R' \prec S'$, that is, the class $\mathfrak{A}(R',S')$ is not empty. It follows from the construction of the matrix $\tilde{A}[N_m, N_{n-1}]$ that it is the canonical matrix of this class. Then by the induction assumption $\tilde{A}[N_m, N_l]$ is the canonical matrix of some nonempty class for any $l \leq n-1$. The theorem is proved. □

COROLLARY 2.1.2. *If for some $i,k \in N_m$, $i < k$, and some $l \in N_n$ a pair of elements of $\tilde{A} = \tilde{A}(R,S) = (\tilde{a}_{ij})$ is such that $\tilde{a}_{il} = 0$, $\tilde{a}_{kl} = 1$, then*

$$\sum_{j=1}^{l} \tilde{a}_{ij} = \sum_{j=1}^{l} \tilde{a}_{kj}.$$

Indeed, as was proved, $\tilde{A}[N_m, N_l]$ is the canonical matrix whose last column is $(\tilde{a}_{1l},\dots,\tilde{a}_{ml})^T$. By the construction of the canonical matrix, the last column can have a coordinate equal to one below a coordinate equal to zero only in the case where the row sums for the rows containing these coordinates coincide.

2.1.2. The interchange operation. The main tool for describing the $(0,1)$-matrices forming a Ryser class is the interchange operation. Consider two $(0,1)$-matrices of order 2:

$$\begin{pmatrix} 1 & 0 \\ 0 & 1 \end{pmatrix}, \quad \begin{pmatrix} 0 & 1 \\ 1 & 0 \end{pmatrix}. \tag{2.1.3}$$

These matrices are complementary to each other (it is natural to call the matrices A and $\mathbf{I} - A$ complementary). If a $(0,1)$-matrix contains one of these matrices as a submatrix, then, replacing it by the complementary matrix, we obtain a matrix having the same row and column sums, that is, a matrix from the same Ryser class. Replacement in a $(0,1)$-matrix of one of the matrices (2.1.3) by the complementary matrix is called the interchange operation. The matrices in (2.1.3) are called interchangeable. The following theorem, proved by Ryser, shows the importance of the interchange operation.

THEOREM 2.1.3. *Let A, B be matrices belonging to the Ryser class $\mathfrak{A}^{m,n}(R,S)$ with $n \geq 2$, $R = (r_1, \dots, r_m)$, $S = (s_1, \dots, s_n)$. Then A can be transformed into B by a finite number of interchange operations.*

PROOF. Let us prove that an arbitrary matrix $A = (a_{ij})$ of a nonempty Ryser class consisting of matrices with $n \geq 2$ columns can be transformed by interchange operations into the canonical matrix $\tilde{A} = (\tilde{a}_{ij})$ of this class. Consider the nth columns of the matrices A and $\tilde{A}$. If they do not coincide, then there exist $i, k \in N_m$ such that $a_{in} = 0$, $a_{kn} = 1$, $\tilde{a}_{in} = 1$, $\tilde{a}_{kn} = 0$. Let r_i, r_k be the corresponding row sums of the matrices of the class $\mathfrak{A}(R,S)$. Let us show that there exists an index $j \in N_{n-1}$ such that $a_{ij} = 1$, $a_{kj} = 0$. Indeed, if for each j such that $a_{ij} = 1$ we have $a_{kj} = 1$, it follows that $r_i < r_k$. But if $i < k$, then this contradicts the monotonicity of the vector R. If $i > k$, then by Corollary 2.1.2 the equalities $\tilde{a}_{in} = 1$, $\tilde{a}_{kn} = 0$ can be true only in the case $r_i = r_k$. We again obtain a contradiction. Thus, A contains an interchangeable submatrix $A[\{i,k\},\{j,n\}]$. Realizing the corresponding interchange operation in A, we arrive at a matrix such that the number of elements of the nth column which do not coincide with elements of the nth column of $\tilde{A}$ is less by 2 than in A. Obviously, repeating the similar operation, we transform A into a matrix A' which has the same nth column as $\tilde{A}$. It is clear that $A'[N_m, N_{n-1}]$ and $\tilde{A}[N_m, N_{n-1}]$ belong to the same class $\mathfrak{A}(R', S')$ described in Theorem 2.1.1, where $S' = (s_1, \dots, s_{n-1})$, the coordinates of R' are given by (2.1.1) or (2.1.2), and $\tilde{A}[N_m, N_{n-1}]$ is the canonical matrix of this class. Proceeding in the same way, we can by interchange operations transform the matrix A' into a matrix A'' such that the columns with numbers n and $n-1$ of A'' coincide with the corresponding columns of $\tilde{A}$. By Theorem 2.1.1 we can continue the process (since $A[N_m, N_l]$, $l = 1, 2, \dots$, is the canonical matrix of the corresponding class) until we obtain the matrix $\tilde{A}$. By the same process the matrix B can be transformed into $\tilde{A}$. The reverse sequence of interchanges transforms $\tilde{A}$ into B, and, combining this reverse sequence with the sequence of interchanges which transforms A to $\tilde{A}$, we obtain the required sequence of operations which transforms A into B. The theorem is proved. □

The process described in the proof of Theorem 2.1.3 can be considered as an algorithm for transforming an arbitrary matrix A in a class $\mathfrak{A}(R,S)$ into an arbitrary matrix B in the same class by a sequence of interchange operations such that the canonical matrix $\tilde{A}$ appears as an intermediate matrix in this process. It is clear that, in general, this sequence is not the shortest one. Moreover, it is not easy to obtain an estimate for the number of interchanges in this algorithm needed for the transition from A to $\tilde{A}$. Later we consider another algorithm which does not use the transition to the canonical form.

2.1.3. Canonical matrix and constructive enumeration. Consider a nonempty Ryser class $\mathfrak{A} = \mathfrak{A}^{m,n}(R,S)$ and let $\tilde{A} = \tilde{A}(R,S)$ be the canonical matrix of $\mathfrak{A}$. According to Theorem 2.1.1, if $\tilde{S}_n$ is the nth column of $\tilde{A}$, that is, $\tilde{A} = [\tilde{A}' \vdots \tilde{S}_n]$, then $\tilde{A}' = \tilde{A}[N_m, N_{n-1}]$ is the canonical matrix of the class $\mathfrak{A}' = \mathfrak{A}^{m,n-1}(R', S')$, where $S' = (s_1, \dots, s_{n-1})$ and the coordinates of R' are defined by (2.1.1) or (2.1.2). Hence the nonemptiness of $\mathfrak{A}$ is equivalent to the nonemptiness of $\mathfrak{A}'$. Moreover, below we show how to construct all matrices of the class $\mathfrak{A}$ if we know all matrices of $\mathfrak{A}'$. First we obtain an auxiliary result.

Let $S_n = (a_{1n}, \dots, a_{mn})^T$ be an m-dimensional column vector, and let $t(S_n)$ denote the number $s_n - (\tilde{S}_n, S_n)$, that is, the number of positions where $\tilde{S}_n$ has the element equal to one and S_n has the zero element. Denote by I the set of all $i \in N_m$ such that $a_{in} = 0$, $\tilde{a}_{in} = 1$, and by K the set of all $k \in N_n$ such that $a_{kn} = 1$, $\tilde{a}_{kn} = 0$, and set $t = |I| = |K|$. We assume that the elements $i_1, \dots, i_t$ and $k_1, \dots, k_t$ of the sets I and K are arranged so that $i_1 < \cdots < i_t$ and $k_1 < \cdots < k_t$.

LEMMA 2.1.4. *If the nonempty class $\mathfrak{A} = \mathfrak{A}^{m,n}(R,S)$ contains a matrix A with the nth column S_n, then the class $\mathfrak{A}'$ contains a matrix $A' = (a'_{ij})$ and a sequence $j_1, \dots, j_t$, $j_q \in N_{n-1}$, such that $a'_{i_1 j_q} = 0$, $a'_{k_q j_q} = 1$, $q = 1, \dots, t$. The converse is also true.*

PROOF. Indeed, in the proof of Theorem 2.1.3 it was shown that in a matrix $A \in \mathfrak{A}$ with a column S_n for any pair of indices (i,k), $i \in I$, $k \in K$, there exists an index $j \in N_{n-1}$ such that $a_{ij} = 1$, $a_{kj} = 0$. Let indices $j_1, \dots, j_t$ be such that $a_{i_q j_q} = 1$, $a_{k_q j_q} = 0$, $q = 1, \dots, t$. In the matrix $A[N_m, Nn-1]$, we replace $2t$ elements a_{ij_q} by $1 - a_{ij_q}$, $i \in I, K$, $q = 1, \dots, t$. The resulting matrix belongs to $\mathfrak{A}'$ and satisfies the conditions of the lemma.

Conversely, let an m-dimensional $(0,1)$-vector $S_m = (a_{1n}, \dots, a_{mn})$ such that $t = s_n - (\tilde{S}_n, S_n)$ be given. As above, we define the subsets I, K of the set N_m. Consider the set $\mathfrak{A}'(S_n)$ of all matrices of $\mathfrak{A}'$ which satisfy the conditions of the lemma. If A is an arbitrary matrix of $\mathfrak{A}'(S_n)$, then first we replace all elements of A in the positions (i, j_q), $i \in I, K$, $q = 1, \dots, t$, by the complements to 1, and then we form the nth column vector, setting $a_{in} = \tilde{a}_{in}$ if $i \notin I, K$ and $a_{in} = 1 - \tilde{a}_{in}$ if $i \in I, K$. Adding this column to the matrix obtained in the first step, we find a matrix of the class $\mathfrak{A}(R,S)$ with the nth column S_n. The lemma is proved. □

Now it is not difficult to formulate an algorithm for constructing all matrices of the class $\mathfrak{A}^{m,n}(R,S)$ from the matrices of the class $\mathfrak{A}' = \mathfrak{A}^{m,n-1}(R', S')$. We consider consecutively all m-dimensional $(0,1)$-vector columns S_n with s_n ones. For each of them, we find in $\mathfrak{A}'$ the subset $\mathfrak{A}'(S_n)$ as in Lemma 2.1.4. For each matrix in $\mathfrak{A}'(S_n)$, we construct, as described above, all matrices $A \in \mathfrak{A}^{m,n}(R,S)$ with the nth column S_n. We denote the set of all matrices obtained in such a way by $\mathfrak{A}(S_n)$. This set is the set of all matrices with the nth column S_n. Clearly, the union of all $\mathfrak{A}(S_n)$ coincides with $\mathfrak{A}^{m,n}(R,S)$.

2.1.4. Interchanges and graph representations of $(0,1)$-matrices. For a $(0,1)$-matrix there exist several representations using graphs. Let $A = (a_{ij})$ be a $(0,1)$-matrix of size $m \times n$. We consider the representation of A using a bigraph: the set of vertices V of the bigraph is divided into two subsets $X = \{x_1, \ldots, x_m\}$ and $Y = \{y_1, \ldots, y_n\}$, and the vertices x_i and y_j are connected by an edge if and only if $a_{ij} = 1$, $i = 1, \ldots, m$, $j = 1, \ldots, n$. It is clear that the row and column sums of A are the degrees of vertices of the bigraph belonging to X and Y, respectively. We can slightly modify this representation. Consider the complete bipartite graph $K_{m,n}$ with $m+n$ vertices. We introduce an orientation of the graph corresponding to the matrix A; namely, we replace an edge $\{x_i, y_j\}$ by the arc (x_i, y_j) if $a_{ij} = 1$ and by the arc (y_j, x_i) if $a_{ij} = 0$, $i = 1, \ldots, m$, $j = 1, \ldots, n$. Denote the resulting digraph by $O(A)$. It is clear that A and $O(A)$ determine each other uniquely, a permutation of rows of A corresponds to a renumbering of vertices of X, and a permutation of columns of A corresponds to a renumbering of vertices of Y in $O(A)$. If A contains an interchangeable submatrix (see (2.1.3)), then $O(A)$ contains a circuit (directed cycle) of minimal possible length equal to 4. The operation of interchange applied to A corresponds to reversing the direction of one such circuit of $O(A)$. In these terms Theorem 2.1.3 can be formulated as follows: If A and B are matrices of the same Ryser class, then by a sequential reorientation of several circuits of length four the graph $O(A)$ can be transformed into the graph $O(B)$.

Let A, B belong to some Ryser class $\mathfrak{A}(R, S)$. We denote by $O(A) - O(B)$ the subgraph of $O(A)$ with the set of arcs consisting of all arcs of $O(A)$ with opposite directions relative to the directions of the corresponding arcs of $O(B)$. It is clear that the number of arcs in $O(A) - O(B)$ is equal to the number $d(A, B)$ of positions where A and B have different elements. It is also obvious that the adjacency matrices of $O(A) - O(B)$ and $O(B) - O(A)$ are mutually complementary. For matrices A and B of a common Ryser class, the matrix $A - B$ with elements 0, ± 1 has zero row and column sums. The nonzero elements of this matrix are located in the positions where A and B differ. Therefore the number of nonzero elements equals the number $d(A, B)$ of arcs of the digraph $O(A) - O(B)$. The number of ones in each row and each column of $A - B$ is equal to the number of ones in the line. Hence the outdegree $d^+(v)$ for any vertex v of $O(A) - O(B)$ equals the indegree $d^-(v)$. A digraph Γ is called even if $d^+(v) = d^-(v)$ for all v in Γ. Thus, the graph $O(A) - O(B)$ is even. Now we need the following well-known fact.

PROPOSITION 2.1.5. *In any even digraph Γ there exist circuits $\gamma_1, \ldots, \gamma_q$ such that the arc set $E(\Gamma)$ of Γ is partitioned into sets $E(\gamma_1), \ldots, E(\gamma_q)$.*

The proposition is easily proved by induction on the number of arcs of the digraph.

If for the set $E(\Gamma)$ and circuits $\gamma_1, \ldots, \gamma_q$ the partition

$$E(\Gamma) = E(\gamma_1) \cup \cdots \cup E(\gamma_q)$$

takes place, then we say that Γ is decomposed (or has the decomposition) into the circuits $\gamma_1, \ldots, \gamma_q$. It is clear that

$$|E(\Gamma)| = \sum_{i=1}^{q} |E(\gamma_i)|. \tag{2.1.4}$$

2.1.5. The representation of the class $\mathfrak{A}(R, S)$ as an interchange graph. For the study of Ryser classes, we define a graph structure on the set of all matrices of a nonempty Ryser class. The vertices of the graph $G(R,S)$ are the matrices of the nonempty Ryser class $\mathfrak{A} = \mathfrak{A}^{m,n}(R,S)$. In $G(R,S)$, vertices A_1, A_2 are connected by an edge if it is possible to transform A_1 into A_2 by one interchange operation. The graph $G(R,S)$ is called an interchange graph. For example, let $m = n$ and $R = S = (1,1,\dots,1)$. Then $\mathfrak{A}$ is the set of all permutation matrices of order n and the number of vertices of $G(R,S)$ equals $n!$; two matrices are connected by an edge if the permutation corresponding to A_2 can be obtained from the permutation corresponding to A_1 by a transposition.

Since it is usually difficult to find the exact number of matrices in a given Ryser class $\mathfrak{A}(R,S)$, we often know neither the number of vertices nor the number of edges in the graph of $G(R,S)$. But we have already obtained one important property of the graph. Namely, Theorem 2.1.3 says that the graph $G(R,S)$ of a nonempty class $\mathfrak{A}(R,S)$ is connected. In this section, we study properties of $G(R,S)$ related to the distance between vertices and the property of connectivity in more detail.

2.1.6. Distance in $G(R,S)$. The distance between two vertices of a connected graph is the length of the shortest path connecting these vertices. The diameter of a connected graph is the greatest distance between the vertices of the graph. We denote by $i(A,B)$ the distance between vertices A and B in $G(R,S)$.

In the study of the distance, we use the graphs $O(A)$, $O(B)$, and $O(A)-O(B)$ defined in Section 2.1.4. By Proposition 2.1.5 the set of arcs of the digraph $O(A)-O(B)$ can be partitioned into the sets $E(\gamma_1),\dots,E(\gamma_q)$, where $\gamma_1,\dots,\gamma_q$ are the circuits of the decomposition of the even graph $O(A)-O(B)$. It is easy to give examples demonstrating that such a partition is not unique.

Let $m = n = 4$, $R = S = (2,2,2,2)$,

$$B = \begin{pmatrix} 1 & 0 & 0 & 1 \\ 1 & 1 & 0 & 0 \\ 0 & 1 & 1 & 0 \\ 0 & 0 & 1 & 1 \end{pmatrix}, \qquad A = \begin{pmatrix} 0 & 1 & 1 & 0 \\ 0 & 0 & 1 & 1 \\ 1 & 0 & 0 & 1 \\ 1 & 1 & 0 & 0 \end{pmatrix}.$$

Then $O(B)-O(A) = O(B)$, and the set E of arcs of the graph has the partitions

$$E = E(\gamma_1) \cup E(\gamma_2) \cup E(\gamma_3) \cup E(\gamma_4),$$
$$E = E(\gamma_1') \cup E(\gamma_2'),$$

where

$$\gamma_1 = (x_1, y_1, x_3, y_2, x_1),$$
$$\gamma_2 = (x_1, y_4, x_3, y_3, x_1),$$
$$\gamma_3 = (x_2, y_1, x_4, y_4, x_2),$$
$$\gamma_4 = (x_2, y_2, x_4, y_3, x_2),$$
$$\gamma_1' = (x_1, y_1, x_4, y_4, x_3, y_3, x_2, y_2, x_1),$$
$$\gamma_2' = (x_1, y_4, x_2, y_1, x_3, y_2, x_4, y_3, x_1).$$

The maximum number q of circuits in the decomposition of an even graph is called the packing circuit number. For matrices A and B of the same Ryser class, we denote by $q(A,B)$ the packing circuit number q of the graph $O(A)-O(B)$. It is clear that in the example above $q(A,B) = 4$.

There exists a close connection between the packing circuit number and the distance between matrices in $G(R,S)$.

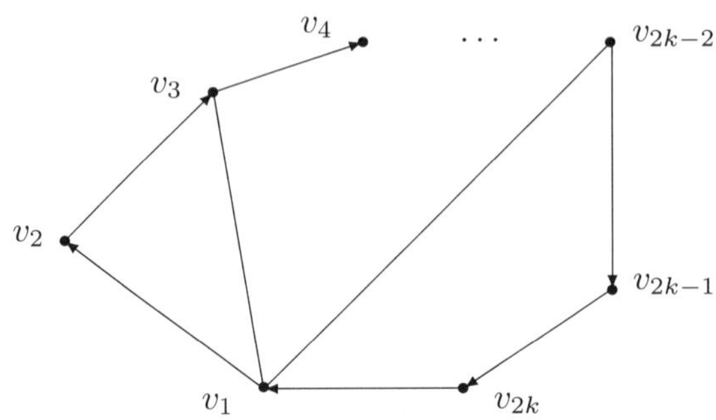

FIGURE 2.1. Circuit γ

THEOREM 2.1.6. *Let $\mathfrak{A}$ be a Ryser class, A, B matrices of the class, and let $O(A) - O(B)$ be a circuit of length $2k$. Then there exists a sequence of $k-1$ interchanges which transforms the matrix A into the matrix B.*

PROOF. For visualization, we use the representation of the circuit γ as a convex polygon on the plane (see Figure 2.1).

We prove the theorem by induction on the length of the circuit. For $k = 2$, the assertion is obvious. Let $k > 2$. We denote the sequential vertices of γ by $v_1, v_2, \ldots, v_{2k}$. Consider the set of arcs in the complete digraph $O(A)$ connecting the vertex v_1 with the vertices $v_4, v_6, \ldots, v_{2k-2}$. These arcs will be referred to as the diagonals from v_1. Either these diagonals are all directed from v_1, or there exists a minimal $l > 0$ such that (v_r, v_1) is an arc of $O(A)$, where $r = 4+2(l-1)$. We reverse the direction of the circuit $(v_1, v_{r-2}, v_{r-1}, v_r, v_1)$ of length four, which is equivalent to an interchange in $O(A)$. As a result the circuit $(v_1, v_{r-4}, v_{r-3}, v_{r-2}, v_1)$ appears in the graph of the new matrix. Then in this matrix we perform an exchange equivalent to reversing the direction of this circuit of length four. Continuing this process, after l interchanges we obtain a matrix A' such that the direction of the arcs of the part of the circuit γ from the vertex v_1 to v_r is reversed. The directions of all diagonals from v_1, except the arc between v_1 and v_r, in $O(A')$ and $O(A)$ coincide, since in the process described above they either did not change, or did change twice. Thus, we find a matrix A' such that $O(A') - O(B)$ consists of the circuit $(v_1, v_r, v_{r+1}, \ldots, v_{2k}, v_1)$ of length

$$2k - r + 2 = 2k - (4 + 2(l-1)) + 2 = 2(k-l).$$

By the induction assumption it is possible to go from A' to B using $k - l - 1$ interchanges. We have transformed A to A' with the use of l interchanges. Thus, we can pass from A to B using $k-1$ interchanges.

If all diagonals in $O(A)$ are directed from v_1, that is, the graph contains the arcs $(v_1, v_4)(v_1, v_6), \ldots, (v_1, v_{2k-2})$, then it contains the circuit $(v_1, v_{2k-2}, v_{2k-1}, v_{2k}, v_1)$ of length 4. Reversing the direction of arcs of this circuit, we obtain a matrix A' such that $O(A') - O(B)$ is a circuit of length $2(k-1)$. Thus, we can apply the induction assumption to this matrix again. The theorem is proved. □

COROLLARY 2.1.7. *If A,B are matrices of the same Ryser class and the digraph $O(A) - O(B)$ is decomposed into circuits $\gamma_1, \ldots, \gamma_s$ of lengths $k_1, \ldots, k_s$, respectively, then A can be transformed into B by $\sum_{i=1}^{s} k_i - s$ interchanges.*

The assertion follows from the fact that by the theorem the matrix A can be transformed by $k_1 - 1$ interchanges to a matrix A_1 such that $O(A_1) - O(B)$ is decomposed into the circuits $\gamma_2, \ldots, \gamma_s$, since the sets of arcs of the circuits $\gamma_1, \ldots, \gamma_s$ are disjoint and the sequence of interchanges converting the circuit γ_1 does not change the direction of the remaining arcs of the graph $O(A) - O(B)$.

COROLLARY 2.1.8. *Let $q(A, B)$ be the packing circuit number of the digraph $O(A) - O(B)$. Then*

$$i(A,B) \leq d(A,B)/2 - q(A,B). \tag{2.1.5}$$

Indeed, by (2.1.4), $d(A, B) = 2(k_1 + \cdots + k_q)$ if $k_1, \ldots, k_q$ are the lengths of the circuits $\gamma_1, \ldots, \gamma_q$ of the decomposition of $O(A) - O(B)$ into the maximum number q of circuits.

We see that the greater $q(A, B)$ is, the smaller is the number of interchanges usually needed for passing from A to B.

2.1.7. A formula for the distance in $G(R, S)$. Theorem 2.1.6 suggests yet another way to find a sequence of interchanges transforming a matrix A into a matrix B of the same Ryser class. This way is connected with a decomposition of the digraph $O(A) - O(B)$ into circuits. If such a decomposition is known, then for each circuit it is possible to construct a sequence of interchanges as it was done in the proof of Theorem 2.1.6. It is clear that during each of these constructions the remaining arcs of $O(A) - O(B)$ do not change, since the circuits forming the decomposition have no common arcs. Therefore this procedure can be repeated until we obtain the matrix B. It is possible to estimate the number of interchanges which have to be performed in this process. Namely, we will show that the equality takes place in (2.1.5). This fact was proved by Walkup in [**116**]. The proof of this result given below follows [**116**].

THEOREM 2.1.9. *Let $A, B \in \mathfrak{A}(R, S)$. Then*

$$i(A,B) = d(A,B)/2 - q(A,B). \tag{2.1.6}$$

PROOF. We prove the theorem by induction on the number $r = i(A, B)$. If $r = 1$, then $d(A, B) = 4$ and $q(A, B) = 1$, that is, (2.1.6) is true. Let $r > 1$. There exists a sequence of matrices $A_0 = A, A_1, \ldots, A_r = B$ such that A_k is obtained from A_{k-1} by some interchange operation, $k = 1, \ldots, r$. Since $i(A, B) = r$, the distance between A_1 and B is exactly $r - 1$, and therefore, by the induction assumption,

$$r - 1 = i(A_1, B) = d(A_1, B)/2 - q(A_1, B). \tag{2.1.7}$$

We have $B - A = (B - A_1) + (A_1 - A)$, where the matrices $B - A$, $B - A_1$, $A_1 - A$ consist of zeros and ± 1. Let $D(A_1, B) = O(A_1) - O(B)$, $q_1 = q(A_1, B)$, and let $\gamma_1, \ldots, \gamma_{q_1}$ be the circuits of the decomposition of $D(A_1, B)$. Consider also the digraph $D(A, A_1) = O(A) - O(A_1)$, which is a circuit of length 4. If $D(A_1, B)$ and $D(A, A_1)$ have common arcs, then they go in the opposite directions, and therefore the corresponding arcs in $O(A) - O(B)$ are absent. We shall keep this remark in mind in what follows. The graph $D(A, A_1)$ can have common arcs with k circuits of the set $\gamma_1, \ldots, \gamma_{q_1}$, $k = 0, 1, 2, 3, 4$. Note that then

$$d(A,B) \leq d(A_1,B) - k + 4 - k. \tag{2.1.8}$$

Consider five possible cases.

If $k = 0$, then

$$d(A, B) = 4 + d(A_1, B), \qquad q(A, B) \geq q_1 + 1. \tag{2.1.9}$$

Let $k = 1$. If some circuit has four common arcs with the circuit of length 4, then it coincides with it. Thus, either

$$d(A, B) = d(A_1, B) - 4, \qquad q(A, B) \geq q_1 - 1, \tag{2.1.10}$$

or

$$d(A, B) \leq d(A_1, B) + 2, \qquad q(A, B) \geq q_1. \tag{2.1.11}$$

If $k = 2$, then

$$d(A, B) \leq d(A_1, B), \qquad q(A, B) \geq q_1 - 1. \tag{2.1.12}$$

If $k = 3$, then

$$d(A, B) \leq d(A_1, B) - 2, \qquad q(A, B) \geq q_1 - 2. \tag{2.1.13}$$

If $k = 4$, then

$$d(A, B) \leq d(A_1, B) - 4, \qquad q(A, B) \geq q_1 - 3. \tag{2.1.14}$$

In each of these cases, (2.1.7) and, respectively, (2.1.9)–(2.1.14) imply that

$$i(A, B) \geq d(A, B)/2 - q(A, B). \tag{2.1.15}$$

For example, from (2.1.7) and (2.1.11) we obtain

$$d(A, B)/2 - q(A, B) \leq \frac{d(A_1, B) + 2}{2} - q_1 = r - 1 + 1 = r.$$

The assertion of the theorem follows from (2.1.15) and (2.1.6). □

COROLLARY 2.1.10. *If $A, B \in \mathfrak{A}(R, S)$ and $O(A) - O(B)$ is a circuit of length $2k$, then $i(A, B) = k - 1$.*

COROLLARY 2.1.11. *The diameter of the graph $G(R, S)$ of the class $\mathfrak{A}^{m,n}(R, S)$ does not exceed $mn/2 - 1$.*

2.1.8. Connectivity of the graph $G(R, S)$ and the minimal degree of vertices. As noted above, Theorem 2.1.3 asserts that the graph $G(R, S)$ is connected if the class $\mathfrak{A}(R, S)$ consists of matrices with at least 2 columns. It is obvious that the graph $G(R, S)$ is also connected in the case $n = 1$, since then the nonempty class consists of a single matrix. We consider the problem of connectivity of $G(R, S)$ in more detail, following the paper [**60**].

Recall that the connectivity (more precisely, vertex connectivity) $\varkappa = \varkappa(G)$ of a graph G is the minimum number of vertices whose removal from G gives a disconnected or trivial graph. Theorem 2.1.3 shows that $\varkappa(G(R, S)) \geq 1$. Let $\delta(G(R, S))$ be the minimal degree of vertices of $G(R, S)$. If $G(R, S)$ consists of a single vertex, then $\varkappa(G(R, S)) = 1$. If $\delta(G(R, S)) = 1$, then, taking a vertex of degree 1, we see that the adjacent vertex also has degree 1, that is, $G(R, S)$ is K_1 and $\varkappa(G(R, S)) = 1$. It is not difficult to give examples of graphs $G(R, S)$ with $\delta(G(R, S)) = 2$. Let, for example, $m = 2$, $n = 3$, $R = (2, 1)$, $S = (1, 1, 1)$. Then $\mathfrak{A}(R, S)$ consists of the matrices

$$\begin{pmatrix} 1 & 1 & 0 \\ 0 & 0 & 1 \end{pmatrix}, \quad \begin{pmatrix} 1 & 0 & 1 \\ 0 & 1 & 0 \end{pmatrix}, \quad \begin{pmatrix} 0 & 1 & 1 \\ 1 & 0 & 0 \end{pmatrix}.$$

Each of them has degree 2 in $G(R,S)$, and $G(R,S) = K_3 = C_3$, that is, $G(R,S)$ is a cycle with three vertices. It is clear that $\varkappa(G(R,S)) = 2$.

Now let $m = n = 4$, $R = S = (3,3,1,1)$. Then $G = G(R,S)$ consists of the matrix

$$A = \begin{pmatrix} 1 & 1 & 0 & 1 \\ 1 & 1 & 1 & 0 \\ 0 & 1 & 0 & 0 \\ 1 & 0 & 0 & 0 \end{pmatrix}$$

and three more matrices which can be obtained from A by interchanges in the submatrices $A[\{1,2\}],\{3,4\}]$ and $A[\{3,4\},\{1,2\}]$. It is clear that $G = C_4$ and $\varkappa(G) = 2$.

Let us show that a graph $G = G(R,S)$ with $\varkappa(G) = 2$ can be either C_3 or C_4. First we prove the following auxiliary result.

LEMMA 2.1.12. *Let $A, B \in \mathfrak{A}(R,S)$, and let the even graph $O(A) - O(B)$ be decomposed into circuits among which there is a circuit of length $t \geq 6$. Then the degree of A in $G(R,S)$ is at least* 3.

PROOF. Indeed, let $\gamma = (v_1, v_2, \dots, v_{2t}, v_1)$. Considering the diagonals connecting v_1 with $v_4, v_6, \dots, v_{2t-2}$ in the proof of Theorem 2.1.6, we proved that $O(A)$ contains a circuit of length 4 such that one of its arcs coincides with one of the diagonals. The same is true for the set of diagonals from v_2 and the set of diagonals from v_3. Thus, we can find three circuits of length 4. It is clear that they are pairwise distinct, since each of them contains a diagonal which is not contained in the other circuits. □

THEOREM 2.1.13. *Let the interchange graph $G = G(R,S)$ of some nonempty Ryser class be such that $\delta(G) = 2$. Then $G = C_3$ or $G = C_4$.*

PROOF. Let A be a vertex of degree 2 in $G(R,S)$. Then, by Lemma 2.1.12, $O(A)$ does not contain a circuit of length $t > 4$. This means that $O(A)$ consists of exactly two different circuits γ_1 and γ_2 of length 4. Consequently, $O(A)$ contains either two (strongly connected) components δ_1 and δ_2 coinciding, respectively, with γ_1 and γ_2, or one component δ containing both circuits γ_1 and γ_2. We denote by $X = \{x_1, \dots, x_a\}$ and $Y = \{y_1, \dots, y_b\}$ the parts of the bipartite graph $O(A)$.

First we consider the case where $O(A)$ consists of a single component. In this case, the strongly connected graph $O(A)$ contains a vertices of X, b vertices of Y and $ab \leq 8$ arcs. Thus, $ab = 4, 6, 8$. If $ab = 4$, then $\gamma_1 = \gamma_2$, and we obtain a contradiction. If $ab = 8$, that is, if $a = 2$, $b = 4$, then let the arcs of $O(A)$ be (x_1, y_1), (x_1, y_2), (x_2, y_3), (x_2, y_4), (y_1, x_2), (y_2, x_2), (y_3, x_1), (y_4, x_1). Then in $O(A)$ there exist 4 distinct contours of length 4, namely,

$$(x_1, y_1, x_2, y_3, x_1), \quad (x_1, y_1, x_2, y_4, x_1), \quad (x_1, y_2, x_2, y_3, x_1), \quad (x_1, y_2, x_2, y_4, x_1).$$

This contradicts the fact that the degree of A is equal to 2. Thus, $a = 2$, $b = 3$, γ_1, γ_2 have two common arcs, and δ is the graph given in Figure 2.2.

Now let $O(A)$ consist of two components δ_1 and δ_2. Then $\gamma_1 = \delta_1$ and $\gamma_2 = \delta_2$. We see that in both the cases after an interchange in the matrix of degree 2 we pass to a matrix A' with the graph $O(A')$ isomorphic (as a digraph) to the graph $O(A)$. Thus, the vertex adjacent to A also has degree 2. Hence $G(R,S)$ is a regular graph of degree 2. In the first and second cases, this graph is C_3 and C_4 respectively. The theorem is proved. □

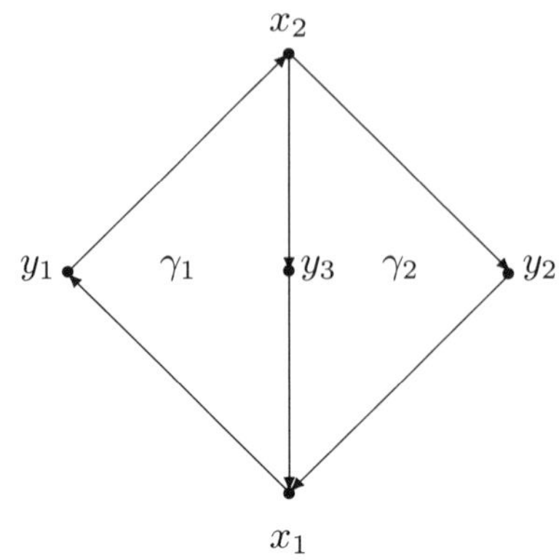

FIGURE 2.2. Graph δ

The examples given above show that classes with graphs C_3 and C_4 really exist.

2.1.9. 3-connectivity of the graphs $G(R, S)$**.** Here we fix an arbitrary nonempty Ryder class $\mathfrak{A}(R, S)$ and consider the graph $G = G(R, S)$. We assume that $\delta(G) \geq 3$, since the graphs with $\delta(G) < 3$ have been described in the previous subsection. The goal is to prove that any graph G with $\delta(G) \geq 3$ is 3-connected. First we prove a result from graph theory. Two paths between vertices u and v of a graph G are called nonintersecting paths if they have no common vertices except u and v.

PROPOSITION 2.1.14. *Let G be a connected graph, and $d(a, b)$ the distance between vertices a, b of G. If u and v are any two vertices of G with $d(u, v) = 2$ and there exist three pairwise nonintersecting paths between u and v, then $\varkappa(G) \geq 3$.*

PROOF. Let S be the set of vertices of the minimal cardinality such that after removing these vertices the graph is disconnected. Suppose that $|S| \leq 2$. If w is a vertex of S, then in view of the minimality of $|S|$ there exist two vertices u and v adjacent to w and belonging to different components of the graphs $G - S$. Thus, $d(u, v) = 2$ and there are no three pairwise nonintersecting paths between u and v. The proposition is proved. $\square$

Now we give a criterion for the existence of three pairwise nonintersecting paths between two vertices A and B of $G(R, S)$.

LEMMA 2.1.15. *Let $A, B \in \mathfrak{A}(R, S)$. If the digraph $O(A) - O(B)$ is a circuit γ of length greater than 4, then there exist three pairwise nonintersecting paths in $G(R, S)$ between A and B.*

PROOF. Let $\gamma = (v_1, v_2, \ldots, v_{2k}, v_1)$ and $k \geq 3$. As we saw in the proofs of Theorem 2.1.6 and Lemma 2.1.12, for each $i = 1, 2, 3$ there exists a sequence of interchanges transforming A into B using the circuits of length 4 constructed using the diagonals from the vertices v_i, $i = 1, 2, 3$. Denote such paths by P_i, and let U_i be an arbitrary point of P_i different from A and B, $i = 1, 2, 3$. Then, by construction, the digraph $O(A) - O(U_i)$ contains some diagonal from v_i, but does not contain the diagonals from v_j, $1 \leq j \leq 3$, $i \neq j$. Thus, the matrices U_1, U_2, and U_3 are pairwise distinct. $\square$

THEOREM 2.1.16. *Let the graph $G = G(R, S)$ be such that $\delta(G) \geq 3$. Then G is a 3-connected graph.*

PROOF. We use the criterion of Proposition 2.1.14. Let A, B be matrices of $\mathfrak{A}(R, S)$ with $i(A, B) = 2$, that is, B can be obtained from A by a sequence of two interchanges. Then for $O(A) - O(B)$ the inequality $q(A, B) \leq 2$ holds. If $q(A, B) = 1$, then $O(A) - O(B)$ is a circuit of length 6 and the result follows from Lemma 2.1.15. Therefore, from here on we assume that $q(A, B) = 2$. We construct three nonintersecting paths from A to B. Denote by γ_1 and γ_2 the pairwise nonintersecting circuits of length 4 forming the decomposition of $O(A) - O(B)$. Let A_i be the matrix obtained from A by the interchange which is equivalent to reversing the directions of the arcs of γ_i, $i = 1, 2$. Then $P_i = A, A_i, B$, $i = 1, 2$, are two nonintersecting paths connecting A and B. Since $\delta(G) \geq 3$, there exists a matrix A_3 adjacent to A which does not coincide with A_1 and A_2. For this matrix,

$$i(A_3, B) \leq i(A, B) + 1 = 3, \qquad q(A_3, B) \leq i(A_3, B) \leq 3.$$

We consider four possible cases.

First suppose that $q(A_3, B) = 1$. Then, depending on the length of the circuit of $O(A_3) - O(B)$, either A_3 and B are adjacent vertices and A, A_3, B is the third required path, or we can apply Lemma 2.1.15.

Let $q(A_3, B) = 2$, $i(A_3, B) = 2$. Then $O(A_3) - O(B)$ can be represented as the union of two circuits γ_1', γ_2' of length 4. Since $A_3 \neq A$, we have $\{\gamma_1, \gamma_2\} \neq \{\gamma_1', \gamma_2'\}$. Let $\gamma_1' \neq \gamma_1, \gamma_2$. Then we can construct the required path P from A_3 to B by changing the directions of arcs, first, of the circuit γ_2' and then of the circuit γ_1'. It is clear that P does not go through A_1 and A_2.

Let $q(A_3, B) = 2$, $i(A_3, B) = 3$. Then $O(A_3) - O(B)$ is the union of nonintersecting circuits γ_1' of length 4 and $\gamma_2' = (v_1, v_2, \ldots, v_6, v_1)$ of length 6. We take a diagonal in γ_2' which does not coincide with an arc of the circuit γ_1'. Without loss of generality we can take $\{v_1, v_4\}$ as such a diagonal. We construct the following three paths P_1, P_2, and P_3 from A_3 to B.

The path P_1 is obtained by first changing the directions of arcs of γ_1', then changing the directions of arcs of γ_2' by sequential changes of the directions of arcs of two circuits formed by the arcs of γ_2' and the diagonal which does not coincide with $\{v_1, v_4\}$.

The path P_2 is obtained by first changing the directions of arcs of γ_2' in the same way as for P_1, and then changing the direction of arcs of γ_1'.

Finally, we obtain the path P_3 as follows. First we change the directions of arcs in one of the circuits of length 4 going through $\{v_1, v_4\}$, then in γ_1', and, finally, in the diagonal $\{v_1, v_4\}$ and the remaining three arcs of γ_2'.

As a result we obtain three pairwise nonintersecting paths from A_3 to B.

Finally, let $q(A_3, B) = 3$. Then the decomposition of $O(A_3) - O(B)$ consists of three different circuits γ_1', γ_2', γ_3' of length 4. We can construct three pairwise nonintersecting paths from A_3 to B sequentially changing the directions of arcs of the contours γ_i', γ_{i+1}', γ_{i+2}', $i = 1, 2, 3$ (the indices are taken modulo 3). At least one of them does not contain A_1, A_2.

Thus, in all cases we have constructed three pairwise nonintersecting paths between two arbitrary vertices with distance 2 in the graph $G(R, S)$. By Proposition 2.1.14 the graph $G(R, S)$ is 3-connected. The theorem is proved. □

The following theorem summarizes the results on the connectivity of the graph $G = G(R, S)$.

THEOREM 2.1.17. *For any nonempty Ryser class $\mathfrak{A}(R, S)$, the graph $G = G(R, S)$ either is 3-connected or coincides with one of the graphs K_1, K_2, C_3 or C_4.*

Problems.

1. Let the matrices
$$A = \begin{pmatrix} 1 & 0 & 1 & 1 & 1 \\ 1 & 1 & 0 & 1 & 0 \\ 0 & 1 & 1 & 0 & 0 \end{pmatrix}, \qquad B = \begin{pmatrix} 1 & 1 & 1 & 1 & 0 \\ 1 & 1 & 1 & 0 & 0 \\ 0 & 0 & 0 & 1 & 1 \end{pmatrix}$$
of the class $\mathfrak{A}^{3,5}((4,3,2),(2,2,2,2,1))$ be given. Find a sequence of interchanges transforming A into B and passing through the canonical matrix of the class. Is this path from A to B minimal?
2. Applying the method used in the proof of Lemma 2.1.4, enumerate the matrices of the class $\mathfrak{A}^{4,3}(R, S)$, where $R = (2, 2, 2, 1)$, $S = (3, 2, 2)$.
3. Let C_1, C_2 be cyclic $(0, 1)$-matrices of order n with generating vectors $(\alpha_1, \dots, \alpha_n)$ and $(\beta_1, \dots, \beta_n)$, where $\alpha_i = \beta_i$, $i \neq 1, n$, $\beta_1 = 1 - \alpha_1$, $\beta_n = 1 - \alpha_n$, $\beta_1 + \beta_n = 1$. Prove that $i(C_1, C_2) = n - 1$.
4. Show that the diameter of the graph $G(R, S)$ for $\mathfrak{A}(R, S) = \mathfrak{A}^n(r)$ does not exceed $n(n - r) - 1$.

2.2. Invariant sets

In this section we study the structure of matrices belonging to the same Ryser class $\mathfrak{A}^{m,n}(R, S)$ from a slightly different point of view as compared with Section 2.1: we find a representation of the matrices of a given monotone class in some standard form. Each matrix of the class can be composed from rectangular blocks whose sizes are the same for all matrices of the class. If the position of a block of size $m' \times n'$ is determined by subsets $I \subseteq N_m$, $J \subseteq N_n$, $|I| = m'$, $|J| = n'$, $m' \le m$, $n' \le n$, then the content of the block is either one of the matrices $\mathbf{J}$ or $\mathbf{0}$ or a matrix $A[I, J]$ belonging to some class $\mathfrak{A}^{m',n'}(R'(I, J), S'(i, J))$, the same for any matrix $A \in \mathfrak{A}^{m,n}(R, S)$. The main tool of the study is the structure matrix T of the class $\mathfrak{A}(R, S)$ introduced in Section 1.2.

2.2.1. Properties of the structure matrix. Let $R = (r_1, \dots, r_m)$, $S = (s_1, \dots, s_n)$ be monotone integer-valued vectors and $\mathfrak{A}^{m,n}(R, S)$ the corresponding monotone class of $(0, 1)$-matrices. The structure matrix $T = (t_{kl})$ is the $(m + 1) \times (n + 1)$ integer-valued matrix with elements

$$t_{kl} = kl + \sum_{i>k} r_i - \sum_{j \le l} s_j, \quad k = 0, 1, \dots, m, \quad l = 0, 1, \dots, n. \tag{2.2.1}$$

The main property of the structure matrix of a nonempty Ryser class was formulated in Theorem 1.2.8: the class $\mathfrak{A}(R, S)$ is not empty if and only if the matrix T is nonnegative. In the first part of the proof of this theorem the following fact was established. Let A be an arbitrary $(0, 1)$-matrix of size $m \times n$, and let k and l be integers such that $0 \le k \le m$, $0 \le l \le n$, $I \subseteq N_m$, $J \subseteq N_n$, $|I| = k$, $|J| = l$ (I and J can be empty). Then

$$t(I, J) = \sigma_0(A[I, J]) + \sigma_1(A[N_m - I, N_n - J]), \tag{2.2.2}$$

where $\sigma_0(X)$, $\sigma_1(X)$ are, respectively, the number of zeros and the number of ones in the $(0,1)$-matrix X (we set $N_0 = \emptyset$).

The following two properties of the structure matrix follow from (2.2.2).

PROPOSITION 2.2.1. *Let $\mathfrak{A}^{m,n}(R,S)$ be a nonempty monotone Ryser class, and $T = (t_{ij})$ the structure matrix of the class. Then the equality $t_{ef} = 0$ for $(e,f) \neq (0,n),(m,0)$ holds if and only if each matrix $A \in \mathfrak{A}^{m,n}(R,S)$ is of the form*

$$A = \begin{pmatrix} \mathbf{J}_{ef} & * \\ * & \mathbf{0} \end{pmatrix},$$

where the matrices $\mathbf{J}_{ef}$ and $\mathbf{0}$ are nonempty, if $ef \neq 0$.

PROPOSITION 2.2.2. *If the structure matrix $T = (t_{ij})$ of a nonempty monotone class is such that $t_{ef} = 0$, then $t_{e'f'} \neq 0$ for any $e' > e$, $f' > f$.*

Indeed, if $t_{e'f'} = 0$, then (2.2.2) implies that $A[N_{e'} - N_e, N_{f'} - Nf] = \mathbf{J}$, but, on the other hand, this nonempty submatrix must be the zero-matrix.

Further, from (2.2.1) it directly follows that

$$t_{k,l+1} - t_{kl} = k - s_{l+1}, \tag{2.2.3}$$

$$t_{k+1,l} - t_{kl} = l - r_{k+1}, \tag{2.2.4}$$

$k = 0,1,\dots,m-1$, $l = 0,1,\dots,n-1$. These relations allow us to construct the structure matrix by going along a row from left to right or along a column from top to bottom. For $k = 0$, it follows from (2.2.3) that

$$t_{0,l+1} = t_{0l} - s_{l+1} = t_{00} - \sum_{j=1}^{l+1} s_j, \qquad l = 0,1,\dots,n-1. \tag{2.2.5}$$

For $l = 0$, it follows from (2.2.4) that

$$t_{k+1,0} = t_{k0} - r_{k+1} = t_{00} - \sum_{i=1}^{k+1} r_i, \qquad k = 0,1,\dots,m-1. \tag{2.2.6}$$

In addition,

$$t_{00} = \sum_{i=1}^{m} r_i = \sum_{j=1}^{n} s_j = \sigma_1(A).$$

Thus, by (2.2.5) the elements of the row with number 0 are

$$t_{0l} = \sum_{j \geq l+1}^{n} s_j, \quad l = 0,1,\dots,n-1, \quad t_{0n} = 0,$$

and by (2.2.6) the elements of the column with number 0 are

$$t_{k0} = \sum_{i \geq k+1}^{m} r_i, \quad k = 0,1,\dots,m-1, \quad t_{m0} = 0.$$

Thus, the elements of the row and column with number 0 of the structure matrix are constructed by using the vectors R and S, and the remaining elements can be found either by the use of (2.2.3) one row after another, or by the use of (2.2.4) one column after another.

Let us give one more representation of the structure matrix of the monotone class $\mathfrak{A}^{m,n}(R,S)$. Consider the matrix

$$T^* = \begin{pmatrix} \tau & -s_1 & -s_2 & \dots & -s_n \\ -r_1 & 1 & 1 & \dots & 1 \\ -r_2 & 1 & 1 & \dots & 1 \\ & & & \dots & 1 \\ -r_m & 1 & 1 & \dots & 1 \end{pmatrix}, \tag{2.2.7}$$

where $r_1, \dots, r_m$, $s_1, \dots, s_n$ are the coordinates of the vectors R and S, respectively, $\tau = r_1 + \dots + r_m$. We denote by E_n the lower triangular $(0,1)$-matrix of order n such that ones are located on and below the principal diagonal and zeros are located above the diagonal.

PROPOSITION 2.2.3 ([**88**]).

$$T = E_m T^* E_n^T. \tag{2.2.8}$$

The proposition can be verified directly.

The construction of the matrix T^* is simpler than the construction of T. The matrix E_m is not singular; therefore (2.2.8) shows that the ranks of T and T^* coincide. Thus, some problems concerning T, which are typical for linear algebra, can be reduced to the corresponding problems for T^*.

2.2.2. Unimodality of the structure matrix along rows and columns. Before we proceed to further investigations of the properties of the structure matrices, we consider several examples.

EXAMPLE 2.2.1. Let $m = n = 4$, $R = S = (3, 2, 2, 1)$. For these vectors, using relations (2.2.3)–(2.2.6), we construct the structure matrix

$$T = \begin{pmatrix} 8 & 5 & 3 & 1 & 0 \\ 5 & 3 & 2 & 1 & 1 \\ 3 & 2 & 2 & 2 & 3 \\ 1 & 1 & 2 & 3 & 5 \\ 0 & 1 & 3 & 5 & 8 \end{pmatrix}.$$

EXAMPLE 2.2.2. Let $m = 5$, $n = 6$, $R = (5, 4, 4, 2, 2)$, $S = (5, 5, 3, 2, 2, 0)$. Then

$$T = \begin{pmatrix} 17 & 12 & 7 & 4 & 2 & 0 & 0 \\ 12 & 8 & 4 & 2 & 1 & 0 & 1 \\ 8 & 5 & 2 & 1 & 1 & 1 & 3 \\ 4 & 2 & 0 & 0 & 1 & 2 & 5 \\ 2 & 1 & 0 & 1 & 3 & 5 & 9 \\ 0 & 0 & 0 & 2 & 5 & 8 & 13 \end{pmatrix}$$

(the role of the upper and lower broken lines will be seen in Section 2.2.6).

We see that the classes corresponding to the vectors R and S in Examples 2.2.1 and 2.2.2 are nonempty.

EXAMPLE 2.2.3. Consider the class $\mathfrak{A}^n(k)$ introduced in Section 1.1. For this class, $R = S = (k, \dots, k)$. It follows from relations (2.2.3)–(2.2.6) that the sequence of elements of the eth row is an arithmetic progression with difference $e - k$ and the sequence of elements of the fth column is an arithmetic progression with difference $f-k$. Further,

$$t_{e0} = k(n-e), \quad t_{0f} = k(n-f), \quad e, f = 0, 1, \dots, n.$$

Consequently,

$$t_{ef} = k(n-e) + f(e-k), \qquad e, f = 0, 1, \dots, n.$$

For example, if $n = 4$, $k = 2$, then

$$T = \begin{pmatrix} 8 & 6 & 4 & 2 & 0 \\ 6 & 5 & 4 & 3 & 2 \\ 4 & 4 & 4 & 4 & 4 \\ 2 & 3 & 4 & 5 & 6 \\ 0 & 2 & 4 & 6 & 8 \end{pmatrix}.$$

From the examples we see that the elements of each row and each column of the structure matrix form a sequence which, generally speaking, is nonincreasing at the beginning and then nondecreasing at the end. Such sequences are called unimodal. In Example 2.2.2, our attention is attracted by the location of zeros: they are located in accordance with rook moves. Later it will be seen that these properties are of the general character (the property which concerns the location of zeros is partially explained by Proposition 2.2.2).

PROPOSITION 2.2.4. *Let $T = (t_{ij})$ be the structure matrix of a nonempty monotone class $\mathfrak{A}^{m,n}(R, S)$, $R = (r_1, \dots, r_m)$, $S = (s_1, \dots, s_n)$. Then the sequence of elements of any row and any column of the matrix T is a unimodal sequence.*

PROOF. Note that the unimodality of a sequence $a_0, a_1, \dots, a_n$ follows from the validity of the inequalities

$$a_k - a_{k+1} \geq a_{k+1} - a_{k+2}, \qquad k = 0, 1, \dots, n-2. \tag{2.2.9}$$

Indeed, if there exists a number q, $0 \leq q \leq n-1$, such that $a_0 \geq a_1, \dots, a_{q-1} \geq a_q$, but $a_q \leq a_{q+1}$, then by (2.2.9) $a_k \leq a_{k+1}$ for all $k = q+1, \dots, n-1$. By (2.2.3) we have

$$t_{ef} - t_{e,f+1} = s_{f+1} - e \geq s_{f+2} - e = t_{e,f+1} - t_{e,f+2}$$

for all e, f, $0 \leq e \leq m$, $0 \leq f \leq n-2$. Thus, for an arbitrary e, the sequence of elements of the eth row is unimodal. For columns the proof is similar. □

COROLLARY 2.2.5. *Let $T = (t_{ij})$ be the structure matrix of a nonempty monotone class $\mathfrak{A}^{m,n}(R, S)$. If for some integers e, f, g, $0 \leq e \leq m$, $0 \leq f < g \leq n$, the equalities $t_{ef} = t_{eg} = 0$ hold, then*

$$t_{e,f+1} = t_{e,f+2} = \cdots = t_{e,g-1} = 0.$$

Similarly, if $t_{df} = t_{ef} = 0$ for some d, e, f, $0 \leq d < e \leq m$, $0 \leq f \leq n$, then

$$t_{d+1,f} = t_{d+2,f} = \cdots = t_{e-1,f} = 0.$$

It is possible to point out the maximum number of the row where the sequence of elements of a given column of the structure matrix attains the minimal value. Let $R = (r_1, \dots, r_m)$ be a integer-valued nonnegative monotone vector and

$$s_f^* = \max\{i \colon r_i \geq f, 0 \leq i \leq m\},$$

where $f \leq r_1$ is a positive integer; for $f > r_1$ we set $s_f^* = 0$. Thus, the numbers $s_1^*, s_2^*, \dots$ can be viewed as parts of the partition of the number $\tau = \sum_{i=1}^m r_i$ which is conjugate to the partition with parts $r_1, \dots, r_m$. Note that here we deviate from the notation accepted in the theory of partitions, where parts of the partition conjugate to the partition with parts $r_1, r_2 \dots$ are usually denoted by $r_1^*, r_2^*, \dots$.

THEOREM 2.2.6. *Let $T = (t_{ij})$ be the structure matrix of a nonempty monotone class $\mathfrak{A}^{m,n}(r, S)$, where $R = (r_1, \dots, r_m)$, $S = (s_1, \dots, s_n)$. Then for any f, $0 \leq f \leq n$, the maximum number q of the row where the sequence $t_{0f}, t_{1f}, \dots, t_{mf}$ attains the minimum is equal to s_j^*. This minimum is equal to*

$$(s_1^* + \dots + s_f^*) - (s_1 + \dots + s_f).$$

PROOF. First let $r_m < f \leq r_1$ and the integer $q \geq 1$ be such that (see (2.2.4))

$$t_{q-1,f} - t_{qf} = r_q - f \geq 0 > r_{q+1} - f = t_{qf} - t_{q+1,f} \tag{2.2.10}$$

for some f, $0 \leq f \leq n$ (such a q exists according to Proposition 2.2.4). But inequalities in (2.2.10) determine q uniquely as the number of rows with sum equal at least to f, that is, $q = s_f^*$. Further, taking into account that $s_1^*, s_2^*, \dots$ are the components of the partition conjugate to the partition $\tau = r_1 + \dots + r_m$, we obtain

$$t_{qf} = fs_f^* + \sum_{i>q} r_i - \sum_{j \leq f} s_j = \sum_{j=1}^{f} s_f^* - \sum_{j=1}^{f} s_j.$$

For $r_m \geq f$, the sequence $t_{0f}, \dots, t_{mf}$ is nonincreasing and $s_f^* = m$. For $r_1 < f$, we have

$$t_{0f} < t_{1f} < \dots, \qquad \sum_{i=1}^{m} r_i = \sum_{j=1}^{f} s_j^*,$$

and the minimal value is equal to

$$t_{0f} = \sum_{j=1}^{f} s_j^* - \sum_{j=1}^{f} s_j.$$

The theorem is proved. □

2.2.3. Invariant sets and invariant positions. Let $\mathfrak{A}$ be a nonempty Ryser class consisting of (0, 1)-matrices of size $m \times n$. Suppose that for some nonempty subsets $I \subseteq N_m$, $J \subseteq N_n$ the number $\sigma_1(A[I, J])$ of elements equal to 1 in the matrix $A[I, J]$ does not depend on the choice of $A \in \mathfrak{A}$. Then the set $I \times J$ is called an invariant set for the class $\mathfrak{A}$. In the case where $I = \{i\}$, $J = \{j\}$, $i \in N_m$, $j \in N_n$, the invariant set $\{i\} \times \{j\}$ is called an invariant position of the class $\mathfrak{A}$. If in any matrix of the class an invariant position is occupied by 1 (by 0), then the position is called an invariant 1-position (invariant 0-position). If the class $\mathfrak{A}$ has an invariant 1-position (0-position), then we say that $\mathfrak{A}$ contains invariant one (invariant zero). Matrices of the form

$$A = \begin{pmatrix} \mathbf{J} & * \\ * & \mathbf{0} \end{pmatrix} \tag{2.2.11}$$

provide evident examples of invariant sets. Let A be an $m \times n$ matrix, and let the size of $\mathbf{J}$ be $e \times f$. Then any set of the form $K \times L$, where $K \subseteq N_e$, $L \subseteq N_f$, is an invariant set in the class containing A. In particular, any position (i, j), $i \in N_e$, $j \in N_f$, gives an invariant one. Similarly, any subset of the set $(N_m - N_e) \times (N_n - N_f)$ is an invariant set; in particular, any zero in the position (i, j), $i \in N_m - N_e$, $j \in N_n - N_f$, is an invariant zero. In fact, the existence of invariant positions in a monotone class means that the class consists of matrices of the form (2.2.11).

THEOREM 2.2.7 (Ryser–Haber). *Let $\mathfrak{A}$ be a monotone class consisting of* $(0,1)$-*matrices of size* $m \times n$. *The class* $\mathfrak{A}$ *has an invariant positions if and only if there exist integers* e, f, $0 \le e \le m$, $0 \le f \le n$, $(e, f) \ne (0, n), (m, 0)$, *such that any matrix of the class* $\mathfrak{A}$ *is of the form* (2.2.11).

PROOF. According to the above, we have to prove only the necessity of the condition. Let $\mathfrak{A}$ contain an invariant one on a position (i, j). Then any position (i', j') for $i' \le i$, $j' \le j$ is an invariant 1-position. Indeed, if this is not the case and the class contains a matrix A with zero in the position (i', j'), then (since the class is monotone) by one interchange (if $i = i'$ or $j = j'$) or by two interchanges (otherwise) we obtain a matrix with zero in the position (i, j), which contradicts the invariance of this 1-position.

Now let (e, f), $e \ge 1$, $f \ge 1$, be the invariant 1-position with maximal value $e + f$. As we have shown, $A[N_e, N_f] = \mathbf{J}_{ef}$ for any $A \in \mathfrak{A}$. If $e = m$ or $f = n$, then the theorem is proved. Let $e < m$, $f < n$. Show that $A[N_m - N_e, N_n - N_f] = \mathbf{0}$. In view of the maximality of $e + f$, the class $\mathfrak{A}$ contains a matrix $A = (a_{ij})$ with $a_{e,f+1} = 0$. Suppose that $A[N_m - N_e, N_n - N_f]$ contains ones in some positions. Let (e', f') be one of these positions, that is, $a_{e'f'} = 1$. If $a_{e'j} = 0$ for $j \le f$, then by one interchange (if $a_{e',f+1} = 1$) or by two interchanges (if $a_{e',f+1} = 0$) we transform the matrix A to a matrix with zero in the position (e, j), which contradicts the invariance of this 1-position. Thus, $a_{ij} = 1$ for all $j \in N_f$ and each $i > e$ such that the matrix $A[N_m - N_e, N_n - N_f]$ contains one in the ith row. This means that no interchange can remove 1 from any such position (i, j); that is, ones in these positions are invariant. This contradicts the maximality of $e + f$. The following assertion can be proved similarly: if $\mathfrak{A}$ contains an invariant zero in the position $(e + 1, f + 1)$, where $0 \le e < m$, $0 \le f < n$ and $e + f$ is minimal, then any matrix of the class $\mathfrak{A}$ is of the form (2.2.11) with the matrix $\mathbf{J}$ of size $e \times f$. The theorem is proved. □

COROLLARY 2.2.8. *A nonempty monotone Ryser class* $\mathfrak{A}^{m,n}(R, S)$ *has invariant positions if and only if* $t_{ef} = 0$ *for an element of the structure matrix* $T = (t_{ij})$ *of the class* $\mathfrak{A}$ *for some* e, f *such that* $(e, f) \ne (m, 0), (0, n)$.

The corollary is obtained by applying Proposition 2.2.1.

Example 2.2.2 shows that the structure matrix can have several zero elements. Thus, the numbers e, f are not determined uniquely.

2.2.4. Composite classes. Consider a nonempty (not necessary monotone) Ryser class $\mathfrak{A} = \mathfrak{A}^{m,n}(R, S)$. The class $\mathfrak{A}$ is called row-composite if there exists a nonempty proper subset $L \subset N_n$ such that the vector of row sums of the submatrix $A[N_m, L]$ is the same for any matrix $A \in \mathfrak{A}$. Similarly, the class $\mathfrak{A}$ is called column-composite if there exists a nonempty proper subset $K \subset N_m$ such that the vector of column sums of the matrix $A[K, N_n]$ is the same for any matrix $A \in \mathfrak{A}$. A class $\mathfrak{A}(R, S)$ is called composite if it is either row-composite or column-composite. A class which is not composite is called elementary.

Consider an example of a composite class. Let a Ryser class $\mathfrak{A}$ contain an $m \times n$ matrix of the form (2.2.11)

$$A = \begin{pmatrix} \mathbf{J}_{ef} & A_1 \\ A_2 & \mathbf{0} \end{pmatrix},$$

where $0 \le e \le m$, $0 \le f \le n$. Then it is clear that each matrix of the class $\mathfrak{A}$ is of the same form. It is easy to see that if $0 < f < n$, then the class $\mathfrak{A}$ is row-composite, and if $0 < e < m$, then $\mathfrak{A}$ is column-composite; under the conditions $0 < f < n$, $0 < e < m$ the class $\mathfrak{A}$ is row-composite and column-composite simultaneously. It is clear that in this example the sets $N_e \times N_f$, $(N_m - N_e) \times N_f$, $N_e \times (N_n - N_f)$, $(N_m - N_e) \times (N_n - N_f)$ are invariant sets. Further it will be seen that the given example of a composite class is a model case; that is, all composite monotone classes have the same structure.

The following result can help us to imagine a concrete structure of composite classes.

LEMMA 2.2.9. *Let $\mathfrak{A} = \mathfrak{A}^{m,n}(R,S)$ be a nonempty row-composite class, and let L_1, L_2 be minimal (with respect to inclusion) subsets of N_n such that the vector of row sums of $A[N_m, L_s]$, $s = 1, 2$, does not depend on the choice of $A \in \mathfrak{A}$. Then either $L_1 = L_2$ or $L_1 \cap L_2 = \emptyset$. Similarly, for a column-composite class $\mathfrak{A}$, minimal subsets $K_1, K_2 \subset N_m$ such that the vector of column sums of $A[K_s, N_n]$, $s = 1, 2$, does not depend on the choice of $A \in \mathfrak{A}$ either coincide or do not intersect.*

PROOF. Let $L = L_1 \cap L_2 \neq \emptyset$, and let L be a proper subset of L_1 and L_2. In view of the minimality of L_1 there exist two matrices $A = [a_{ij}], B = [b_{ij}] \in \mathfrak{A}$ and $k \in N_m$ such that $\sigma_1(A[\{k\}, L]) > \sigma_1(B[\{k\}, L])$. Therefore $a_{kl} = 1$, $b_{kl} = 0$ for some $l \in L$. Then by the conditions of the lemma there exists a position (k, j) such that $a_{kj} = 0$, $b_{kj} = 1$, $j \in L_1 - L$. We denote by s_j, s_l the sums of elements of the jth and lth columns respectively. Let $s_j \le s_l$. Then, as is easily seen, the jth and lth columns of the matrix A contain an interchangeable submatrix. By the corresponding interchange we transform the matrix A into a matrix $A' = (a'_{st})$ such that $a'_{kl} = 0$, $a'_{kj} = 1$ and the remaining elements of the kth row coincide with the corresponding elements of A. Consequently,

$$\sigma_1(A'[\{k\}, L_2]) = \sigma_1(A'[\{k\}, L]) + \sigma_1(A[\{k\}, L_2 - L]) < \sigma_1(A[\{k\}, L_2]).$$

This inequality contradicts the identity of corresponding row sums for all matrices in $\mathfrak{A}$. Similarly, if $s_j < s_l$, then the jth and lth columns of the matrix B contain an interchangeable submatrix and by the corresponding interchange the matrix B is transformed into a matrix B' such that

$$\sigma_1(B'[\{k\}, L_2]) > \sigma_1(B[\{k\}, L_2]).$$

The assertion of the lemma for column-composite classes is proved similarly. The lemma is proved. □

Lemma 2.2.9 helps us make a step to the representation of matrices of a given Ryser class as a composition of blocks which we mentioned above. Indeed, let $K_1, \dots, K_p$ be a partition of the set N_m into minimal disjoint subsets such that for each matrix $A = (a_{ij})$ of a given class $\mathfrak{A}$ the column sums

$$\sum_{i \in K_r} a_{ij}, \qquad r = 1, \dots, p,$$

are equal to each other, and let $L_1, \dots, Lq$ be a similar partition of the set N_n. Then the vectors $R^{(i)}$ and $S^{(i)}$ of the row and column sums, respectively, in any matrix $A[K_i, L_j]$ do not depend on the choice of $A \in \mathfrak{A}$, $i = 1, \dots, p$, $j = 1, \dots, q$. Thus, for any matrix $A \in \mathfrak{A}$ the submatrix $A[K_i, K_j]$ belongs to a fixed class

$\mathfrak{A}^{m_i,n_j}(R^{(i)}, S^{(j)})$, where $m_i = |K_i|$, $n_j = |L_j|$. Note that such a representation is of no value for elementary classes, since in that case $p = q = 1$.

2.2.5. A criterion for composite classes. In this subsection, we give a criterion for determining on the base of the structure matrix whether a class is composite, and clarify the structure of composite classes by obtaining information about the content of some blocks.

THEOREM 2.2.10. *Let* $\mathfrak{A} = \mathfrak{A}^{m,n}(R, S)$ *be a monotone Ryser class and* $T = (t_{ij})$ *the structure matrix of* $\mathfrak{A}$. *Then* $\mathfrak{A}$ *is row-composite if and only if* $t_{ef} = 0$ *for some integers* e, f, $0 \le e \le m$, $0 < f < n$. *The class* $\mathfrak{A}$ *is column-composite if and only if* $t_{ef} = 0$ *for some* $0 < e < m$, $0 \le f \le n$.

PROOF. Let $t_{ef} = 0$ and $(e, f) \ne (0, n), (m, 0)$. Then by Proposition 2.2.1 all matrices of the class are of the form (2.2.11). As shown above, if $0 < f < n$, then the class $\mathfrak{A}$ is row-composite, and if $0 < e < m$, then it is column-composite.

Now let $\mathfrak{A}$ be a row-composite class, and $N_n = L_1 \cup L_2$ a partition of the set of indices of columns into subsets with the same vectors of row sums for each subset. Let $j \in L_1$, $l \in L_2$ and $j < l$. Note that the columns with numbers j and l do not contain an interchangeable matrix, since the interchange modifies the vectors of row sums of the submatrices $A[N_m, L_u]$, $u = 1, 2$. Thus, interchanges can be performed only in the columns which both belong either to L_1 or to L_2. Now let a matrix $A = (a_{ij}) \in \mathfrak{A}$ not contain zero columns. Then there exists $a_{il} = 1$ in the lth column. The element a_{ij} also equals 1, since otherwise the jth and lth columns would contain an interchangeable submatrix by the monotonicity of the class. For the same reason the one in the position (i, j) must be invariant. Then by Theorem 2.2.7 all matrices in $\mathfrak{A}$ are of the form (2.2.11), and by Corollary 2.2.8 there are integers e, f such that $t_{ef} = 0$ and $(e, f) \ne (0, n), (m, 0)$.

Now let a matrix of the class $\mathfrak{A}$ contain $n - f > 0$ zero columns. Then, obviously, $t_{ef} = 0$ (see, for example, (2.2.2)). Thus, the assertion of the theorem for row-composite classes is proved. The proof for the column-composite classes is carried out in perfect analogy. □

COROLLARY 2.2.11. *If the matrices of a Ryser class do not contain lines consisting only of zeros or only of ones, then the class is row-composite if and only if it is column-composite.*

COROLLARY 2.2.12. *Let* $\mathfrak{A}^{m,n}(R, S)$ *be a nonempty Ryser class. If* $mn \ge 2$, *then it contains an invariant position if and only if it is composite.*

We see that all matrices of row-composite and column-composite classes are in fact of the form (2.2.11).

Thus, invariant positions necessarily exist in composite classes and only in them. But, maybe, invariant sets of greater cardinality can be found in elementary classes as well. This assumption is wrong: Brualdi and Ross [**27**] proved that if a Ryser class contains an invariant set, then it necessarily contains an invariant position. We do not give the proof of this interesting result here because of its technical complexity, but instead refer the reader to [**27**].

2.2.6. The standard representation of composite classes. The results obtained above show that the structure of composite classes depends on the location of zeros in the corresponding structure matrices T. Proposition 2.2.2 and

Corollary2.2.5 give the following picture of the location of positions occupied by zeros in $T = (t_{ij})$.

In each row and each column containing zeros, they form a unique interval (see Corollary 2.2.5).

If $t_{ef} = t_{e'f'} = 0$, then either $e \le e'$ or $f \le f'$ (see Proposition 2.2.2).

Thus, positions of zeros are sequences of horizontal and vertical intervals, which can form broken lines. These sequences can be ordered from left bottom to right top. Some properties of the location of zeros in the structure matrix are given in the following lemmas.

LEMMA 2.2.13. *Let $T = (t_{ij})$ be the structure matrix of a nonempty monotone Ryser class. Let $t_{ef} = 0$ and either $t_{e'f} \ne 0$ for some $e' < e$ or $t_{ef'} \ne 0$ for some $f' > f$. In this case, if $t_{e-1,g} = 0$, where $g > f$, then*

$$t_{e-1,f} = t_{e-1,f+1} = \cdots = t_{e-1,g-1} = 0;$$

similarly, if $t_{d,f+1} = 0$, where $d < e$, then

$$t_{df} = t_{d+1,f} = \cdots = t_{e-1,f} = 0.$$

The proof is easily carried out with the use of formula (2.2.2), and we leave it to the reader (see Problem 3 in this section).

LEMMA 2.2.14. *If (e, f) and (e', f') are, respectively, the positions corresponding to the end of one broken line in the sequence of zeros of the structure matrix and to the beginning of the other, then $|e - e'|, |f - f'| \ge 2$.*

The assertion follows from Lemma 2.2.13.

Let $\mathfrak{A} = \mathfrak{A}^{m,n}(R, S)$ be a nonempty Ryser class with monotone vectors $R = (r_1, \ldots, r_m)$, $S = (s_1, \ldots, s_n)$ and the structure matrix $T = (t_{ij})$. Suppose that zeros of T form v connected components, which we denote by $P_{i-1} \to Q_i$, where P_{i-1} is the initial point and Q_i is the end point of the ith component, $i = 1, \ldots, v$. Let

$$P_0 = (m, 0), \quad P_i = (e_i, f_i), \quad Q_i = (g_i, h_i), \quad i = 1, \ldots, v-1, \quad Q_v = (0, n),$$
$$m = e_0 \ge g_1 > e_1 \ge g_2 > \cdots > e_{v-1} \ge g_v = 0,$$
$$0 = f_0 \le h_1 < f_1 \le h_2 < \cdots < f_{v-1} \le h_v = n.$$

In particular, for an elementary class, $v = 2$, $P_0 = Q_1 = (m, 0)$, $P_1 = Q_2 = (0, n)$. We draw two broken lines Λ_0 and Λ_1 in the structure matrix. Let the lines be parallel and follow the sequences $P_0 \to Q_1, \ldots, P_{v-1} \to Q_v$ and border them from below and from the right. Between these intervals, the broken line Λ_0 consists of the horizontal interval from the point $Q_i = (g_i, h_i)$ to the point (g_i, f_i) and the vertical interval from (g_i, f_i) to $(e_i, f_i) = P_i$, and Λ_1 consists of the vertical interval from Q_i to the point (e_i, h_i) and the horizontal interval from (e_i, h_i) to P_i, $i = 1, \ldots, v-1$. The broken lines Λ_0 and Λ_1 are drawn in the structure matrix T in Example 2.2.2. For an arbitrary matrix $A \in \mathfrak{A}$, we slightly change the broken lines Λ_0 and Λ_1, cutting off the initial horizontal interval corresponding to zero in T in the position $(m, 0)$ and the end vertical interval corresponding to zero in T in the position $(0, n)$. Then

- to the left of Λ_1 and above Λ_1 in A, a set of invariant ones is located,
- to the right of Λ_0 and below Λ_0 in A, a set of invariant zeros is located,
- and $m_i = |g_i - e_i| \ge 2$, $n_i = |f_i - h_i| \ge 2$, $i = 1, \ldots, v-1$ (see Lemma 2.2.14).

Let us clarify what is located in the remaining positions of T. It is clear that all sets of the form $(N_{g_i} - N_{e_i}) \times (N_{f_i} - N_{h_i})$ are invariant. Moreover, if

$$\begin{aligned} R^{(i)} &= (r_{e_i+1} - h_i, \ldots, r_{g_i} - h_i), \\ S^{(i)} &= (s_{h_i+1} - e_i, \ldots, s_{f_i} - e_i) \end{aligned} \tag{2.2.12}$$

are integer-valued nonnegative vectors of dimension m_i and n_i, respectively, then the matrix $A[N_{g_i} - N_{e_i}, N_{f_i} - N_{h_i}]$ corresponding to the ith interval belongs to the class $\mathfrak{A}^{(i)} = \mathfrak{A}^{m_i,n_i}(R^{(i)}, S^{(i)})$, $i = 1, \ldots, v-1$, for any matrix $A \in \mathfrak{A}$. On the other hand, if we take arbitrary matrices $A^{(i)} \in \mathfrak{A}^{(i)}$ and put them on the corresponding positions in A and leave intact zeros and ones in the remaining positions, then we obtain a matrix of the class $\mathfrak{A}$.

Let us show that all classes $\mathfrak{A}^{(i)}$, $i = 1, \ldots, v-1$, are elementary. If on the contrary some class $\mathfrak{A}^{(i)}$ is composite, then by Theorems 2.2.10 and 2.2.7 each matrix $A^{(i)} \in \mathfrak{A}^{(i)}$ is of the form (2.2.11) with $\mathbf{J} = \mathbf{J}_{e'f'}$, $(e', f') \neq (0, n'), (m', 0)$. Then by (2.2.2) $t_{ef} = 0$, where $e = e_i + e'$, $f = h_i + f'$, and $(e, f) \neq (g_i, h_i), (e_i, f_i)$. We obtain a contradiction.

Thus, we obtain the following result.

THEOREM 2.2.15. *Let $\mathfrak{A} = \mathfrak{A}^{m,n}(R, S)$ be a monotone Ryser class with $R = (r_1, \ldots, r_m)$ and $S = (s_1, \ldots, s_n)$, and let T be the structure matrix of the class. Also, let the sequence of zeros in T consist of v connected components $P_{i-1} \to Q_i$, where P_{i-1} denotes the initial point and Q_i the end point of the ith component,*

$$\begin{gathered} P_0 = (m, 0), \quad P_i = (e_i, f_i), \quad Q_i = (g_i, h_i), \quad i = 1, \ldots, v-1, \quad Q_v = (0, n), \\ m = e_0 \geq g_1 > e_1 \geq g_2 > \cdots > e_{v-1} \geq g_v = 0, \\ 0 = f_0 \leq h_1 < f_1 \leq h_2 < \cdots < f_{v-1} \leq h_v = n. \end{gathered}$$

In $A \in \mathfrak{A}$ let the broken lines Λ_0 and Λ_1 be drawn as described above. Then any matrix of the class $\mathfrak{A}$ consists of the set of invariant ones located from the left and above Λ_1, the set of invariant zeros located from the right and below Λ_0, and the submatrices $A^{(i)} = A[N_{g_i} - N_{e_i}, N_{f_i} - N_{h_i}]$ belonging to the elementary classes $\mathfrak{A}^{m_i,n_i}(R^{(i)}, S^{(i)})$, where the vectors $R^{(i)}$ and $S^{(i)}$ are defined in (2.2.12) and $m_i, n_i \geq 2$, $i = 1, \ldots, v-1$.

Theorem 2.2.15 reduces the study of composite monotone Ryser classes to the study of several elementary classes of less dimension. The structure of a nonmonotone class can be easily derived from the structure of the corresponding monotone class.

2.2.7. Rank of the structure matrix. The rank of the structure matrix T of the class $\mathfrak{A}^{m,n}(R, S)$, the trace and the eigenvalues (for $m = n$) of this matrix are of interest in theory and applications. These characteristics are investigated with the use of Proposition 2.2.3 and the matrix T^* defined in (2.2.7) (see [**100**] and [**37**]). Here we present one of the main results of this type.

THEOREM 2.2.16 ([**86**]). *Let T be the structure matrix of a nonempty Ryser class $\mathfrak{A}^{m,n}(R, S)$, where $R = (r_1, \ldots, r_m)$, $S = (s_1, \ldots, s_n)$ are integer-valued monotone vectors, and let $r(T)$ be the rank of T. Then*

$$r(T) \leq 3.$$

In addition, $r(T) = 3$ *if and only if* $m, n \geq 2$, $r_1 > r_m$, $s_1 > s_n$; $r(T) = 2$ *if and only if* $m, n \geq 1$ *and either* $1 \leq r_m = \cdots = r_1 < n$ *or* $1 \leq s_n = \cdots = s_1 < m$; $r(T) = 1$ *if and only if either* $r_m = \cdots = r_1 = n$, *or* $r_1 = 0$, $s_1 = 0$.

PROOF. Since $r(T) = r(T^*)$, we will prove all assertions for the matrix T^*.

If $m, n \geq 3$, then any minor of the fourth order of T^* equals zero. Therefore $r(T^*) \leq 3$.

Let $m, n \geq 2$. The minor

$$\Delta = \Delta(T^*[\{0, 1, m\}, \{0, 1, n\}])$$

is equal to $(s_1 - s_n)(r_m - r_1)$. It takes the value zero if and only if $s_1 = s_n$ or $r_1 = r_m$. Therefore the rank of the matrix T^* is 3 if $s_1 \neq s_n$ and $r_1 \neq r_m$.

For $s_n = s_1$ or $r_m = r_1$ any minor of order 3 of T^* equals zero. Consider the minor

$$\Delta_1 = \Delta(T^*[\{0, 1\}, \{0, n\}]) = \tau - r_1 s_n,$$

where $\tau = r_1 + \cdots + r_m = s_1 + \cdots + s_n$. It is clear that $\Delta_1 = (n - r_1)s_n$ for $r_m = r_1$ and $\Delta_1 = (m - s_n)r_1$ for $s_n = s_1$. Therefore $\Delta_1 \neq 0$ if $r_1 < n$ in the first case, and $\Delta_1 \neq 0$ if $s_n < m$ in the second case. Thus, if the conditions of the second assertion are fulfilled, then $r(T^*) = 2$.

Finally, now it is obvious that the equality $r(T^*) = 1$ is equivalent to the validity of one of the conditions: $r_1 + \cdots + r_m = n$ and $r_1 = 0$, $s_1 = 0$, which are the conditions of the third assertion of the theorem. □

Note that Theorem 2.2.16 seems to be useful in the problem of classification of mathematical models in economics (see [**100**]).

Problems.

1. Let $\mathfrak{A} = \mathfrak{A}^{9,13}(R, S)$, where
$$R = (13, 11, 10, 10, 9, 7, 4, 4, 1),$$
$$S = (9, 8, 7, 7, 7, 7, 6, 5, 4, 4, 3, 1, 1).$$
Find the structure matrix T for the class $\mathfrak{A}$.
2. Let $\mathfrak{A} = \mathfrak{A}^{m,n}(R, S)$ be a nonempty monotone class of $(0, 1)$-matrices, where $R = (r, r, \ldots, r)$, $S = (s_1, s_2, \ldots, s_n)$ are integer-valued positive vectors, and $s_1 < m$. Prove that the class $\mathfrak{A}$ is elementary.
3. Prove Lemma 2.2.13.
4. For the class $\mathfrak{A}$ introduced in Problem 1, find the representation given in Theorem 2.2.15.
5. Let $\mathfrak{A} = \mathfrak{A}^{n,n}(R, S)$ be a nonempty monotone class of square $(0, 1)$-matrices, $R = (r_1, r_2, \ldots, r_n)$, $S = (s_1, s_2, \ldots, s_n)$, $\tau = \sum_{i=1}^n r_i$, and let T be the structure matrix of the class and T^* the matrix defined by (2.2.7). Show that the eigenvalues of T^* satisfy the equation
$$(-\lambda)^{n-2}(\lambda^3 - p_1\lambda^2 + p_2\lambda - p_3) = 0,$$
where
$$p_1 = n + \tau,$$
$$p_2 = n\tau - \sum_{i=1}^n r_i s_i,$$
$$p_3 = \sum_{i<j} (r_i - r_j)(s_i - s_j) = n \sum_{i=1}^n r_i s_i - \tau^2.$$

Using Proposition 2.2.3, derive from this equation the following formula for the trace of T:

$$\operatorname{tr} T = \sum_{k=1}^{n} k(k + r_k - s_{n-k+1}).$$

2.3. Estimates of the term rank

The notion of the term rank was introduced in Section 1.2. The term rank $\rho(A)$ is an important global characteristic of a $(0,1)$-matrix A. In this section, we show that it is convenient to study this characteristic with the use of the structure matrix of the Ryser class containing the matrix A. In particular, we obtain the estimate of $\rho(A)$ mentioned at the end of Section 1.2. The term rank of a $(0,1)$-matrix does not change if we add or remove a zero line. Therefore in this section we consider the classes $\mathfrak{A}(R,S)$ with positive integer-valued monotone vectors R, S.

2.3.1. The term rank and interchanges. Let A be a $(0,1)$-matrix of size $m \times n$, and k a positive integer, $k \leq \min\{m,n\}$. Let $P_1, \dots, P_k$ be a set of positions in A containing ones. If any two of these positions do not belong to a common row or a common column, then the ones located in these positions are called independent. Let a $(0,1)$-matrix have the term rank $\rho(A) = \rho$. Then any ρ independent ones are called basis ones and the set of positions occupied by these ones are called a basis set of positions.

The aim of the next few subsections is to derive the formula for the maximal term rank $\bar{\rho}(\mathfrak{A})$ of a class $\mathfrak{A}$ of $(0,1)$-matrices obtained by Ryser. This formula expresses $\bar{\rho}(\mathfrak{A})$ in terms of the structure matrix of $\mathfrak{A}$. To this end, we have to prove a series of auxiliary results.

LEMMA 2.3.1. *An interchange changes the term rank of a $(0,1)$-matrix by at most one.*

If A_1 is obtained from A by an interchange, then the covering of ones of A_1 can be obtained from the covering of ones of A by adding at most one line. Therefore, by the König Theorem $\rho(A_1) \leq \rho(A) + 1$. Similarly, $\rho(A) \leq \rho(A_1)+1$.

THEOREM 2.3.2. *Let $\mathfrak{A}$ be a nonempty Ryser class, $\tilde{\rho} = \tilde{\rho}(\mathfrak{A})$ the minimal and $\bar{\rho} = \bar{\rho}(\mathfrak{A})$ the maximal term rank of the class. Then for any integer ρ, $\tilde{\rho} \leq \rho \leq \bar{\rho}$, there is a matrix $A \in \mathfrak{A}$ such that $\rho(A) = \rho$.*

PROOF. Indeed, by Lemma 2.3.1 any interchange modifies the term rank at most by one. But it is possible to transform a matrix $B \in \mathfrak{A}$ with $\rho(B) = \tilde{\rho}$ to a matrix $C \in \mathfrak{A}$ with $\rho(C) = \bar{\rho}$ by a sequence of interchanges (see Theorem 2.1.3). Therefore among the matrices which are obtained by this sequence of interchanges there is a matrix with term rank equal to ρ for any ρ such that $\tilde{\rho} \leq \rho \leq \bar{\rho}$. □

2.3.2. Matrix $A_{\bar{\rho}}$. Let $\mathfrak{A}$ be a nonempty Ryser class and $\bar{\rho}(\mathfrak{A}) = \bar{\rho}$. In what follows the matrices of the class $\mathfrak{A}$ with some special location of the positions of $\bar{\rho}$ basis ones will be used.

Before proving the existence of such matrices, we formulate a technical assertion which, in fact, has been used above.

LEMMA 2.3.3. *Let a $(0,1)$-matrix $A = (a_{pq})$ belong to the class $\mathfrak{A}^{m,n}(R,S)$ with $R = (r_1, \dots, r_m)$, $S = (s_1, \dots, s_n)$. If $a_{ij} = 0$, $a_{kj} = 1$, and $r_i \geq r_k$ for some $i, k \in N_m$ and $j \in N_n$, then the rows with numbers i and k contain an interchangeable* *submatrix with elements* a_{ij} *and* a_{kj}*, and by the corresponding interchange the matrix* A *is transformed into a matrix* $B = (b_{pq})$ *with* $b_{ij} = 1$, $b_{kj} = 0$*. Similarly, if* $a_{ij} = 0$, $a_{il} = 1$*, and* $s_j \geq s_l$ *for some* $j, l \in N_n$ *and* $i \in N_m$*, then the columns with numbers* j *and* l *contain an interchangeable submatrix, and by the corresponding interchange the matrix* A *is transformed into a matrix* $C = (c_{pq})$ *with* $c_{ij} = 1$, $c_{il} = 0$.

Below, performing interchanges, we often use this assertion, sometimes without reference to the lemma.

THEOREM 2.3.4. *Let* A *be a matrix of a monotone Ryser class* $\mathfrak{A} = \mathfrak{A}^{m,n}(R, S)$ *with* $R, S > 0$*, and let* $\rho = \rho(A)$ *be the term rank of* A*. Then* $\mathfrak{A}$ *contains a matrix* B *with* $\rho(B) \geq \rho$ *such that* ρ *independent ones are located in* B *in the positions* $(1, \rho), (2, \rho - 1), \dots, (\rho, 1)$*. The matrix* A *can be transformed into* B *by a sequence of interchanges such that no interchange gives a matrix with term rank less than* ρ*.*

PROOF. Let $A = (a_{ij})$. First we use interchanges to obtain from A a matrix A' such that $\rho(A') \geq \rho$ and ρ independent ones are located in the first ρ columns. Let $\mathcal{B}$ be some set of basis positions in A. Denote by J the subset of N_ρ consisting of the indices of all columns which do not contain positions of $\mathcal{B}$, and by L the set of all indices l such that $(i, l) \in \mathcal{B}$ for some $i \in N_m$ and $l > \rho$. Then $|J| = |L|$. Consider an arbitrary pair (j, l), $j \in J$, $l \in L$. There exists an index i such that $(i, l) \in \mathcal{B}$, that is, $a_{il} = 1$. Since $j < l$, performing an interchange if required (see Lemma 2.3.3), we transform A to a matrix with one in the position (i, j). It is clear that $\mathcal{B}' = (\mathcal{B} - \{(i, l)\}) \cup \{(i, j)\}$ is a set of positions of independent ones. Proceeding in the same way with each pair (j, l), $j \in J$, $l \in L$, we obtain the required matrix $A' = (a'_{ij})$ with ρ independent ones in the first ρ columns.

Now consider the first row and the ρth column of the matrix A'. By the construction, there exists $i \in N_m$ such that $Q = (i, \rho) \in \mathcal{B}'$ and, consequently, $a'_{i\rho} = 1$. It is clear that if the first row does not contain a position of $\mathcal{B}'$, then, performing an interchange if required, we can obtain a matrix with one in the position $(1, \rho)$. This one and the ones in the positions of $\mathcal{B}' - \{Q\}$ form a set of ρ independent ones in the new matrix.

Now let $P = (1, j) \in \mathcal{B}'$ for some $j \in N_\rho$. Consider the matrix $A'[\{1, i\}, \{j, \rho\}]$. Show that A' can be transformed into a matrix in $\mathfrak{A}$ with ones in the positions (i, j) and $(1, \rho)$ and in all positions of $\mathcal{B}' - \{P, Q\}$ as well. Indeed, if $a'_{1\rho} = a'_{ij} = 0$, then this can be done by one interchange. If $a'_{1\rho} = 1$, $a'_{ij} = 0$, then we transfer the one from the position (i, ρ) to the position (i, j) by the use of an interchangeable submatrix in the columns with numbers j and ρ (see Lemma 2.3.3). If $a'_{i\rho} = 0$, $a_{ij} = 1$, then we obtain one in the position $(1, \rho)$ using an interchangeable matrix in the rows with numbers 1 and i. It is clear that the ones in the positions of the set $(\mathcal{B}' - \{P, Q\}) \cup \{(1, \rho), \{(i, j)\}$ are independent. We proceed in the same way with the second row and the $(\rho - 1)$th column, and so on up to the ρth row and the first column. Finally, we obtain a matrix B with ones in the positions $(1, \rho), (2, \rho - 1), \dots, (\rho, 1)$. The theorem is thus proved. □

COROLLARY 2.3.5. *Let* $\bar{\rho}$ *be the maximal term rank of a nonempty Ryser class* $\mathfrak{A}$*. Then in* $\mathfrak{A}$ *there is a matrix* $A_{\bar{\rho}} = (a_{ij})$ *with basis set* $(1, \bar{\rho}), (2, \bar{\rho} - 1), \dots, (\bar{\rho}, 1)$*, that is,* $a_{1\bar{\rho}} = a_{2,\bar{\rho}-1} = \cdots = a_{\bar{\rho}1} = 1$.

It is clear that any matrix $A_{\bar\rho}$ in the corollary is of the form

$$A_{\bar\rho} = \begin{pmatrix} A_1 & A_2 \\ A_3 & \mathbf{0} \end{pmatrix}, \tag{2.3.1}$$

where A_1 is a matrix of order $\bar\rho$ such that $\rho(A_1) = \bar\rho$.

2.3.3. The standard representation of matrices of an arbitrary class. The following theorem gives a representation of matrices of an arbitrary monotone class generalizing the representation (2.2.11). This representation implies a formula for the maximal term rank.

THEOREM 2.3.6. *Let $\mathfrak{A} = \mathfrak{A}^{m,n}(R,S)$ be a monotone Ryser class with positive vectors $R = (r_1, \dots, r_m)$, $S = (s_1, \dots, s_n)$, and let $\bar\rho$ be the maximal term rank of $\mathfrak{A}$. There exist integers e, f, $0 \le e \le m$, $0 \le f \le n$, such that any matrix A of the class $\mathfrak{A}$ is of the form*

$$A = \begin{pmatrix} {}^{(1)}A & * \\ * & {}^{(0)}A \end{pmatrix}, \tag{2.3.2}$$

where ${}^{(1)}A = A[N_e, N_f]$, ${}^{(0)}A = A(N_e, N_f)$ and

$$\sigma_0({}^{(1)}A) + \sigma_1({}^{(0)}A) = \bar\rho - (e+f). \tag{2.3.3}$$

In addition, for any matrix $A_{\bar\rho}$ introduced in Corollary 2.3.5 the equalities

$$\sigma_0({}^{(1)}A_{\bar\rho}) = 0, \qquad \sigma_1({}^{(0)}A_{\bar\rho}) = \bar\rho - (e+f) \tag{2.3.4}$$

hold.

PROOF. We prove the theorem by induction on the number of rows m. First of all, note that according to (2.2.1) for any e, f such that $0 \le e \le m$, $0 \le f \le n$ we have

$$\sigma_0(A[N_e, N_f]) + \sigma_1(A(N_e, N_f)) = t_{ef},$$

where t_{ef} is an element of the structure matrix T of the class $\mathfrak{A}$. Therefore the validity of equalities (2.3.3) and (2.3.4) does not depend on the choice of the matrix $A \in \mathfrak{A}$. Thus, it is sufficient to prove the assertion for an arbitrary matrix of the class. We prove relations (2.3.4) for the matrix $A_{\bar\rho}$ from Corollary 2.3.5.

Note that the validity of (2.3.4) for the degenerate matrix ${}^{(1)}A$ of size $e \times f$ is equivalent in our case to the validity of either the equality $\bar\rho = m$ or the equality $\bar\rho = n$. Indeed, in view of $S > 0$, for $e = 0$ equalities (2.3.4) imply that $\bar\rho - f \ge n - f$ (that is, $\bar\rho = n$), and since $R > 0$, for $f = 0$ we obtain $\bar\rho = m$. On the other hand, for $\bar\rho = m$ or $\bar\rho = n$ equalities (2.3.4) are obviously fulfilled if we take $(e,f) = (m,0)$ and $(e,f) = (0,n)$, respectively. Thus, the assertion of the theorem is true for $\bar\rho = \min\{m,n\}$. For $m = 1$, this provides the first step of induction.

So from here on we assume that $\bar\rho < m, n$. The proof consists of three steps.

Step 1. Transform $A_{\bar\rho} = (a_{ij})$ into a matrix $B = (b_{ij})$ by a sequence of interchanges applied to the elements of the first row. It is required that the matrix B be such that for $j, l \in N_n$

- the equalities $b_{1j} = 1$ and $b_{1l} = 0$ are not fulfilled for $s_j > s_l$, and
- the equalities $b_{1j} = 1$ and $b_{1l} = 0$ are not fulfilled for $s_j = s_l$.

It is clear that the first requirement can be achieved by interchanges which do not affect the basis ones of the matrix $A_{\bar\rho}$ located in the positions $(2, \bar\rho - 1), \dots, (\bar\rho, 1)$ (see Lemma 2.3.3). The same condition has to be fulfilled when we want to satisfy the second requirement. Here only the situation where $s_j = s_l$ for $j < l < \bar{\rho}$ and $a_{1j} = 1, a_{1l} = 0$, is not clear. If the needed interchange affects the row with number $\bar{\rho} - l + 1$, then, evidently, it moves $a_{\bar{\rho}-l+1,l} = 1$. Then we perform two interchanges: by using first the interchangeable submatrix $A[\{1, \bar{\rho} - l + 1\}, \{j, l\}]$, and then an interchange in the rows with numbers $\bar{\rho} - l + 1$, $\bar{\rho} - j + 1$ and in the lth column such that the one returns to the position $(\bar{\rho} - l + 1, l)$.

Thus, as a result we obtain the matrix

$$B = \begin{pmatrix} \delta_1 & \delta_2 \\ B' & \mathbf{0} \end{pmatrix}, \tag{2.3.5}$$

where B' is an $(m-1) \times n'$ matrix belonging to a monotone class $\mathfrak{A}'$ with positive vectors of row and column sums, δ_1 is an n'-dimensional $(0,1)$-vector, δ_2 is an $(n-n')$-dimensional vector consisting of ones, $n' \le n$. The matrix B' contains ones in the positions $(1, \bar{\rho} - 1), (2, \bar{\rho} - 2), \dots, (\bar{\rho} - 1, 1)$.

Step 2. Let us show that $\bar{\rho}(\mathfrak{A}') = \bar{\rho} - 1$. Assuming the contrary, let $\bar{\rho}(\mathfrak{A}') = \bar{\rho}$. Then by Theorem 2.3.4 in $\mathfrak{A}'$ there is a matrix A' with ones in the positions $(1, \bar{\rho}), (2, \bar{\rho} - 1), \dots, (\bar{\rho}, 1)$. We insert it instead of B' into B defined in (2.3.5) and obtain a matrix of the class $\mathfrak{A}$ as well. Since $\rho(B) = \bar{\rho}$, the equality $b_{1n} = 1$ should be true. Now using at most two interchanges, we can (without affecting the basis ones in the columns with numbers $1, \dots, \bar{\rho} - 1$) first move one from the position $(2, \bar{\rho})$ to the position $(1, \bar{\rho})$ and then one of the nth column to the position $(2, n)$. Thus, we obtain a matrix of the class $\mathfrak{A}$ with $\bar{\rho}+1$ independent ones in the positions

$$(\bar{\rho} + 1, 1), (\bar{\rho}, 2), \dots, (3, \bar{\rho} - 1), (2, n), (1, \bar{\rho}),$$

which contradicts the maximality of $\bar{\rho}$. Therefore B' is a matrix of the form $A'_{\bar{\rho}-1}$ for the class $\mathfrak{A}'$ similar to the matrix A in Corollary 2.3.5 for the class $\mathfrak{A}$.

Step 3. The induction assumption is applicable to the class $\mathfrak{A}'$. Therefore, B' contains the matrix $\mathbf{J}_{e'f'}$, $0 \le e' \le m - 1$, $0 \le f' \le n'$, in the upper left corner. It is possible that $\bar{\rho} - 1 = n'$, so that the matrix $\mathbf{J}_{e'f'}$ can be degenerate. Thus,

$$B' = \begin{pmatrix} \mathbf{J}_{e'f'} & Z \\ Y & {}^{(0)}B' \end{pmatrix}, \tag{2.3.6}$$

and $\sigma_1({}^{(0)}B') = \bar{\rho} - 1 - (e' + f')$. If such a representation is valid for several pairs (e', f'), then we choose the pair with maximal f'.

Now suppose that $b_{1j} = 0$ for some $j \le f'$. Show that then only zeros can be located above the basis one in the matrix Y in (2.3.6). Suppose that this is not the case. Then the jth column of B contains at least two ones, and by the construction of Step 1 the vector δ_2 is empty and $n' = n$. In the case where $\mathbf{J}_{e'f'}$ in (2.3.6) is degenerate, that is, in the case $\bar{\rho} - 1 = n'$, we obtain a contradiction. If $\mathbf{J}_{e'f'}$ is not degenerate, that is, if $e', f' > 0$, then, comparing the jth and lth columns for $l > f'$, we see that $b_{1,f'+1} = \dots = b_{1n} = 0$, since the jth column contains at least $e' + 2$ ones and the lth column contains at most $e' - 1 + 2 = e' + 1$ ones, and the equality $b_{1l} = 1$ would contradict the construction of Step 1. Then we find that $r_1 < r_2$, contradicting the monotonicity of the vector R. Thus, we conclude that there exists a number f, $0 \le f \le f'$, such that the matrix B contains a submatrix $\mathbf{J}_{e'+1,f}$. If it is needed, we can locate one in the position $(1, \bar{\rho})$ by a single interchange. Thereby we obtain a matrix $C \in \mathfrak{A}$ of the same form as in Corollary 2.3.5 such that

$$\begin{aligned}\sigma_0(C[N_{e'+1}, N_f]) &= 0,\\ \sigma_1(C(N_{e'+1}, N_f)) &= \sigma_1({}^{(0)}B') + f' - f\\ &= \bar{\rho} - 1 - (e' + f') + f' - f = \bar{\rho} - (e' + 1 + f).\end{aligned}$$

The theorem is proved. □

2.3.4. Maximal term rank. Theorem 2.3.6 yields a formula for the maximal term rank of a Ryser class.

THEOREM 2.3.7. *Let $\mathfrak{A} = \mathfrak{A}^{m,n}(R, S)$ be a Ryser class with $R = (r_1, \dots, r_m)$, $S = (s_1, \dots, s_n)$, $r_1 \geq \dots \geq r_m > 0$, $s_1 \geq \dots \geq s_n$, and let T be the structure matrix of the class. Then*

$$\bar{\rho}(\mathfrak{A}) = \min_{ij}(t_{ij} + i + j). \tag{2.3.7}$$

PROOF. Let i, j be arbitrary integers such that $0 \leq i \leq m$, $0 \leq j \leq n$. Then for any matrix $A \in \mathfrak{A}$ the equality

$$t_{ij} = \sigma_0(A[N_i, N_j]) + \sigma_1(A(N_i, N_j)) \tag{2.3.8}$$

holds (see(2.2.1)). It is clear that $i + j + \sigma_1(A(N_i, N_j))$ lines cover all ones of any matrix $A \in \mathfrak{A}$. Therefore, taking (2.3.8) into account, we see that $t_{ij} + i + j \geq \bar{\rho}$ for any $i \in N_m$, $j \in N_n$. If we take the same e, f as in Theorem 2.3.6, then, obviously, the function $t_{ij} + i + j$ attains its minimum, equal to $\bar{\rho}$, at $i = e$, $j = f$. The theorem is proved. □

2.3.5. Minimal term rank. For the minimal term rank $\tilde{\rho}$ of a class $\mathfrak{A}(R, S)$ there exist formulas which express $\tilde{\rho}$ in terms of the structure matrix of the class. The first such formula was obtained by Haber [**47**]. A simpler expression for the minimal term rank was found by Brualdi. Here we give the last formula without proof, which can be found in [**20**].

Let $T = (t_{ij})$ be the structure matrix of a nonempty monotone class $\mathfrak{A}$ of $(0, 1)$-matrices of size $m \times n$. For any integers e, f such that $0 \leq e \leq m$, $0 \leq f \leq n$, we consider the function

$$\varphi(e, f) = \min\{t_{k_1, f+l_2} + t_{e+k_2, l_1} + (e - k_1)(f - l_1)\}, \tag{2.3.9}$$

where the minimum is taken over all integers k_1, k_2, l_1, l_2 such that

$$0 \leq k_1 \leq e + k_2 \leq m, \qquad 0 \leq l_1 \leq f + l_2 \leq n.$$

THEOREM 2.3.8. *The minimal term rank $\bar{\rho}(\mathfrak{A})$ is equal to*

$$\min\{e + f \colon \varphi(e, f) \geq t_{ef}\},$$

where the minimum is taken over all pairs (e, f), $0 \leq e \leq m$, $0 \leq f \leq n$.

The theorem is proved by using the demand–supply theorem in the theory of flows in networks.

Thus, for a given $(0, 1)$-matrix A, we can find upper and lower estimates for the term rank. For this purpose we consider the structure matrix of the Ryser class containing A. Then Theorem 2.3.7 gives the upper estimate, and the lower estimate is provided by Theorem 2.3.8. Note that for constructing matrices with minimal and maximal term rank in a given Ryser class there are algorithms based on the theory of flows in networks [**47**].

2.3.6. Conditions for coincidence of $\tilde{\rho}$ and $\bar{\rho}$. Can $\tilde{\rho}$ and $\bar{\rho}$ coincide? Obviously, they can, since there are Ryser classes consisting of a single matrix. Below this question will be examined more closely. First we obtain a simple but useful result. We denote by $\mathfrak{A}(X)$ the Ryser class containing a $(0,1)$-matrix X and, consequently, determined by this matrix.

LEMMA 2.3.9. *Let a $(0,1)$-matrix A of size $m \times n$ be represented in the form*

$$A = \begin{pmatrix} \mathbf{J}_{kl} & X_1 \\ X_2 & \mathbf{0} \end{pmatrix}. \tag{2.3.10}$$

Then

$$\rho(A) = \min\{k + \rho(X_1), l + \rho(X_2)\}.$$

Let $\mathfrak{A} = \mathfrak{A}(A)$, $\mathfrak{A}_i = \mathfrak{A}(X_i)$, $i = 1, 2$. Then

$$\bar{\rho}(\mathfrak{A}) = \min\{k + \bar{\rho}(\mathfrak{A}_2), l + \bar{\rho}(\mathfrak{A}_1), \tag{2.3.11}$$

$$\tilde{\rho}(\mathfrak{A}) = \min\{k + \tilde{\rho}(\mathfrak{A}_2), l + \tilde{\rho}(\mathfrak{A}_1). \tag{2.3.12}$$

PROOF. It is obvious that

$$\begin{aligned}\rho(A) &\geq \rho(X_1) + \rho(X_2) + \min\{k - \rho(X_1), l - \rho(X_2)\} \\ &= \min\{k + \rho(X_2), l + \rho(X_1)\}.\end{aligned}$$

The reverse inequality follows from the König Theorem. The lemma is proved. □

The following results show that the answer to the question on the coincidence of $\tilde{\rho}$ and $\bar{\rho}$ depends on whether or not the Ryser class contain invariant elements 1.

THEOREM 2.3.10 ([**86**]). *Let $\mathfrak{A} = \mathfrak{A}^{m,n}(R, S)$ be a monotone Ryser class such that $R, S > 0$ and $\bar{\rho}(\mathfrak{A}) < \min\{m, n\}$. If $\mathfrak{A}$ has no invariant 1-positions, then $\tilde{\rho}(\mathfrak{A}) \neq \bar{\rho}(\mathfrak{A})$.*

PROOF. Let $T = (t_{ij})$ be the structure matrix of the class $\mathfrak{A}$. By Theorem 2.3.7 there exist integers e, f, $0 \leq e \leq m$, $0 \leq f \leq n$, such that

$$\bar{\rho} = \bar{\rho}(\mathfrak{A}) = t_{ef} + e + f.$$

Since $\bar{\rho} < m, n$, we see that $e < m$, $f < n$, and since $R, S > 0$, the inequalities $e, f > 0$ hold. In $\mathfrak{A}$, we take the matrix $A_{\bar{\rho}}$ considered in Theorem 2.3.4. By Theorem 2.3.6, $A_{\bar{\rho}}[N_e, N_f] = \mathbf{J}_{ef}$ and $A_{\bar{\rho}}(N_e, N_f)$ contains t_{ef} ones. The position $(1,1)$ is not invariant; therefore there exists a matrix $B = (b_{ij}) \in \mathfrak{A}$ with $b_{11} = 0$. Then $\sigma_1(B(N_e, N_f)) < t_{ef}$. But, obviously, all ones of the matrix B can be covered by $e + f + \sigma_1(B(N_e, N_f))$ lines. Therefore, $\tilde{\rho} = \tilde{\rho}(\mathfrak{A}) \leq \rho(B) < e + f + t_{ef} = \bar{\rho}$, that is, $\tilde{\rho} \neq \bar{\rho}$. □

If a class $\mathfrak{A}$ of $(0,1)$-matrices of size $m \times n$ does not contain invariant ones, but $\bar{\rho}(\mathfrak{A}) = \min\{m, n\}$, then a possibility of coincidence of $\tilde{\rho}$ and $\bar{\rho}$ appears. Let, for example, $\mathfrak{A} = \mathfrak{A}^n(k)$. As noted in Section 1.2, each matrix A of this class contains a positive diagonal, that is, $\rho(A) = n$, and $\tilde{\rho}(\mathfrak{A}) = \bar{\rho}(\mathfrak{A}) = n$. On the other hand, it is obvious that there exist classes of matrices with $\bar{\rho} = \min\{m, n\}$ such that $\tilde{\rho} \neq \bar{\rho}$. Consider, for example, the class $\mathfrak{A}(R, S)$ consisting of the square matrices of order 3 with $R = S = (2, 1, 1)$. This class contains the matrices

$$A = \begin{pmatrix} 0 & 1 & 1 \\ 1 & 0 & 0 \\ 1 & 0 & 0 \end{pmatrix}, \qquad B = \begin{pmatrix} 1 & 1 & 0 \\ 0 & 0 & 1 \\ 1 & 0 & 0 \end{pmatrix}$$

such that $\rho(A) = 2$ and $\rho(B) = 3$; that is, $\tilde{\rho} \neq \bar{\rho}$ in this class.

Now consider the Ryser classes which can have invariant ones. In this case, the following theorem of Brualdi [**19**] reduces the question on coincidence of $\tilde{\rho}$ and $\bar{\rho}$ to the similar problem for classes of smaller matrices.

THEOREM 2.3.11. *Let $\mathfrak{A} = \mathfrak{A}^{m,n}(R,S)$ be a Ryser class with monotone (not necessarily positive) vectors R, S. Then the equality $\tilde{\rho}(\mathfrak{A}) = \bar{\rho}(\mathfrak{A})$ is true if and only if there exist integers e, f, $0 \le e \le m$, $0 \le f \le n$, such that any matrix $A \in \mathfrak{A}$ is of the form*

$$A = \begin{pmatrix} \mathbf{J}_{ef} & A_1 \\ A_2 & \mathbf{0} \end{pmatrix} \tag{2.3.13}$$

and $\tilde{\rho}(\mathfrak{A}(A_1)) = e$, $\tilde{\rho}(\mathfrak{A}(A_2)) = f$.

PROOF. Let each matrix $A \in \mathfrak{A}$ be of the form (2.3.13), and let the conditions on the minimal term ranks be fulfilled. Then it follows from equalities (2.3.11) and (2.3.12) that $\tilde{\rho}(\mathfrak{A}) = \bar{\rho}(\mathfrak{A})$.

Now suppose that $\tilde{\rho}(\mathfrak{A}) = \bar{\rho}(\mathfrak{A})$. We show that in this case each matrix $A \in \mathfrak{A}$ has a representation (2.3.13) of the required form.

First let $\mathfrak{A}$ not contain invariant ones. Then Theorem 2.3.10 shows that the equality $\tilde{\rho} = \bar{\rho}$ is possible only if $\bar{\rho} = \min\{m', n'\}$, where m', n' are the numbers of nonzero coordinates of the vectors R, S respectively. If $m' \le n'$, then we take $e = m'$, $f = 0$, and if $n' \le m'$, then we take $e = 0$, $f = n'$. In both cases, we obtain the required representation (2.3.13) with degenerate matrix $\mathbf{J}_{ef}$.

Now let $\mathfrak{A}$ contain invariant ones. Then by Theorem 2.2.7 there exist $k \in N_m$, $l \in N_n$, $(k,l) \neq (0,n), (m,0)$, such that each matrix $A \in \mathfrak{A}$ is of the form

$$A = \begin{pmatrix} \mathbf{J}_{kl} & X_1 \\ X_2 & \mathbf{0} \end{pmatrix}.$$

Let $\tilde{\rho}_i = \tilde{\rho}(\mathfrak{A}(X_i))$, $\bar{\rho}_i = \bar{\rho}(\mathfrak{A}(X_i))$, $i = 1,2$ (it is clear that $\tilde{\rho}_1$, $\tilde{\rho}_2$, $\bar{\rho}_1$, $\bar{\rho}_2$ do not depend on the choice of $A \in \mathfrak{A}$). By Lemma 2.3.9

$$\tilde{\rho}(\mathfrak{A}) = \min\{k + \tilde{\rho}_2, l + \tilde{\rho}_1\},$$
$$\bar{\rho}(\mathfrak{A}) = \min\{k + \bar{\rho}_2, l + \bar{\rho}_1\}.$$

Consider all possible cases.

First let $\tilde{\rho}(\mathfrak{A}) = k + \tilde{\rho}_2 \le l + \tilde{\rho}_1$ and $\bar{\rho}(\mathfrak{A}) = k + \bar{\rho}_2 \le l + \bar{\rho}_1$. Then, since $\tilde{\rho} = \bar{\rho}$, we have $\tilde{\rho}_2 = \bar{\rho}_2$. Applying induction to the class $\mathfrak{A}(X_2)$, we find that there exist integers p, q such that

$$X_2 = \begin{pmatrix} \mathbf{J}_{pq} & X_3 \\ X_4 & \mathbf{0} \end{pmatrix}$$

and $\tilde{\rho}_3 = \tilde{\rho}(\mathfrak{A}(X_3)) = p$, $\tilde{\rho}_4 = \tilde{\rho}(\mathfrak{A}(X_4)) = q$. Then by Lemma 2.3.9 $\tilde{\rho}_2 = p + q$. We set $e = k + p$, $f = q$, and

$$Y = \begin{pmatrix} \mathbf{J}_{k,l-q} & X_1 \\ X_3 & \mathbf{0} \end{pmatrix}.$$

Then

$$A = \begin{pmatrix} \mathbf{J}_{ef} & Y \\ X_4 & \mathbf{0} \end{pmatrix}.$$

In this case, $\tilde{\rho} = k + \tilde{\rho}_2 = k + p + q$ and, consequently, $\rho(Y) = k + p$. Therefore $\tilde{\rho}(\mathfrak{A}(Y)) = e$, $\tilde{\rho}(\mathfrak{A}(X_4)) = f$. Thus, in this case the assertion is proved.

Now let $\bar{\rho}(\mathfrak{A}) = k + \bar{\rho}_2 \le l + \bar{\rho}_1$ and $\tilde{\rho}(\mathfrak{A}) = l + \tilde{\rho}_1 \le k + \tilde{\rho}_2$. Then

$$\tilde{\rho}(\mathfrak{A}) = l + \tilde{\rho}_1 \le k + \tilde{\rho}_2 \le k + \bar{\rho}_2 = \bar{\rho}(\mathfrak{A}) = \tilde{\rho}(\mathfrak{A}).$$

Hence $\bar{\rho}_2 = \tilde{\rho}_2$ and $\tilde{\rho}(\mathfrak{A}) = k + \tilde{\rho}_2$. Thus, we get the case considered above.

The remaining two cases for $\tilde{\rho}$ and $\bar{\rho}$ are treated in the similar way. The theorem is proved. □

2.3.7. Classes with prescribed values of $\tilde{\rho}$ and $\bar{\rho}$. Here we estimate the difference between $\tilde{\rho}$ and $\bar{\rho}$ in a Ryser class consisting of $(0,1)$-matrices of a given size $m \times n$. As shown by Brualdi [**19**], the difference $\bar{\rho} - \tilde{\rho}$ can vary within wide limits. First we prove a relation between $\tilde{\rho}$ and $\bar{\rho}$ for an arbitrary class $\mathfrak{A}$.

THEOREM 2.3.12. *Let $\tilde{\rho} = \tilde{\rho}(\mathfrak{A})$ and $\bar{\rho} = \bar{\rho}(\mathfrak{A})$. Then*

$$\bar{\rho} \le \tilde{\rho} + (\tilde{\rho}/2)^2$$

for even $\tilde{\rho}$, and

$$\bar{\rho} \le \tilde{\rho} + \frac{\tilde{\rho}+1}{2}\frac{\tilde{\rho}-1}{2}$$

for odd $\tilde{\rho}$.

PROOF. Consider $A \in \mathfrak{A}$ with $\rho(A) = \tilde{\rho}$. By König's Theorem, in A there exist a set I consisting of e rows and a set J consisting of f columns such that the lines of $I \cup J$ contain all ones of the matrix A and $e + f = \rho(A) = \tilde{\rho}$. Let $T = (t_{ij})$ be the structure matrix of the class $\mathfrak{A}$. Then by (2.2.1)

$$t_{ef} \le t(I,J) \le ef + 0 = ef.$$

By Theorem 2.3.7

$$\bar{\rho} \le t_{ef} + e + f \le ef + \tilde{\rho}.$$

Hence we obtain the assertion of the theorem. □

THEOREM 2.3.13. *Let m, n, k, l be integers, $m, n > 0$. There exist nonnegative integer-valued vectors R, S of sizes m and n, respectively, such that $\mathfrak{A} = \mathfrak{A}(R,S)$ is not empty and $\tilde{\rho}(\mathfrak{A}) = k$, $\bar{\rho}(\mathfrak{A}) = l$ if and only if*

$$0 \le k \le l \le \begin{cases} \min\{k + k^2/4, m, n\}, & k \text{ is even}, \\ \min\{k + (k^2-1)/4, m, n\}, & k \text{ is odd}. \end{cases} \tag{2.3.14}$$

PROOF. The necessity of the condition follows directly from Theorem 2.3.12.

Suppose now that condition (2.3.14) is fulfilled. Let k be even. We set $p = k/2$, $q = l - k$. The existence of a class $\mathfrak{A}$ with $\tilde{\rho}(\mathfrak{A}) = k$, $\bar{\rho}(\mathfrak{A}) = l$ will be proved by a direct construction. Let X be a $(0,1)$-matrix of size $p \times q$ with monotone row sum vector, and let each column of X have exactly one zero and each row have at most p zeros (the last requirement can be realized by virtue of (2.3.14)). Let Y be a $(0,1)$-matrix of size $q \times p$ such that there is exactly one zero in each row, the column sum vector of Y is monotone, and each column has at most p zeros. Consider the matrix

$$A = \begin{pmatrix} \mathbf{J}_{pp} & \mathbf{J}_{pp} & X & \mathbf{J}_{pv} \\ \mathbf{J}_{pp} & \mathbf{0} & \mathbf{0} & \mathbf{0} \\ Y & \mathbf{0} & \mathbf{I}_q & \mathbf{0} \\ \mathbf{J}_{up} & \mathbf{0} & \mathbf{0} & \mathbf{0} \end{pmatrix}, \tag{2.3.15}$$

where $u = m - k - q$, $v = n - k - q$. We denote by R and S the monotone vectors of row and column sums of A, respectively, and by $T = (t_{ij})$ the structure matrix of the class $\mathfrak{A} = \mathfrak{A}(R, S)$. It is clear that A has at least $2p + q = l$ independent ones. On the other hand, all ones of A can be covered by l lines. By König's Theorem, $\rho(A) = l$. Note that

$$\bar{\rho}(\mathfrak{A}) \leq t_{pp} + p + p = q + 2p = l.$$

Hence $\bar{\rho}(\mathfrak{A}) = l$.

Now we show that $\tilde{\rho}(\mathfrak{A}) \geq k$. Let there exist, on the contrary, a matrix B in $\mathfrak{A}$ with $\rho(B) < k$. Then there exist e rows and f columns in B containing all ones, and $e + f = \rho(B) < k = 2p$. Let, for example, $e < p$. Each row sum of B is less than p; therefore $f = n$. This contradicts the condition $k \leq n$. Thus, $\rho(B) \geq k$ and $\tilde{\rho} \geq k$. Further, there exists a sequence of q interchanges which transforms the matrix A defined by (2.3.15) into a matrix of the form

$$C = \begin{pmatrix} Z & \mathbf{J}_{p.n-p} \\ \mathbf{J}_{m-p,p} & \mathbf{0} \end{pmatrix}.$$

By König's Theorem, $\rho(C) \leq 2p = k$, and we obtain that $\rho(C) = k$. Thus, $\tilde{\rho}(\mathfrak{A}) = k$. The case of odd k is treated similarly.

The theorem is proved. □

2.3.8. Paths in the graph $G(R, S)$. In Section 2.1, we considered the interchange graph $G(R, S)$ whose vertices are matrices of a nonempty class $\mathfrak{A}(R, S)$ and two vertices are connected by an edge if they are obtained one from the other by an interchange. Let $\bar{\rho}$ be the maximal term rank of the class $\mathfrak{A}(R, S)$. We denote by $G_{\bar{\rho}}(R, S)$ the subgraph of $G(R, S)$ induced by all the matrices of $\mathfrak{A}$ of maximal term rank. Brualdi (see [**19**, Chapter 6]) asked whether this subgraph is connected. In other words, for $(0, 1)$-matrixes $A, B \in \mathfrak{A}(R, S)$ such that $\rho(A) = \rho(B) = \bar{\rho}$, is it possible to transform A into B by a sequence of interchanges, obtaining a matrix of maximal rank after each interchange? Below we give a positive answer to this question.

First we prove several auxiliary assertions concerning paths of a special form in the interchange graph.

LEMMA 2.3.14. *Any matrix A with $\rho(A) = \bar{\rho}$ can be transformed by interchanges into a matrix $A_{\bar{\rho}}$ defined in Section 2.3.2, obtaining a matrix of maximal term rank after each interchange.*

Indeed, performing interchanges in the proof of Theorem 2.3.4, we obtained at each step a matrix with term rank no less than the rank of the initial matrix.

The next assertion concerns the method, described in Section 2.1, for obtaining the matrix B from a matrix of the same class. We denote by P_p the permutation matrix corresponding to the cycle $(1\,2\ldots p)$.

LEMMA 2.3.15. *Let A, B be $(0, 1)$-matrices of size $m \times n$ of a common Ryser class, and let $A - B = \mathbf{I}_p - P_p$ for some p, $2 \leq p \leq \min\{m, n\}$. Then A can be transformed into B by $p - 1$ interchanges involving each time only two elements of one arbitrarily chosen row and two positions where A and B differ from one another.*

Indeed, the required sequence is precisely the sequence of interchanges described in Section 2.1.6 and illustrated in Fig. 2.2.

Lemma 2.3.15 can be refined. We introduce some notation. If for a $(0, 1)$-matrix of order n the submatrix $X[\{1, i\}, \{i, i + 1\}]$ is interchangeable, then we denote this submatrix by K_i, $i = 2, 3, \dots, n$ (the indices are considered modulo n). The diagonal in a matrix of order n on the positions $(1,2), (2,3), \dots, (n,1)$ is called the second diagonal.

Let an n-dimensional $(0,1)$-vector of the form $(0, \dots, 1)$ be given. It can be written as a sequence of $2t$ intervals $0_1 1_1 0_2 1_2 \dots 0_t 1_t$, where 0_i and 1_i are runs, respectively, of zeros and ones, $t \geq 1$, $i = 1, \dots, t$. If A, B are $(0,1)$-matrices, then we denote by $A \triangle B$ the set of all positions where A and B do not coincide.

LEMMA 2.3.16. *Let*

$$A = \begin{pmatrix} 1 & 0 & & \delta & & \\ \vdots & 1 & 0 & & * & \\ \vdots & & \ddots & \ddots & & \\ \vdots & & & \ddots & \ddots & \\ \vdots & & * & & \ddots & 0 \\ 0 & \dots & \dots & \dots & \dots & 1 \end{pmatrix}, \qquad B = \begin{pmatrix} 0 & 1 & & \delta & & \\ \vdots & 0 & 1 & & * & \\ \vdots & & \ddots & \ddots & & \\ \vdots & & & \ddots & \ddots & \\ \vdots & & * & & \ddots & 1 \\ 1 & \dots & \dots & \dots & \dots & 0 \end{pmatrix} \tag{2.3.16}$$

be $(0,1)$-matrices of order p, where δ is an arbitrary $(p-2)$-dimensional $(0,1)$-vector. If $A \triangle B$ is the set consisting of the principal and the second diagonals, then there exists a sequence $A_1 = A, A_2, \dots, A_p = B$ such that each matrix A_i is obtained from A_{i-1} by the use of some interchangeable matrix of the form K_k, $2 \leq k \leq p$, and $\rho(A_i) = p-1$ or $\rho(A_i) = p$, $i = 2, 3, \dots, p$. The equality $\rho(A') = p-1$ for some member $A' = (a'_{ij})$ of this sequence is possible only if there exists an index $l \in N_p$ such that $a'_{1l} = 0$, $a'_{l-1,l} = a'_{ll} = 1$, $a'_{1,l-1} = 1$, and the lth column takes part in any covering of ones of A' by $p-1$ lines.

PROOF. We represent the sequence of p elements of the first row of the matrix A in the cyclic order beginning with 0 in the position $(1,2)$ and ending by 1 in the position $(1,1)$ in the form of intervals of zeros and ones:

$$(0\delta 1) = 0_1 1_1 0_2 1_2 \dots 0_t 1_t, \qquad t \geq 1,$$

where 0_k consists of $i_k \geq 1$ zeros and 1_k consists of $j_k \geq 1$ ones, $k = 1, \dots, t$,

$$(i_1 + \dots + i_t) + (j_1 + \dots + j_t) = p.$$

The matrices $A_2, A_3, \dots$ will be constructed in the following way.

Step 1. By interchanges with the use of interchangeable matrices of the form K_k, $2 \leq k \leq p$, we move the ones of the interval 1_t of the first row to one position to the left. As a result we obtain a matrix X with j_t ones in the last positions of the first row and j_t ones on the second diagonal in the positions

$$(p+1-j_t, p+2-j_t), \dots, (p-1, p), (p, 1).$$

If as a result we do not obtain the matrix B, then we go to the next step.

Step 2. We move the element one in position $(1, p+1-j_t)$ in the matrix X $i_t - 1$ positions to the left by the interchanges described in the statement of the lemma.

If as a result we do not obtain B, then we repeat alternatively Step 1 and Step 2, each time moving the resulting interval of j_t+1 ones one position to the left and then moving its extreme left element one $i_l - 1$ positions to the left, $l = t-1, t-2, \dots, 1$. As a result we obtain the matrix B using

$$j_t + i_t - 1 + \sum_{r=1}^{t-1} (j_r + 1 + i_r - 1) = p - 1$$

interchanges.

The matrix X_k obtained after k realizations of Step 1 and $k-1$ realizations of Step 2 has

$$s = j_t + \sum_{r=t-k+1}^{t-1} (j_r + 1 + i_{r+1} - 1)$$

ones on the second diagonal in the positions $(p+1-s, p+2-s), \dots, (p, 1)$, it has $p-s-1$ ones on the principal diagonal beginning with the position $(2,2)$ up to the position $(p-s, p-s)$, and in addition it has a one in the position $(1, p+1-s)$. These ones are independent; therefore $\rho(X_k) = p$. Consider now the matrices which appear after the kth realization of Step 1, $k = 1, \dots, t$. In such a matrix $Y = (y_{ij})$, the principal diagonal and the second diagonal in the $(p+1-s)$th column contain zeros and contain ones in some column with number l, where l belongs to the interval $[p+2-s, p-s+j_k]$. In this case, $y_{1l} = 0$, $y_{1,l-1} = 1$, and $y_{1,p+1-s} = 1$. Then it is clear that $Y[N_p - \{1\}, N_p]$ has $p-2$ independent ones. On the other hand, any covering of ones of this submatrix by $p-2$ lines contains the lth column and does not contain the $(p+1-s)$th column. But in this case, the one in the position $(1, p+1-s)$ of Y remains uncovered. Consequently, by König's Theorem, $\rho(Y) \geq p-1$.

By realization of Step 2 we move $i_k - 1$ ones from the principal diagonal to the second diagonal, $k = 1, \dots, t$. As is easily seen, we always obtain a matrix with p independent ones: on the principal and the second diagonals in the last $p-1$ rows and also the moving one in the first row. The lemma is proved. □

LEMMA 2.3.17. *Let $Q = (q_{ij})$ be the matrix obtained from some permutation matrix $P = (p_{ij})$ of order p by moving the element one in the first row to some other position of this row. Then $\rho(Q) = p-1$. If $q_{1l} = 1$, $p_{1j} = 1$, then a covering of ones in Q by $p-1$ lines contains the lth column and does not contain the jth column.*

The assertion is obvious.

2.3.9. Connectivity of the graph $G_{\bar\rho}$. Let $\mathfrak{A} = \mathfrak{A}^{m,n}(R,S)$ be a class of $(0,1)$-matrices and $\bar\rho$ be the maximal term rank of $\mathfrak{A}$. Suppose that $A \in \mathfrak{A}$ and $\rho(A) = \bar\rho$. We have seen (Theorem 2.3.4) that if $\mathfrak{A}$ is a monotone class, then we can transform A by a finite number of interchanges to some matrix $A_{\bar\rho}$ of a special form with ones in the positions $(1, \bar\rho), (2, \bar\rho - 1), \dots, (\bar\rho, 1)$. In this case (see Lemma 2.3.14), the corresponding path in the interchange graph goes through matrices with maximal term rank. By Theorem 2.3.6, for the monotone class $\mathfrak{A}$ there exist integers e, f, $0 \leq e \leq m$, $0 \leq f \leq n$, such that the matrix $A_{\bar\rho}$ is of the form

$$A_{\bar\rho} = \begin{pmatrix} \mathbf{J}_{ef} & V \\ W & U \end{pmatrix}, \tag{2.3.17}$$

where $\mathbf{J}_{ef}$ is the $e \times f$ matrix consisting of ones and the number of ones $\sigma = \sigma(U)$ in the matrix U equals $\bar{\rho} - (e + f)$. Thus, the elements of U are zeros, except for the independent ones located in the positions

$$(\bar{\rho} - f, f + 1), (\bar{\rho} - f - 1, f + 2), \ldots, (e + 1, f + \sigma).$$

THEOREM 2.3.18. *Let $\mathfrak{A} = \mathfrak{A}^{m,n}(R, S)$ be a nonempty Ryser class with maximal term rank $\bar{\rho}$, and let $G = G(R, S)$ be the interchange graph of $\mathfrak{A}$. The subgraph $G_{\bar{\rho}}(R, S)$ of G is a connected graph.*

PROOF. First of all, note that it is sufficient to prove the theorem for a monotone class. Indeed, any matrix in $\mathfrak{A}$ can be transformed into a matrix with monotone vectors of row and column sums by the use of a pair of permutations of rows and columns common for all matrices in $\mathfrak{A}$.

By Theorem 2.3.4 and Lemma 2.3.14 any matrix of a monotone class with maximal term rank can be transformed into a matrix of the form (2.3.17) by interchanges preserving the value of the term rank. Thus, it is sufficient to prove the theorem for an arbitrary pair of matrices of the form (2.3.17).

The set of matrices $A_{\bar{\rho}} = (a_{ij})$ is characterized by the following properties. Let $R = (r_1, \ldots, r_m)$, $S = (s_1, \ldots, s_n)$, $r_1 \geq \cdots \geq r_m$, $s_1 \geq \cdots \geq s_n$. Then:

1. The matrix V belongs to the class $\mathfrak{A}_1 = \mathfrak{A}^{e,n-f}(R_1, S_1)$, where

$$R_1 = (r_1 - f, \ldots, r_e - f),$$

$$S_1 = \begin{cases} (s_{f+1}, \ldots, s_n), & \sigma = 0, \\ (s_{f+1} - 1, \ldots, s_{f+\sigma} - 1, s_{f+\sigma+1}, \ldots, s_n), & \sigma \neq 0. \end{cases}$$

The matrix W belongs to the class $\mathfrak{A}_2 = \mathfrak{A}^{m-e,f}(R_2, S_2)$, where

$$R_2 = \begin{cases} (r_{e+1}, \ldots, r_m), & \sigma = 0, \\ (r_{e+1} - 1, \ldots, r_{e+\sigma} - 1, r_{e+\sigma+1}, \ldots, r_m) & \sigma \neq 0, \end{cases}$$

$$S_2 = (s_1 - e, \ldots, s_f - e).$$

2. $\rho(V) = \bar{\rho}(\mathfrak{A}_1) = e$, $\rho(W) = \bar{\rho}(\mathfrak{A}_2) = f$.
3. $a_{1\bar{\rho}} = a_{2,\bar{\rho}-1} = \cdots = a_{\bar{\rho}1} = 1$.

We derive the assertion of the theorem from the following fact concerning arbitrary (not necessarily monotone) classes of $(0,1)$-matrices.

THEOREM 2.3.19. *Let A and B be $(0,1)$-matrices of size $g \times h$ of a common Ryser class, and let $\mathcal{B}$ be a set of ρ_0 pairwise noncollinear positions in a matrix of size $g \times h$. Suppose that in A and B the positions of $\mathcal{B}$ are filled by ones. Then the interchange graph of the class contains a path from A to B passing only through matrices having ρ_0 independent ones located in the same rows and columns as $\mathcal{B}$.*

PROOF. Let

$$A \triangle B = C_1 \cup C_2 \cup \cdots \cup C_q, \tag{2.3.18}$$

where C_i is the set of positions containing the arcs of some simple circuit of the digraph $O(A) - O(B)$, $i = 1, \ldots, q$ (see Section 1.2.1). Since (2.3.18) is a partition, it is sufficient to consider the case where $O(A) - O(B)$ is a simple circuit Γ of length $2p$, that is, the case $q = 1$.

Let I, J be the sets of vertices of the circuit Γ which are the indices of the rows and, respectively, the columns, $|I| = |J| = p$. We denote by $\mathcal{B}_1$ the positions of $\mathcal{B}$

located in the submatrix $A[I, J]$, and let $\mathcal{B}_2 = \mathcal{B} - \mathcal{B}_1$. We perform in A and B the same permutations of rows and columns so that the arcs of Γ directed as in $O(A)$ are on the principal diagonal and the remaining arcs are on the second diagonal. Thus, A and B are transformed into matrices A' and B' such that the submatrices $A'[N_p, N_p]$ and $B'[N_p, N_p]$ are a pair of matrices of the form (2.3.16). In this case, the positions of the sets $\mathcal{B}$, $\mathcal{B}_1$, $\mathcal{B}_2$ turn into the sets of pairwise noncollinear positions $\mathcal{B}' = \mathcal{B}_1' \cup \mathcal{B}_2'$, with $\mathcal{B}_1'$, $\mathcal{B}_2'$ in the matrices A', B', respectively. It is clear that ones in the positions in $\mathcal{B}_1'$ form a set of independent ones in $A'[N_p, N_p]$ and $B'[N_p, N_p]$ located outside the principal and the second diagonals of these matrices.

For $p = 2$, the assertion is obvious. Let $p > 2$. Consider the two possible cases.

Let $A'[N_p, N_p]$ have a row which does not contain a position of $\mathcal{B}_1'$. Then by Lemma 2.3.15 it is possible to transform A' into B' by $p - 1$ interchanges which do not touch elements in the positions of $\mathcal{B}_1'$. Thus, each of these interchanges gives a matrix with at least ρ_0 independent ones.

Now let each row of $A'[N_p, N_p]$ contain a position of $\mathcal{B}_1'$; that is, ones of A' and B' located in the positions of $\mathcal{B}_1'$ correspond to a permutation matrix $P = (p_{ij})$ of order p such that $P < A'[N_p, N_p]$, $B'[N_p, N_p]$.

Let us construct a sequence transforming A' into B' by the rule described in Lemma 2.3.16 with the use of the first row of $A'[N_p, N_p]$. Let us show that in this case after each interchange the resulting matrix has at least ρ_0 independent ones. In fact we prove that the term rank of the submatrix $C[N_p, N_p]$ is equal to p for each matrix C obtained during the interchanges. Let $p_{1j} = 1$. Until the interchanges do not touch one in the position $(1, j)$, we have $P < C$ and therefore $\rho(C) = p$. Now let us perform an interchange that moves this one to the position $(1, j - 1)$ (see Step 1 in the proof of Lemma 2.3.17). Suppose, on the contrary, that $\rho(C) = p - 1$ for the resulting matrix C. By Lemma 2.3.16 this is possible only if there exists a covering by $p - 1$ lines containing the jth column. Let Q be the matrix which is obtained from P by transferring one of the first row to the position $(1, j - 1)$. It is clear that $Q < C$. Therefore a covering of ones of C by $p - 1$ lines covers ones of Q. However by Lemma 2.3.17 this is impossible, since no covering of ones of Q by $p - 1$ lines can include the jth column. After the next interchange we again obtain one in the position $(1, j)$, which remains in this position after all further interchanges. Thus, after each interchange we obtain a matrix C with $|\mathcal{B}_1'| = p$ independent ones in $C[N_p, N_p]$ and altogether with at least $\rho_0 = p + |\mathcal{B}_2'|$ independent ones in C. Thus, Theorem 2.3.19 is proved. □

We return to the proof of Theorem 2.3.18. Let $A_{\bar{\rho}}$, $B_{\bar{\rho}}$ be matrices of the initial Ryser class $\mathfrak{A}^{m,n}(R, S)$ such that

$$A_{\bar{\rho}} = \begin{pmatrix} \mathbf{J}_{ef} & V_1 \\ W_1 & U \end{pmatrix}, \qquad B_{\bar{\rho}} = \begin{pmatrix} \mathbf{J}_{ef} & V_2 \\ W_2 & U \end{pmatrix},$$

where $\rho(V_1) = \rho(V_2) = \bar{\rho}(\mathfrak{A}_1) = e$, $\rho(W_1) = \rho(W_2) = \bar{\rho}(\mathfrak{A}_2) = f$. In this case, e independent ones in V_1 and V_2 are located in the same positions in the columns with numbers $f + \sigma + 1, \ldots, f + \sigma + e$ and f independent ones in W_1 and W_2 are located in the rows with numbers $e + \sigma + 1, \ldots, e + \sigma + f$. Theorem 2.3.19 shows that V_1 can be transformed into V_2 by interchanges so that each matrix obtained during these transformations has e independent ones in the columns with the same numbers $f + \sigma + 1, \ldots, f + \sigma + e$. Similarly, W_1 can be transformed into W_2 by interchanges so that each intermediate matrix has f independent ones in the rows

$e+\sigma+1, \ldots, e+\sigma+f$. Thus, performing by the corresponding interchanges first the transformation of V_1 into V_2 and then the transformation of W_1 into W_2, we obtain a path from $A_{\bar{\rho}}$ to $B_{\bar{\rho}}$ going through matrices with term rank equal to $e+f+(\bar{\rho}-(e+f))=\bar{\rho}$. Theorem 2.3.18 is thus proved. □

In conclusion, note that a similar question on the connectivity of the subgraph $G_{\tilde{\rho}}$ of the graph $G(R,S)$ induced by the matrices of the class $\mathfrak{A}(R,S)$ with minimal term rank remains open.

Problems.

1. Let a $(0,1)$-matrix A with $\rho(A)=\rho$ be such that for any interchangeable matrix $A[\{i,k\},\{j,l\}]$ there exists a covering of ones in A by ρ lines containing either the rows i,k or the columns j,l. Show that $\rho(A')\leq\rho$ for any matrix A' adjacent to A in the interchange graph.
2. Let $\mathfrak{A}$ be a class of $(0,1)$-matrices of order n containing a matrix with positive diagonal. What is the minimal value of the term rank of matrices in $\mathfrak{A}$?
3. Let $A=(a_{ij})$ and $B=(b_{ij})$ be $(0,1)$-matrices of size $m\times n$ such that the directed bipartite graph $O(A)-O(B)$ is a circuit
$$i_1\to j_1\to i_p\to j_p\to i_{p-1}\to j_{p-1}\to\cdots\to i_2\to j_2\to i_1$$
of length $2p$, where
$$1\leq i_1\leq\cdots\leq i_p\leq m,\qquad 1\leq j_1\leq\cdots\leq i_p\leq n.$$
Suppose also that there exists $s\in N_p$ such that $a_{i_s,j_{s-1}}=b_{i_s,j_{s-1}}=1$. Show that A can be transformed into B by $p-1$ interchanges such that each interchange uses two elements of the row with number i_s and the last interchange uses elements of the columns with numbers j_{s-1} and j_s.
4. Suppose that $\mathfrak{A}$ is a Ryser class of $(0,1)$-matrices of size $m\times n$, $\tilde{\rho}=\tilde{\rho}(\mathfrak{A})$ is the minimal term rank of $\mathfrak{A}$, and matrices $A,B\in\mathfrak{A}$ possess the following properties:
$$\rho(A)=\rho(B)=\tilde{\rho},\quad A-B=\mathbf{I}-P_p,\quad p\leq\min\{m,n\},$$
and one stands in the position $(1,p)$ in A and B. Suppose also that there exists a covering of ones of A by $\tilde{\rho}$ lines including the rows $1,2,\ldots,k$ and the columns $k+1,k+2,\ldots,p$, where k is one of the numbers $0,1,\ldots,p$. Using the result of Problem 3, show that A can be transformed into B by interchanges, obtaining after each interchange a matrix with the minimal term rank.

CHAPTER 3

Nonnegative Matrices and Extremal Combinatorial Problems

Introduction

In this chapter, we consider a series of extremal combinatorial problems related to nonnegative matrices. In some cases, matrices provide a convenient means for analysis of such extremal problems, although from the beginning these problems are formulated in other terms. In other cases, these are problems on the extremum of functions directly connected with the matrices themselves.

The problem of forbidden configurations is a rather general combinatorial problem. From the very beginning, this problem was considered from two different points of view: as a problem on $(0,1)$-matrices, and as a purely set-theoretic problem about the maximal density for systems of sets. In Section 3.1, we give one of the most general results in this direction, formulating it in terms of density but with the proof based on a matrix construction. In the problem on forbidden configurations, an extremal matrix is a simple matrix (that is, a matrix without identical rows) which has the maximum number of rows among the matrices not containing the forbidden configurations. We present an inductive construction of such a matrix in the problem where the forbidden configurations are all k-dimension vectors with exactly s nonzero coordinates. In this section, we follow [**7, 9, 10, 61**].

One of the best known extremal combinatorial problems is the covering problem considered in Section 3.2. This problem consists in determining the minimum number of subsets taken from a given system of subsets of a finite set which, taken together, cover the set and, moreover, in finding the covering itself. This number, called the covering number, is interpreted in matrix terms as the depth of the incidence matrix of the given system of subsets. In the literature on combinatorics, the same extremal problem is often formulated in terms of hypergraphs as the problem on the transversal number of a hypergraph (see, for example, [**14, 113**]). A natural generalization of the covering problem is the α-covering problem, where α is a given positive integer. In matrix terms, this is the problem about the α-depth of a nonnegative matrix. We derive some properties of the α-depth. For an arbitrary $(0,1)$-matrix A belonging to some Ryser class $\mathfrak{A}$, the α-depth $\varepsilon(A;\alpha)$ is a number in the interval $[\tilde{\varepsilon}(\mathfrak{A};\alpha), \bar{\varepsilon}(\mathfrak{A};\alpha)]$, where $\tilde{\varepsilon}$ and $\bar{\varepsilon}$ are, respectively, the minimal and the maximal depth of the matrices in the class $\mathfrak{A}$. It is shown that any number in this interval can be a value of the α-depth of some matrix of this class. An effective algorithm for finding the minimal depth of a Ryser class is suggested. Then a more difficult problem on the maximal depths is considered. We describe some properties of the maximal depth and, in particular, derive relations between the maximal depths of the classes $\mathfrak{A}^n(k)$ for various n, k. Next, in the matrix terminology, we develop a general approach to the covering problem. This approach is related to a construction in a $(0,1)$-matrix and a series of numerical characteristics of the matrix. One of these characteristics is the number of pairwise distinct orthogonal rows with maximal row sums in a matrix (the number of matchings in a hypergraph). A lower bound for this number is given. The use of this bound in the construction mentioned above leads to an upper bound for the covering number of an arbitrary system of subsets. First this bound is expressed in terms of parameters of the matrix related to the numbers of matchings, and then it is presented in a form containing only the parameters of the size of the $m \times n$ matrix, the maximum r of the row sums, and the minimum s of the column sums. This upper bound allows us to estimate the asymptotic behaviour of the maximal depth as $m, n, r, d \to \infty$ so that both m and n and, respectively, r and d have the same orders. In conclusion we deduce the general upper bound for the maximal α-depth, $\alpha \geq 1$, which is very useful for obtaining a number of asymptotic results on the maximal α-depth. In Section 3.2, we use the methods and results from [**104, 103, 105, 57, 71**].

The famous van der Waerden problem on the minimum of the permanent of a doubly stochastic matrix, which remained unsolved for more than half a century, is a combinatorial problem on the extremum of a function. The solution of this problem was obtained simultaneously in [**35**] and [**38**] (in the last paper, without proof of the uniqueness of the minimizing matrix). In Section 3.3, we give both solutions following, along with the original papers [**35**] and [**38**], an elegant presentation by van Lint [**115**].

3.1. Forbidden configurations

3.1.1. Formulation of the problem. Here we consider a typical extremal combinatorial problem admitting a convenient interpretation in terms of nonnegative matrices. In various combinatorial problems, one must find an upper bound for the size of a given combinatorial configuration which does not contain certain forbidden configurations of a special form. Consider, in particular, the following question: how many edges can there be in an n-vertex graph that does not contain, as an induced subgraph, the complete graph K_r with r vertices? This question can easily be formulated in matrix terms. A matrix is called simple if it does not contain two identical rows. We denote by $A(G)$ and $B(G)$, respectively, the adjacent matrix and the incidence matrix of a graph G (see Section 1.1). Then a graph G with n vertices contains the graph K_r if and only if $B(G)$ contains as a submatrix a simple matrix F belonging to the class

$$\mathfrak{A}^{m,r}(2, r-1), \quad m = \binom{r}{2}, \quad r \leq n$$

(see Section 1.1). Thus, the initial question can be formulated as follows. Let B be a simple $(0,1)$-matrix with n columns and all row sums equal to 2. What is the maximum number m of rows of B such that B does not contain as a submatrix a matrix of the class

$$\mathfrak{A}^{m',n'}(2, r-1), \quad m' = \binom{r}{2}, \quad n' = r,$$

for a given $r \geq 3$? The same problem can be formulated in terms of the adjacent matrix $A(G)$: what is the maximum number of ones in a symmetric $(0,1)$-matrix A of order n with zero trace which contains neither a zero row nor a submatrix

$A[N', N']$, $N' \subset N$, of the form

$$C_r^{r-1} = \begin{pmatrix} 0 & 1 & 1 & \dots & 1 \\ 1 & 0 & 1 & \dots & 1 \\ & & & \dots & 1 \\ 1 & 1 & 1 & \dots & 0 \end{pmatrix}, \tag{3.1.1}$$

where r is the order of the submatrix. Note that the problem on the maximum number of edges in the graph without triangles (that is, for $r = 3$) was solved by Mantel in [**71**] and for an arbitrary r by Turan in [**112**].

In the theory of matroids, the problems on the existence and the construction of matroids with forbidden minors (see [**10**]) are similar to the above-described extremal problems.

Let us formulate the general problem on forbidden configurations. First we give several definitions. Let $\mathcal{S} = (S, \mathcal{E}, \varphi)$ be a configuration (see Section 1.1). A configuration $\mathcal{S}' = (S', \mathcal{E}', \varphi')$ is called a subconfiguration of $\mathcal{S}$ if the incidence matrix $B(\mathcal{S}')$ is permutationally equivalent to some submatrix of the incidence matrix $B(\mathcal{S})$. In this case, we write $\mathcal{S}' \subseteq \mathcal{S}$. It is clear that we can identify $\mathcal{S}'$ with some subset of $\mathcal{S}$ and $\mathcal{E}'$ with some subset of $\mathcal{E}$; and since $\varphi'(E', s') = \varphi(E', s')$ if $s' \in \mathcal{S}'$, $E' \in \mathcal{E}'$, in this case we denote φ' by φ.

We will consider configurations with weights from the set $Z_q = \{0, 1, \dots, q-1\}$. Let n, k be positive integers and $k \leq n$. Consider a configuration $\mathcal{S}_0 = (N_n, \mathcal{E}_0, \varphi)$ and its subconfiguration $\mathcal{F} = (N_k, \mathcal{E}_1, \varphi)$ with a simple incidence matrix $A(\mathcal{F})$. We identify $E \in \mathcal{E}_0$ with a row vector of the incidence matrix $A(\mathcal{E}_0)$. Consider the following question: what is the maximum number of different sets in the configuration $\mathcal{S} = (N_n, \mathcal{E}, \varphi) \subseteq \mathcal{S}_0$ which does not contain $\mathcal{F}$ as a subconfiguration? In this case, $\mathcal{F}$ is called a forbidden configuration and the problem itself is called the problem on forbidden configurations. The configuration $\mathcal{S}_0$ is called the base configuration. The most studied case is the case where $\mathcal{S}_0$ includes the family of all subsets of N_n with weights in Z_q, and just this case will be considered in this section. However, other base configurations are also of significant interest. For example, in the problem on graphs with forbidden subgraphs the role of $\mathcal{S}_0$ is played by the family of all 2-subsets of N_n (or all subsets with at most two elements, if we consider graphs with loops).

Note that, in the matrix interpretation, as the forbidden configuration we often consider not $\mathcal{F}$ itself but the family of permutationally equivalent nonnegative matrices which represent this configuration. These incidence matrices (or adjacent matrices for graphs) can be imagined as one of these matrices in the standard form. For example, in the problem on forbidden complete subgraphs such a matrix can be chosen in the form (3.1.1).

3.1.2. Examples. Connection with the problem on the density for a set of vectors. The statement of the problem given above is usually made more explicit. As the base configuration, the configuration consisting of all subsets of the set N_n with all possible weights from Z_q, $q \geq 2$, is often used. In this case, the subsets with weights are represented as row vectors with coordinates in Z_q. The following two configurations were most intensively studied as forbidden configurations.

One of them is the configuration $\mathcal{C}_k^s$ consisting of all k-dimensional vectors with coordinates in Z_q each containing exactly s nonzero coordinates, $0 \leq s \leq k \leq n$.

For $q = 2$, this configuration can be viewed as the family of all s-subsets of some k-set. Then the question on the forbidden configuration can be formulated as follows: what is the maximum number m of rows in a simple $(0,1)$-matrix A with n columns which does not contain the configuration $\mathcal{C}_k^s$? For $k = 3$, $s = 2$, this is the question on the maximum number of edges in a simple hypergraph, that is, in a hypergraph without multiple edges which does not contain "triangles".

The second is the configuration $\mathcal{C}_{(k)}$ consisting of all k-dimensional vectors with coordinates in Z_q. The forbidden configurations of such a form appear to be the simplest ones for investigation. On the other hand, $\mathcal{C}_{(k)}$ is of interest in the general context of this section. Indeed, if we find an upper bound in the problem with the forbidden configuration $\mathcal{C}_{(k)}$, then this bound is an upper bound in the problem with any forbidden configuration $\mathcal{F}$ determined on k-subsets, since $\mathcal{F}$ is contained in $\mathcal{C}_{(k)}$ as a subconfiguration.

It is interesting to note that the problem under consideration was simultaneously studied for two systems of notions, namely in terms of $(0,1)$-matrices, as presented above, and in terms of dense families of sets and, later, families of vectors. The latter formulation led to further generalizations of the problem. Namely, suppose that in the problem the following restrictions on the weights of the elements $1, 2, \dots, n$ are imposed: the weight of an element i cannot be greater than some number $q_i - 1$, $i = 1, 2, \dots, n$, where $2 \leq q_1, \dots, q_n \leq q$. Such vectors are called $(n; q_1, \dots, q_n)$-vectors, and a matrix with $(n; q_1, \dots, q_n)$-vectors as its row vectors is called an $(n; q_1, \dots, q_n)$-matrix. Let $S = \{i_1, \dots, i_k\} \subseteq N_n$. If $x = (x_1, \dots, x_n)$ is an $(n; q_1, \dots, q_n)$-vector, denote by $x|_S$ the vector $(x_{i_1}, \dots, x_{i_k})$, that is, the projection of x on S. Consider a family $\mathcal{S}$ of subsets of N_n. A family M of $(n; q_1, \dots, q_m)$-vectors is called S-dense for $S = \{i_1, \dots, i_k\}$ if for any $(n; q_{i_1}, \dots, q_{i_k})$-vector $x' = (x'_{i_1}, \dots, x'_{i_k})$ there is an $x \in M$ such that $x|_S = x'$. If this condition is fulfilled, then a matrix $A = A(M)$ with the set of row vectors coinciding with M is called S-dense. A family M of $(n; q_1, \dots, q_n)$-vectors (a matrix $A(M)$) is called $\mathcal{S}$-dense if it is S-dense for some $S \in \mathcal{S}$. If $\mathcal{S}$ consists of all k-subsets of N_n, then an $\mathcal{S}$-dense family M of $(n; q_1, \dots, q_n)$-vectors (a matrix $A(M)$) is called k-dense. For $q = 2$ and $\mathcal{S}$ consisting of all t-subsets of N_n, we obtain the notion of a family of sets of density at least t, that is, a family containing together with a t-set also all subsets of this t-set including the empty set (see [**97**]). The maximum number t such that M has this property was called in [**97**] the density of the family M. Now we extend the definition of the configuration $\mathcal{C}_k^s$. Namely, we say that an $(n, q_1, \dots, q_n)$-matrix A contains $\mathcal{C}_k^s$ if for some $J = \{i_1, \dots, i_k\} \subseteq N_n$ each of the $(k; q_{i_1}, \dots, q_{i_k})$-vectors having exactly s nonzero coordinates occurs as a row vector in the submatrix of A formed by the columns with numbers in J, $0 \leq s \leq k \leq n$.

3.1.3. The Alon Theorem. The most general result on forbidden configurations was obtained by Alon in [**7**] as a theorem on the density of a set of vectors. We prove this theorem using $(n; q_1, \dots, q_n)$-matrices, as was done in [**9**].

THEOREM 3.1.1. *Let A be a simple $(n; q_1, \dots, q_n)$-matrix with m rows, where $m, n, q_1, \dots, q_n$ are positive integers, $q_i \geq 2$, $i = 1, 2, \dots, n$. Let $\mathcal{S}$ be a family of subsets of N_n. Suppose that A is not $\mathcal{S}$-dense. Then*

$$m \leq f(n, \mathcal{S}), \tag{3.1.2}$$

where $f(n,\mathcal{S})$ is the number of all $(n;q_1,\dots,q_n)$-vectors such that the simple matrix formed by these vectors as the rows does not contain a row x such that

$$x|_S = (0,\dots,0)$$

for any $S \in \mathcal{S}$. In other words, any simple $(n;q_1,\dots,q_n)$-matrix of size $m \times n$ is $\mathcal{S}$-dense for $m > f(n,\mathcal{S})$.

PROOF. We prove the theorem by induction on the number of columns n. For $n = 1$, the assertion is obviously true. For $n > 1$, we introduce two families of subsets

$$\begin{aligned} \mathcal{S}' &= \{S - \{n\}\colon S \in \mathcal{S}, n \in S\}, \\ \mathcal{S}'' &= \{S \in \mathcal{S}, n \notin S\}. \end{aligned} \tag{3.1.3}$$

Let us show that

$$f(n,\mathcal{S}) = f(n-1,\mathcal{S}') + (q_n - 1) f(n-1,\mathcal{S}''). \tag{3.1.4}$$

Indeed, consider $f(n,\mathcal{S})$ pairwise distinct row vectors having the property from the statement of the theorem. Let the nth coordinate of one such row x be nonzero. By the conditions of the theorem there is no subset $S \in \mathcal{S}''$ such that $x|_S = (0,\dots,0)$. The number of such rows is, obviously, $(q_n - 1)f(n-1,\mathcal{S}'')$. If we take a row with the zero nth coordinate, then $x|_{S\setminus\{n\}} \neq (0,\dots,0)$ for any $S \in \mathcal{S}$, $n \in S$, and there are $f(n-1,\mathcal{S}')$ such rows.

Take an $m \times n$ matrix A which is not $\mathcal{S}$-dense. We exchange the rows of A so that first we have the row vectors with the nth coordinate equal to 0, then to 1, and so on, and finally we have the rows with the nth coordinate equal to $q_n - 1$. We set $B = A[N_m, N_{n-1}]$ and denote by B_i the submatrix of B whose rows correspond to the rows of A with the nth coordinate equal to i, $i = 0,1,\dots,q_n - 1$. Then, turning to the transpose matrix for convenience of writing, we see that

$$A^T = \begin{pmatrix} B_0^T & \vdots & B_1^T & \vdots & \dots & \vdots & B_{q_n-1} \\ 0\dots0 & \vdots & 1\dots1 & \vdots & \dots & \vdots & q_n-1\dots q_n-1 \end{pmatrix}. \tag{3.1.5}$$

It is clear that $B = [B_0^T B_1^T \dots B_{q_n-1}^T]^T$; that is, the rows of B are the columns of the matrices $B_0^T, B_1^T, \dots, B_{q_n-1}^T$.

By the induction hypothesis the matrix B has at most $f(n-1,\mathcal{S}'')$ distinct rows. Consider the set $\mathcal{R}$ consisting of all distinct row vectors of B_0 (that is, distinct column vectors of B_0^T) which appear among the rows in each of the matrices $B_1,\dots,B_{q_n-1}$ (among the columns in each of the matrices $B_1^T,\dots,B_{q_n-1}^T$). We show that $|\mathcal{R}| \le f(n-1,\mathcal{S}')$. Indeed, otherwise, by the induction hypothesis, B_0 is $\mathcal{S}'$-dense, and there exists $S = \{i_1,\dots,i_k\} \in \mathcal{S}$, $i_1 < \dots < i_k = n$, such that for any $x' = (x_{i_1},\dots,i_k)$ the set $\mathcal{R}$ contains a row vector $x = (x_1,\dots,x_n)$ with $x|_{S-\{n\}} = x'$. Let $x_{i_k} = a$. In B_a there exists a row x. It is clear that $(x_1,\dots,x_{n-1},a)|_S = (x_{i_1},\dots,x_{i_k}) = x'$. Thus, the matrix A is $\mathcal{S}$-dense, in contradiction with the assumption.

Thus, B has at most $f(n-1,\mathcal{S}'')$ distinct rows, at most $f(n-1,\mathcal{S}')$ of them can appear q_n times and the remaining rows at most $q_n - 1$ times. Consequently,

$$\begin{aligned} m &\le (q_n - 1)(f(n-1,\mathcal{S}'') - f(n-1,\mathcal{S}')) + q_n f(n-1,\mathcal{S}') \\ &= f(n-1,\mathcal{S}') + (q_n - 1) f(n-1,\mathcal{S}'') = f(n,\mathcal{S}) \end{aligned}$$

(see (3.1.4)); that is, we obtain (3.1.2). The theorem is thus proved. □

A simple $(n; q_1, \dots, q_n)$-matrix with $f(n, \mathcal{S})$ rows which is not $\mathcal{S}$-dense is called an extremal matrix with respect to the family $\mathcal{S}$.

COROLLARY 3.1.2. *Let M be an extremal $m \times n$ matrix with respect to some family $\mathcal{S}$ of subsets of N_n, and let $m = f(n, \mathcal{S})$. If $B = M[N_m, N_{n-1}]$, then B contains exactly $f(n-1, \mathcal{S}'')$ distinct rows, and exactly $f(n-1, \mathcal{S}')$ of them are included in each of the submatrices $B_0, B_1, \dots, B_{q_n - 1}$ defined in (3.1.5), where $\mathcal{S}'$, $\mathcal{S}''$ are the families of subsets defined for $\mathcal{S}$ by (3.1.3).*

The assertion of the corollary follows directly from (3.1.2) and (3.1.4).

Let $\mathcal{S}$ be a family of subsets of N_n and $S \in \mathcal{S}$. We denote by $\mathcal{C}_S$ the configuration consisting of all subsets of S with weights, including the empty set which corresponds to the vector $(0, \dots, 0)$. Recall that we identify subsets of the base configuration and the corresponding row vectors of its incidence matrix.

COROLLARY 3.1.3. *Let $\mathcal{S}$ be a family of subsets of N_n and let w be an integer such that $0 < w < \min\{|S| \colon S \in \mathcal{S}\}$. For a simple $(n; q_1, \dots, q_n)$-matrix A, let A' be the submatrix formed by all row vectors of A with at least w nonzero coordinates. If the number of rows in A is at least $f(n, \mathcal{S}) + 1$, then for some $S \in \mathcal{S}$ the matrix A' contains the configuration consisting of all rows of $\mathcal{C}_S$ with at least w nonzero coordinates.*

PROOF. The matrix A can be obtained from A' by adding row vectors with at most $w - 1$ nonzero coordinates each. By Theorem 3.1.1 the matrix A is $\mathcal{S}$-dense and therefore contains the configuration $\mathcal{C}_S$ for some $S \in \mathcal{S}$. Then all rows of A containing rows of $\mathcal{C}_S$ with at least w nonzero coordinates should belong to A'. □

COROLLARY 3.1.4. *Let $\mathcal{S}(r)$ consist of all r-subsets of N_n. Then for any k, $0 < k \le n$, there exists a simple $(n; q_1, \dots, q_n)$-matrix with*

$$\sum_{r=n-k+1}^{n} \sum_{\{i_1, \dots, i_r\} \in \mathcal{S}(r)} (q_{i_1} - 1) \dots (q_{i_r} - 1) \tag{3.1.6}$$

rows which does not contain the subconfiguration $\mathcal{C}_{(k)}$.

Indeed, it is obvious that the number (3.1.6) coincides with $f(n, \mathcal{C}_{(k)})$.

3.1.4. The extremal matrix for the configuration $\mathcal{C}_k^s$. Corollary 3.1.2 and the proof of Theorem 3.1.1 suggest the idea of the inductive construction of extremal matrices for problems on forbidden configurations. We will show how this idea is realized in the case of the forbidden configuration $\mathcal{C}_k^s$ defined in Section 3.1.2.

Here we consider the family $\mathcal{S} = \mathcal{S}(k)$ consisting of all k-subsets of N_n, $0 < k \le n$. Observe that by the definition of the function $f(n, \mathcal{S})$ in this case there exists a simple $(n; q_1, \dots, q_n)$-matrix with $f(n, \mathcal{S})$ rows which does not contain $\mathcal{C}_k^0$. For this $\mathcal{S}$, the number $f(n, \mathcal{S})$ is denoted by $f(n, k)$.

THEOREM 3.1.5. *Let $\mathcal{S}$ be the family of all k-subsets of N_n, and let s be an integer such that $0 \le s \le k$. For each n there exists a simple $(n; , q_1, \dots, q_n)$-matrix of size $f(n, k) \times n$ which does not contain $\mathcal{C}_k^s$.*

PROOF. We prove the theorem by induction on the number of columns. For $s = 0$, the assertion is true according to the above remark. Therefore we assume that $0 < s \le k$. We denote the required matrix by $A_n(k, s)$. The case $k > n$ is allowed, in which case for $A_n(k,s)$ one should take the matrix with all $q_1q_2\dots q_n$ possible row vectors. We will construct $A_n(k,s)$ so that in the case $k>s$ this matrix contains $A_n(k-1,s)$ as a submatrix.

First we set $A_1(1,1)=[0\ 2\ 3\ \dots\ (q_1-1)]^T$ and $A_1(k,s)=[0\ 1\ 2\ \dots\ (q_1-1)]^T$ for $k>1$. Then suppose that all $A_t(k,s)$ for $0<s\le k$ and $t<n$ have been constructed, and if $k>s$, then $A_t(k,s)$ contains $A_t(k-1,s)$ as a submatrix.

The matrix $A_n(k,s)$ will be constructed as follows. If $k=s$, then for $A_n(k,k)$ we take the matrix with n columns whose rows form the set of all pairwise distinct $(n;q_1,\dots,q_n)$-rows with k elements equal to 1. Then it is clear that $A_n(k,k)$ is a simple matrix with $f(n,k)$ rows, and it does not contain the configuration $\mathcal{C}_k^k$. If $k>s$, then, as will be proved below, for $A_n(k,s)$ we can take the matrix

(3.1.7)

$$A=\begin{pmatrix} A_{n-1}(k-1,s)^T & \vdots & A_{n-1}(k,s)^T & \vdots & \dots & \vdots & A_{n-1}(k,s)^T \\ 0\dots0 & \vdots & 1\dots1 & \vdots & \dots & \vdots & (q_n-1)\dots(q_n-1)\end{pmatrix}^T.$$

First of all, observe that A is simple; indeed, by the induction assumption the matrices $A_{n-1}(k-1,s)$ and $A_{n-1}(k,s)$ are simple. Further, by (3.1.3) and (3.1.4) the matrix A has

$$f(n-1,k-1)+(q_n-1)f(n-1,k)=f(n,k)$$

rows. This follows from the fact that in this case $\mathcal{S}'$ is the set of all $(k-1)$-subsets and $\mathcal{S}''$ is the set of all k-subsets of N_{n-1}. Suppose that A contains the configuration $\mathcal{C}_k^s$. If the incidence matrix $B(\mathcal{C}_k^s)$ of this configuration had zeros in the last column of the matrix A, then it would follow that $A_{n-1}(k-1,s)$ contains the configuration $\mathcal{C}_{k-1}^s$, which is impossible by the induction assumption. But for $k>s$ the configuration $\mathcal{C}_k^s$ contains zeros in each column. Consequently, $\mathcal{C}_k^s$ is contained in the first $n-1$ columns of the matrix A. All the rows of the matrix $B(\mathcal{C}_k^s)$ are pairwise distinct. In addition, take into account that by the induction assumption the rows of $A_{n-1}(k-1,s)$ appear among the rows of $A_{n-1}(k,s)$. Therefore we obtain that $A_{n-1}(k,s)$ contains $\mathcal{C}_k^s$, and we arrive at a contradiction. Thus, A does not contain $\mathcal{C}_k^s$.

It remains to show that $A_n(k,s)$ contains $A_{n-1}(k-1,s)$ if $k>s$. Consider two possible cases.

Let $k>s+1$. Then the matrix $A_{n-1}(k,s)$ contains $A_{n-1}(k-1,s)$, which in turn contains $A_{n-1}(k-2,s)$ by the induction assumption. For $A_n(k-1,s)$, it is possible to construct the representation similar to (3.1.7). Comparing it with A, we obtain the desired result.

Let $k=s+1$. Again consider (3.1.7). By construction, the matrix $A_n(s,s)$ is such that all its rows are $(n;q_1,\dots,q_n)$-rows with at most $s-1$ elements equal to 1. Note that $A_{n-1}(s,s)$ contains all possible $(n-1)$-dimensional row vectors with at most $s-1$ elements equal to 1. Therefore A contains all row vectors with at most $s-1$ elements equal to 1 and with the nth component equal to zero. In the same way $A_{n-1}(s+1,s)$ contains (by the induction assumption) $A_{n-1}(s,s)$, and therefore A contains all such row vectors with the nth coordinate equal to $1,2,\dots,q_n-1$. Thus, $A_n(s+1,s)$ contains $A_n(s,s)$.

Thus, each step of the induction construction has been realized, and we see that the matrix A given by (3.1.7) can be taken as the required matrix $A_n(k,s)$. The theorem is proved. □

COROLLARY 3.1.6. *For any positive integer n there exists a simple $(0,1)$-matrix of size $m \times n$ which does not contain the configuration C_k^s, where*

$$m = \sum_{i=0}^{k-1} \binom{n}{i}, \qquad s = 0, 1, \dots, k.$$

This corollary follows from Theorem 3.1.5 if we set $q_1 = \cdots = q_n = 2$ (see (3.1.6)).

Note that, in particular, the matrix $A_n(k,s)$ constructed in Theorem 3.1.5 does not contain the $(0,1)$-matrix of size $\binom{k}{s} \times k$ consisting of all rows with exactly s ones. It is clear that the elements 0 and 1 in this matrix can be replaced by any pair of distinct elements $\{i,j\}$, $i,j \le \min_{1\le r\le n}(q_r - 1)$.

3.1.5. Some remarks. In the first proof of Theorem 3.1.1, a shift operator defined on the set of $(n; q_1, \dots, q_n)$-matrices was used. Here we describe this operator, which was also used in the proofs of some other results on forbidden configurations.

Let $n, q_1, \dots, q_n$ be positive integers, $q_i \ge 2$, $i = 1, \dots, n$, and let x be an $(n; q_1, \dots, q_n)$-vector with coordinates $(a_1, \dots, a_n)$. For $1 \le i \le n$, $j+1 \le q_i - 1$, we define the operator $\bar{T}_{ij}$, setting

$$\bar{T}_{ij}(x) = \begin{cases} (a_1, \dots, a_n), & a_i \ne j, \\ (a_1, \dots, a_{i-1}, a_i + 1, a_{i+1}, \dots, a_n, & a_i = j. \end{cases}$$

Let A be an $(n; q_1, \dots, q_n)$-matrix with row vectors $x_1, \dots, x_m$. We define the shift operator T_{ij} for any $j < q_i - 1$, $i \in N_n$, setting

$$T_{ij} = (y_1, \dots, y_m)^T,$$

where for $k = 1, 2, \dots, m$

$$y_k = \begin{cases} \bar{T}_{ij}(x_k) & \text{if } \bar{T}_{ij}(x_k) \ne (x_1, \dots, x_m), \\ x_k & \text{otherwise.} \end{cases}$$

Thus, the action of the operator T_{ij} is the replacement of the element j in the ith column of the $(n; q_1, \dots, q_n)$-matrix A by the element $j+1$ if the resulting row is not contained in A. The following property is the main property of the shift operator T_{ij} used in the proof of Theorem 3.1.1: if A is a simple $(n; q_1, \dots, q_n)$-matrix with m rows, then for any $J \subseteq N_n$ the number of distinct rows in $A[N_m, J]$ is no less than in $T_{ij}(A)[N_m, J]$.

Note that in **[10]** the proof of Theorem 3.1.1 in the case $q_1 = \cdots = q_n = 2$ is carried out by methods of linear algebra; that paper includes also a survey of open questions related to the problem on forbidden configurations.

3.2. Covering problem

3.2.1. α-depth of a nonnegative matrix. In this section, we consider functions on matrices related to the covering problem, one of the most interesting extremal combinatorial problems. Let A be an $m \times n$ nonnegative matrix with positive vector of column sums, ${}^sA$ its support, and d the minimum number of positive elements of a column of A. For a positive integer α, $1 \le \alpha \le d$, the α-depth of the matrix A is the minimum number of rows of ${}^sA$ such that, in the submatrix formed by these rows, the sum of elements in each column is at least α. We denote the

α-depth of the matrix A by $\varepsilon(A;\alpha)$. The number $\varepsilon(A;1)$ will be called the depth of A and denoted by $\varepsilon(A)$. Let A be a nonnegative $m \times n$ matrix with a positive vector of row sums, ${}^sA$ its support, and r the minimum number of positive elements of a row of A. For a positive integer α, $1 \leq \alpha \leq r$, the α-width of the matrix A is the minimum number of columns of ${}^sA$ such that, in the submatrix formed by these columns, the sum of elements in each row is at least α. It is clear that the α-width of the matrix A coincides with the α-depth of the transpose matrix A^T, and therefore we do not introduce a special notation for the α-width. The 1-width $\varepsilon(A^T)$ of A will be referred to as the width of A. The α-depth (α-width) of a nonnegative matrix is determined by the properties of its support ${}^sA$; therefore in this section we consider only $(0,1)$-matrices.

3.2.2. The depth of a matrix and the covering problem. Let X be a finite set consisting of n elements, and $\mathcal{X} = \{X_1, \ldots, X_m\}$ a family of its subsets. Suppose that the sets in $\mathcal{X}$ form a covering of X, that is, $X = \bigcup_{i=1}^m X_i$.

The following questions arise:

(a) What is the minimum number of subsets in a subfamily $\mathcal{X}'$ of $\mathcal{X}$ such that $\mathcal{X}'$ covers X?

(b) How to find such a subfamily $\mathcal{X}'$?

These questions constitute one of the most interesting combinatorial problems, known as the covering problem. In this section, we consider problem (a) and problems closely related to it. The number introduced in (a) is called the covering number for the family $\mathcal{X}$. We consider questions related to estimating the covering number and some related constants.

One more aspect related to coverings should be mentioned, namely, how to list all coverings of a given set which possess some properties. Problems on the enumeration of coverings will be considered in Chapter 4.

Let a set X consist of n elements and let $\mathcal{X} = (X_1, \ldots, X_m)$ be a family of m subsets of X; that is, let the configuration $\mathcal{C} = (X, \mathcal{X}, \chi)$ be given, where χ is the characteristic function of $\mathcal{X}$. Consider the incidence matrix $A = A(\mathcal{C})$ of the configuration $\mathcal{C}$. This is a $(0,1)$-matrix of size $m \times n$. If $X_1, \ldots, X_m$ form a covering of X, then $A(\mathcal{C})$ does not contain zero columns. It is obvious that in this case the covering number for $\mathcal{X}$ is exactly the depth $\varepsilon(A)$ of A defined in Section 3.2.1. Thus, the solution of problem (a) on the covering number will be obtained if we give a method for finding the depth of $(0,1)$-matrices. Note that in this book, in accordance with the common practice, the term "covering" in connection with a nonnegative matrix is used in two different senses, namely, in the case considered just above, where we deal with covering the columns of a matrix by rows; and in the case where we deal with covering all positive elements of a matrix by lines. These meanings of the term covering concern different questions, so the use of the same term usually does not lead to misunderstanding.

There is a natural generalization of the covering problem, when the problems similar to (a) and (b) are stated for a subfamily $\mathcal{X}'$ of the family $\mathcal{X}$ which is an α-covering of the set X. By this is meant a covering such that each element $x \in X$ is α-covered, that is, belongs to at least α subsets of $\mathcal{X}$, where α is a positive integer. The corresponding α-covering number for this family coincides with the α-depth $\varepsilon(A;\alpha)$ of the incidence matrix of the configuration formed by the subsets of $\mathcal{X}$. For a family $\mathcal{X} = \{X_1, \ldots, X_m\}$ of nonempty subsets of a set $X = \{x_1, \ldots, x_n\}$, a similar problem on the minimum number of representatives can be formulated as follows: what is the minimum number of elements in a set $T \subseteq X$ such that any subset of $\mathcal{X}$ contains an element of T? This number is called the transversal number for $\mathcal{X}$. It is clear that if $A = A(\mathcal{C})$ is the incident matrix of a configuration $\mathcal{C} = (X, \mathcal{X}, \chi)$, then the transversal number is equal to the width $\varepsilon(A^T)$ of the matrix A. A similar interpretation is valid for the α-width $\varepsilon(A^T; \alpha)$.

The covering problem and a series of closely related problems have clear analogues in extremal problems for hypergraphs, problems whose modern analysis started in the late 1970s (see [**14, 1, 44, 113**]).

A hypergraph H is a family of subsets (hyperedges or simply edges) of a finite set X, which is the set of vertices or vertex set (see Section 1.1). The maximum number of vertices in an edge (over all edges of a hypergraph H) is called the rank of the hypergraph, and is denoted by $r = r(H)$. The number of subsets containing a given vertex $x \in X$ is called the degree of x, and is denoted by $\delta(x)$. If each of the edges contains the same number of vertices r, then the hypergraph is called r-uniform, or an r-hypergraph, or simply an r-graph. In a hypergraph, multiple edges (that is, identical subsets in the family of subsets) and edges consisting of one vertex (loops) are usually permitted. Thus, an ordinary graph without isolated vertices is a 2-graph (but not a hypergraph of rank 2). The transversal number $\tau(H)$ for a hypergraph H on a set X is defined as the minimum number of vertices in a set $T \subseteq X$ such that T has nonempty intersection with each of the edges of H. The transversal number is sometimes called the covering number for H (see [**14**]). The transversal number $\tau(H)$ coincides with the width $\varepsilon(A^T)$ of the incidence matrix A of the family of subsets composing the hypergraph H.

3.2.3. Properties of the depth. The maximal and minimal depth of the class $\mathfrak{A}(R, S)$. First of all, note that the α-depth of a matrix is invariant under permutations of the rows and columns of the matrix. This property will often be used below.

A set C of different rows of a $(0,1)$-matrix A of size $m \times n$ is called its α-covering (or an α-covering of its columns) if all column sums of the submatrix $A[I(C), N_n]$ are at least α, where $I(C) \subseteq N_m$ is the set of indices of the rows in C. An α-covering of C is minimal if it stops being an α-covering after removing any row. If for a $(0,1)$-matrix $A = (a_{ij})$ of size $m \times n$, a subset $I \subseteq N_m$, and some $j \in N_n$

$$\sum_{i \in J} a_{ij} \geq \alpha, \tag{3.2.1}$$

then we say that the column j is α-covered by the rows with numbers from I. In particular, for $a_{ij} > 0$, we say that the row i covers the column j. If (3.2.1) is an equality, then we say that the column j is α-multiply covered (if $\alpha = 1$, then it is singly covered) by the set of rows C with numbers from I. If an α-covering is minimal, then for each row of this covering there exists at least one column covered by this row, which is α-multiply covered.

LEMMA 3.2.1. *Under the interchange operation, the α-depth of a $(0,1)$-matrix A can change by at most one.*

PROOF. Let C be a minimal α-covering of the columns of A. Let the matrix A turn into a matrix A' by the interchange operation. Let the related interchangeable submatrix be $A[\{i,k\},\{j,l\}]$. If $i,k \in I(C)$ or $i,k \notin I(C)$, then the rows with numbers from $I(C)$ form a covering in A' as well. If $i \in I(C)$, $k \notin I(C)$, and $a_{ij} > 0$, then the column j in A' need not be α-covered by the rows from C.

However the rows with numbers in $I(C) \cup \{k\}$ form an α-covering in A'. Now let $|I(C)| = \varepsilon(A;\alpha)$. Then, according to the aforesaid, $\varepsilon(A';\alpha) \le \varepsilon(A;\alpha) + 1$, and in the same way $\varepsilon(A;\alpha) \le \varepsilon(A';\alpha) + 1$. The lemma is thus proved. □

Any $(0,1)$-matrix A of size $m \times n$ belongs to the class $\mathfrak{A}^{m,n}(R,S)$ determined by the vector R of its row sums and by the vector S of its column sums. The maximal (minimal) α-depth of the Ryser class is the maximal (minimal) α-depth over all matrices of the class. We denote the maximal α-depth of the class $\mathfrak{A}$ by $\bar{\varepsilon}(\mathfrak{A};\alpha)$ and the minimal α-depth of the class by $\tilde{\varepsilon}(\mathfrak{A};\alpha)$. Then for any matrix $A \in \mathfrak{A}$,

$$\tilde{\varepsilon}(\mathfrak{A};\alpha) \le \varepsilon(A;\alpha) \le \bar{\varepsilon}(\mathfrak{A};\alpha). \tag{3.2.2}$$

THEOREM 3.2.2. *Let $\mathfrak{A}$ be a Ryser class, and $\bar{\varepsilon}(\mathfrak{A};\alpha)$, $\tilde{\varepsilon}(\mathfrak{A};\alpha)$ its maximal and minimal α-depth, respectively. For any integer ε from the interval $[\tilde{\varepsilon}(\mathfrak{A};\alpha), \bar{\varepsilon}(\mathfrak{A};\alpha)]$ there exists a matrix A in $\mathfrak{A}$ such that $\varepsilon(A;\alpha) = \varepsilon$.*

The theorem follows directly from Lemma 3.2.1 and the theorem on the connectivity of the interchange graph for the class $\mathfrak{A}$ (see Theorem 2.1.3).

Finding the α-depth of a $(0,1)$-matrix is a complicated problem. Therefore estimates of the α-depth, and, first of all, investigations of the maximal and minimal depths of classes of $(0,1)$-matrices, are of interest.

THEOREM 3.2.3. *Let $\mathfrak{A} = \mathfrak{A}^{m,n}(R,S)$ be the nonempty class of $(0,1)$-matrices with vector of row sums $R = (r_1,\ldots,r_m)$ and vector of column sums $(d,d,\ldots,d)$. Let $\bar{r}_1 \ge \bar{r}_2 \ge \cdots \ge \bar{r}_m$ be the coordinates of R arranged in nonincreasing order, and let*

$$m_0 = \min_{1 \le k \le m} \left\{ k \colon \sum_{i=1}^{k} \bar{r}_i \ge n \right\}.$$

Then

$$\tilde{\varepsilon}(\mathfrak{A}) = m_0.$$

PROOF. It is obvious that $\tilde{\varepsilon}(\mathfrak{A}) \ge m_0$. Let us show that in $\mathfrak{A}(\bar{R},S)$, where $\bar{R} = (\bar{r}_1,\ldots,\bar{r}_m)$ (and consequently, in $\mathfrak{A}(R,S)$), there exists a matrix A with $\varepsilon(A) = m_0$. The total number of ones in any matrix of the class $\mathfrak{A}$ equals

$$\sum_{i=1}^{m} r_i = dn.$$

Consider the sequence $s = (1,1,\ldots,1)$ of dn ones. We construct the matrix A in the following way (as if we are developing the sequence s). In the first row, beginning with the position $(1,1)$, we place in succession $\bar{r}_1$ ones and put the other elements equal to zero. Then, in the second row, we place $\bar{r}_2$ ones in the positions from $(2, \bar{r}_1 + 1)$ to $(2, \bar{r}_1 + \bar{r}_2)$ and put the other elements equal to zero. In the same way, we sequentially fill the rows $3,\ldots,m$: in the kth row we place ones in succession in the positions from $(k, \sum_{i=1}^{k-1} \bar{r}_i + 1)$ to $(k, \sum_{i=1}^{k} \bar{r}_i)$, and put the other elements equal to zero (the column numbers are taken modulo n). The matrix A obtained in such a way belongs to the class $\mathfrak{A}(\bar{R},S)$, since the column with number j contains, obviously, those and only those ones of the sequence s which are placed in the positions whose numbers are congruent to j modulo n, that is, all column sums of A are equal to d. On the other hand, it is clear that $\varepsilon(A) = m_0$. The theorem is thus proved. □

COROLLARY 3.2.4. *For the classes* $\mathfrak{A}^{m,n}(R,S)$, *where*

$$R = (r, r \dots, r), \quad S = (d, d, \dots, d), \quad rm = dn,$$

the equality $\tilde{\varepsilon}(\mathfrak{A}) = \lceil n/r \rceil$ *holds.*

In particular, $\tilde{\varepsilon}(\mathfrak{A}) = \lceil n/r \rceil$ *for any* $\mathfrak{A} = \mathfrak{A}^n(r)$.

To find $\tilde{\varepsilon}(\mathfrak{A}; \alpha)$ effectively, where $\mathfrak{A}$ is an arbitrary monotone class $\mathfrak{A}^{m,n}(R,S)$, the following approach exists (see [**87**, Chapter VI, Section 5] and [**43**]). First we pass to the class $\mathfrak{A}' = \mathfrak{A}^{n,m}(S,R)$ consisting of the transposed matrices of the class $\mathfrak{A}$. In $\mathfrak{A}'$, we take a canonical matrix $\tilde{A}'$ (see Section 2.1). If $S = (s_1, \dots, s_n)$, then for any $\alpha = 1, 2, \dots, s_n$ the value $\tilde{\varepsilon}_\alpha = \tilde{\varepsilon}(\mathfrak{A}; \alpha)$ coincides with the number of the column where the αth from the left 1 of the last row of the matrix $\tilde{A}'$ is located.

The problem of finding or estimating the maximal depth of a Ryser class is significantly more complicated. This problem is considered in this section. First we list some simple properties of the function $\bar{\varepsilon}$.

3.2.4. Relations between the maximal depths of classes with different parameters. Let m, n, r, d be positive integers, with $r \leq m$ and $d \leq n$. We denote by $\mathfrak{A}^{m,n}(\leq r, \geq d)$ the family of all $(0,1)$-matrices of size $m \times n$ such that the row sums are at most r and the column sums are at least d. Clearly $\mathfrak{A}^{m,n}(\leq r, \geq d)$ is the union of all Ryser classes with the corresponding restrictions on the row and column sums. For a positive integer k, we denote by $\mathcal{R}^{(k)}(A)$ the set of the rows of a $(0,1)$-matrix A with row sums equal to k.

Let $A = (a_{ij})$ be a $(0,1)$-matrix of size $m \times n$, and let $P = (p_{ij})$ be the permutation matrix of order p, $p \leq \min\{m,n\}$, corresponding to a permutation π, that is,

$$p_{ij} = 1 \iff j = \pi(i), \qquad i, j = 1, \dots, p.$$

Suppose that there exist ordered sets

$$I = (i_1, \dots, i_p), \qquad i_1 < \dots < i_p,$$
$$J = (j_1, \dots, j_p), \qquad j_1 < \dots < j_p,$$

such that $a_{i_s j_t} = 1$ for $t = \pi(s)$, $s, t = 1, \dots, p$. Then we say that A includes P, or P can be separated from A, and write $P \subseteq A$. The corresponding sets of indices will be denoted by $I(P)$, $J(P)$. If a $(0,1)$-matrix A of size $m \times n$ includes the permutation matrix P of order p corresponding to a permutation π, $p \leq \min\{m,n\}$, then we define the matrix $A^+ = (a^+_{ij})$ of size $(m+1) \times (n+1)$ such that

$$a^+_{ij} = 0 \qquad \text{if } i \in I(P),\ j \in J(P),$$

$$a^+_{i,n+1}, a^+_{m+1,j} = \begin{cases} 1 & \text{if } i \in I(P),\ j \in J(P), \\ 0 & \text{if } i \notin I(P),\ j \notin J(P), \end{cases}$$

$$a^+_{ij} = a_{ij} \qquad \text{in all other cases.}$$

Such a matrix is called the extension of the matrix A with the use of the permutation matrix P. It is clear that the matrix A^+ has p ones in the $(m+1)$th row and p ones in the $(n+1)$th column. It is also clear that for a $(0,1)$-matrix $B = (b_{ij})$ of size $(m+1) \times (n+1)$ having exactly p ones in some row with number i and exactly p ones in some column with number j located in rows and columns of sets I and J, respectively, it is possible to define the operation of contraction to a matrix of size $m \times n$ inverse to the operation of extension if $b_{ij} = 0$ and $B[I,J]$ contains p pairwise noncollinear zeros.

LEMMA 3.2.5. *If a $(0,1)$-matrix A^+ of size $(m+1)\times(n+1)$ is obtained by extension of a matrix A of size $m\times n$, then*

$$\varepsilon(A^+)\le\varepsilon(A)+1,$$

and if in A^+ some row R with number $i\le m$ and the row R' with number $m+1$ satisfy the inequality $R\ge R'$, then

$$\varepsilon(A)\le\varepsilon(A^+).$$

PROOF. Let A^+ be obtained from A by the use of a permutation matrix P, and let $I=I(P)$. Any covering of the matrix A is a covering of A^+ if it contains at least one row with number in I. Therefore $\varepsilon(A^+)\le\varepsilon(A)+1$.

Now let the condition of the second assertion of the lemma be fulfilled. Then for any covering C of the matrix A^+ there exists a covering of a matrix A with the same number of rows: if C contains R', then we replace R' by the row R. Therefore $\varepsilon(A)\le\varepsilon(A^+)$. □

PROPOSITION 3.2.6. *Let A be a matrix of the class $\mathfrak{A}^{m,n}(\le r,\ge r)$, and let the set of rows $\mathcal{R}^{(r)}(A)$ be nonempty. If A contains a row $R\in\mathcal{R}(A)$ with a number i such that the submatrix $A_0=A[N_m-\{i\},J_i]$ of size $(m-1)\times r$, where J_i is the set of numbers of the columns covered by R, contains a nonzero row not equal to the ith row, then A includes some permutation matrix of order r.*

PROOF. Denote by s the minimum of positive row sums in A_0, and let R_1 be a row in A_0 with s ones. By the condition, $1\le s\le r-1$. We consider s columns in A_0 covered by R_1 and choose in each of them an element equal to one, complying with the following conditions:

(a) no two of these elements lie in a common row, and
(b) one of them belongs to R_1.

Since $s\le r-1$ and each column in A_0 has at least $r-1$ ones, such a choice is possible. We denote the s rows containing the chosen elements by $R_1,\dots,R_s$. Each of the remaining $r-s$ columns in A_0 has at least $(r-1)-(s-1)=r-s$ ones outside the rows $R_1,\dots,R_s$, since it has no ones in R_1. Therefore in each of them it is possible to choose one so that condition (a) is satisfied and, in addition, each of these ones does not lie in the rows $R_1,\dots,R_s$.

Thus, we have found a permutation matrix P of order r such that $P\subseteq A_0$. □

COROLLARY 3.2.7. *Any matrix A of the class $\mathfrak{A}^n(r)$ which is not equivalent to the direct sum of the matrices $\mathbf{J}_{rr}$ can be extended to a matrix A^+ of the class $\mathfrak{A}^{n+1}(r)$ with two identical rows.*

THEOREM 3.2.8. *For any $n,r\ge 1$ with $n\ge r$,*

$$\bar{\varepsilon}(\mathfrak{A}^n(r))\le\bar{\varepsilon}(\mathfrak{A}^{n+1}(r)).$$

PROOF. Consider in $\mathfrak{A}=\mathfrak{A}^n(r)$ the family $\overline{\mathfrak{A}}$ of matrices with maximal depth. Suppose that any matrix $A\in\overline{\mathfrak{A}}$ possesses the property that any two rows of A are either orthogonal or identical. Then A is permutationally equivalent to the direct sum of the matrices $\mathbf{J}_{rr}$ and by Corollary 3.2.4

$$\bar{\varepsilon}(\mathfrak{A})=\varepsilon(A)=n/r=\lceil(n+1)/r\rceil=\tilde{\varepsilon}(\mathfrak{A}^{n+1}(r))\le\bar{\varepsilon}(\mathfrak{A}^{n+1}(r)).$$

Thus, in this case the theorem is proved.

Now suppose that in $\overline{\mathfrak{A}}$ there exists a matrix A with two distinct and nonorthogonal rows. By Corollary 3.2.7 the matrix A can be extended to a matrix $A^+ \in \mathfrak{A}^{n+1}(r)$ with two identical rows. Then by Lemma 3.2.5

$$\bar{\varepsilon}(\mathfrak{A}) = \varepsilon(A) \leq \varepsilon(A^+) \leq \bar{\varepsilon}(\mathfrak{A}^{n+1}(r))).$$

□

It is easy to see that for the classes $\mathfrak{A}^n(r)$ another property of monotonicity is fulfilled.

THEOREM 3.2.9. *For any $n, r, \alpha \geq 1$ with $n \geq r \geq \alpha$,*

$$\bar{\varepsilon}(\mathfrak{A}^n(r); \alpha) \geq \bar{\varepsilon}(\mathfrak{A}(r+1); \alpha).$$

We leave the proof of this assertion to the readers (see Problem 2 in this section).

The following fact is established by analyzing the conditions under which a matrix $A \in \mathfrak{A}^n(r)$ allows a contraction and by applying Lemma 3.2.5.

THEOREM 3.2.10. *Let n, r be positive integers. If $n \geq r(r+1)/3$, then*

$$\bar{\varepsilon}(\mathfrak{A}^n(r)) - \bar{\varepsilon}(\mathfrak{A}^{n-1}(r)) \leq 1.$$

We omit the proof, since it contains significant technical difficulties (see [**103**]).

3.2.5. A method for finding an upper bound for an arbitrary matrix. Let a matrix A belong to the class $\mathfrak{A}^{m,n}(R, S)$, where $R = (r_1, \ldots, r_m)$ and $S = (s_1, \ldots, s_n)$. We assume that

$$r = \max_{1 \leq i \leq m} r_i \geq 2, \qquad d = \min_{1 \leq j \leq n} s_j \geq 2$$

and the set $\mathcal{R}^{(r)}(A)$ is nonempty. We construct a sequence of submatrices

$$A^{(r-1)}, A^{(r-2)}, \ldots, A^{(1)}$$

such that $A^{(i)}$ is embedded in $A^{(i+1)}$ for $i \leq r-1$ and the row sums in $A^{(i)}$ do not exceed i for $i = 1, \ldots, r-1$. We set $A^{(r)} = A$, $m_r = m$, $n_r = n$. Suppose that we have constructed the submatrices $A^{(r)}, A^{(r-1)}, \ldots, A^{(r-i)}$, $0 \leq i \leq r-2$, such that the submatrix $A^{(r-s)}$ of size $m_{r-s} \times n_{r-s}$ is embedded in $A^{(r-s+1)}$, $1 \leq s \leq i$, and its row sums do not exceed $r-s$, $s = 0, 1, \ldots, i$. Choose in $A^{(r-i)}$ the maximum number μ_{r-i} of pairwise orthogonal rows with sums equal to $r-i$. If there is no such row, that is, $\mu_{r-i} = 0$, then we set $A^{(r-i-1)} = A^{(r-i)}$. If $\mu_{r-i} > 0$, then we move the chosen μ_{r-i} rows and the columns covered by them to the first positions in $A^{(r-i)}$. Then the submatrix

$$A^{(r-i)}[N_{\mu_{r-i}}, N_{n_{r-i}} - N_{(r-i)\mu_{r-i}}]$$

is the zero matrix and, in view of the maximality of the number μ_{r-i}, the submatrix $A^{(r-i)}(N_{\mu_{r-1}}, N_{(r-i)\mu_{r-i}})$ has at most $r-i-1$ ones in each row. We denote this matrix by $A^{(r-i-1)}$. It is obvious, that $A^{(r-i-1)}$ is a matrix of size $m_{r-i-1} \times n_{r-i-1}$, where $m_{r-i-1} = m_{r-i} - \mu_{r-i}$, $n_{r-i-1} = n_{r-i} - (r-i)\mu_{r-i}$. For example, after the first two steps we obtain a matrix of the form

$$A = A^{(r)} = \begin{pmatrix} * & \mathbf{0} & \mathbf{0} \\ * & * & \mathbf{0} \\ * & * & A^{(r-2)} \end{pmatrix},$$

where the first part of the matrix consists of μ_r rows, the second of μ_{r-1} rows, and the submatrix $A^{(r-2)}$ is of the size

$$(m - \mu_r - \mu_{r-1}) \times (n - r\mu_r - (r-1)\mu_{r-1}).$$

After $r-1$ steps we arrive at a submatrix $A^{(1)}$ of the following form:

$$A^{(1)} = \begin{pmatrix} 1 & & & \\ 1 & & & \\ \vdots & & \mathbf{0} & \\ 1 & & & \\ & 1 & & \\ & 1 & & \\ & \vdots & & \\ & 1 & & \\ & & \ddots & \\ & & & 1 \\ & & & 1 \\ & & \mathbf{0} & \vdots \\ & & & 1 \end{pmatrix}$$

It is obvious that

$$m_{r-i} = m - \sum_{s=0}^{i-1} \mu_{r-s},$$
$$n_{r-i} = n - \sum_{s=0}^{i-1} (r-s)\mu_{r-s}, \qquad i = 0, 1, \ldots, r-1.$$

It is clear from the construction that the first

$$\sum_{i=0}^{r-1} \mu_{r-i} = \sum_{i=1}^{r} \mu_i$$

rows of the matrix A rearranged in such a way form a covering of this matrix. Note that for the matrix A whose row sums are at least r and whose column sums do not exceed a positive integer d, the similar construction can be performed for columns (instead of rows). In particular, such a construction can be realized for the incidence matrix of a uniform r-hypergraph (r-graph) with the maximal degree of a vertex equal to Δ for estimating the transversal number of this hypergraph.

The algorithm for finding a covering belongs to the class of so-called gradient algorithms. The $(0,1)$-matrix constructed by this algorithm is called the standard form of the initial matrix A.

3.2.6. A lower bound for the number of orthogonal rows of a $(0,1)$-matrix (the number of matchings in a hypergraph). In order to estimate the maximal depth of a class of $(0,1)$-matrices, we need to find lower bounds for $\mu_r, \mu_{r-1}, \ldots, \mu_1$, which appear in the algorithm of the preceeding subsection. Note that μ_r is simply the number of matchings in the hypergraph of rank at most r corresponding to the initial matrix A. The remaining numbers $\mu_{r-1}, \ldots, \mu_1$ have a similar sense.

THEOREM 3.2.11 ([**104**]). *Let B be a $(0,1)$-matrix of size $m \times n$, whose row sums are at most r and whose column sums are at least d, where r, d are positive integers. Then B contains at least*

$$t = \max\{0, \lceil (dn - (r-1)m/(rd - r + 1)\rceil\} \tag{3.2.3}$$

pairwise orthogonal rows with row sums equal to r.

PROOF. Let t be the maximum number of orthogonal rows in B with row sums equal to r, and assume $t > 0$. We move these rows and the columns covered by them to the first positions of B and obtain the $m \times n$ matrix

$$\begin{pmatrix} * & \mathbf{0} \\ * & B' \end{pmatrix},$$

where B' is a matrix of size $(m-t) \times (n - rt)$. The matrix B' has at least $d(n-rt)$ ones located in $m - t$ rows, and the row sums in B' are at most $r - 1$. Therefore

$$d(n - rt) \le (r-1)(m-t);$$

that is, $t(rd - r + 1) \ge dn - (r-1)m$. Since $d \ne (r-1)/r$, we obtain the required estimate

$$t \ge (dn - (r-1)m)/(rd - r + 1).$$

□

It is clear that inequality (3.2.3) can be formally extended to the case of the degenerate matrix with $n = 0$.

Theorem 3.2.11 can be formulated in terms of hypergraphs. We denote by $\nu(H)$ the number of matchings in a hypergraph H.

THEOREM 3.2.12. *Let H be a hypergraph with n vertices and m edges such that $r(H) = r$. Then*

$$\nu(H) \ge (\delta n - (r-1)m)/(r\delta - r + 1), \tag{3.2.4}$$

where δ is the minimal degree of a vertex in H.

We denote by $\beta(H)$ the maximum number of vertices of maximal degree in the set M of independent vertices of the hypergraph H (that is, no two vertices in M belong to a common edge).

THEOREM 3.2.13. *Let H be a hypergraph with n vertices and m edges. If each edge of H contains at least r vertices, then*

$$\beta(H) \ge (rm - (\Delta - 1)m)/(r\Delta - \Delta + 1), \tag{3.2.5}$$

where Δ is the maximal degree of a vertex in H.

Theorem 3.2.13 is true, in particular, for r-uniform hypergraphs.

There exists an infinite series of nontrivial examples of attainability of estimates (3.2.3), (3.2.4), (3.2.5) of Theorems 3.2.11, 3.2.12, 3.2.13. Consider, for example, the class of matrices $\mathfrak{A}^n(r+1)$. For $n > r^2 + r + 1$, Theorem 3.2.11 shows that the number t of pairwise orthogonal rows in any matrix of this class is at least $\lceil n/(r^2+r+1)\rceil$, that is, at least 2; however, for $n = r^2 + r + 1$, the theorem asserts only that $t \ge 1$. Let there exist a finite projective plane of order r (r can be an arbitrary power of a prime number). The incidence matrix A of this projective plane is a matrix of the class $\mathfrak{A}^n(r+1)$, where $n = r^2 + r + 1$. The matrix A does not contain two orthogonal rows, since any pair of distinct lines has a nonempty intersection. Thus, for $n = r^2 + r + 1$ estimate (3.2.4) is attainable.

3.2.7. An upper bound for the maximal depth of a matrix (the covering number of a family of sets, the transversal number of a hypergraph). Any $(0,1)$-matrix belongs to some class of matrices $\mathfrak{A}^{m,n}(\le r, \ge d)$ (see Section 3.2.4).

THEOREM 3.2.14 ([**76, 61**]). *Let a $(0,1)$-matrix A belong to $\mathfrak{A}^{m,n}(\le r, \ge d)$, $r, d \ge 2$. Then*

$$\varepsilon(A) \le n - L(A), \tag{3.2.6}$$

where

$$L(A) = \sum_{i=2}^{r}(i-1)\mu_i, \tag{3.2.7}$$

and for $i = 2, 3, \ldots, r$

$$\mu_i \ge \left\lceil \frac{1}{id - i + 1}\left(dn - (i-1)m - \sum_{s=i+1}^{r}(sd - i + 1)\mu_s\right)\right\rceil. \tag{3.2.8}$$

PROOF. Let μ_{r-i} denote the maximum number of pairwise orthogonal rows in the submatrix $A^{(r-i)}$, $i = 0, 1, \ldots, r-1$, which appear in the algorithm in Section 3.2.5. Let us show that these numbers satisfy inequality (3.2.8). Indeed, for any i, $0 \le i \le r-1$, the matrix $A^{(r-i)}$ has

$$m_{r-i} = m - \sum_{s=0}^{i-1}\mu_{r-s}$$

rows and

$$n_{r-i} = n - \sum_{s=0}^{i-1}(r-s)\mu_{r-s}$$

columns. Applying Theorem 3.2.11 to $A^{(r-i)}$, replacing n in (3.2.3) by n_{r-i}, m by m_{r-i}, and r by $r-i$, we obtain

$$\mu_{r-i} \ge \frac{dn - (r-i-1)m - \sum_{s=0}^{i-1}((r-s)d - (r-i) + 1)\mu_{r-s}}{(r-i)d - (r-i) + 1},$$

and, replacing $r-i$ by i, $r-s$ by s, we obtain inequality (3.2.8). By the construction, the first $\mu_1 + \cdots + \mu_r$ rows of the matrix A in the standard form constitute a covering of the matrix. It is clear that

$$\mu_1 = n - \sum_{i=0}^{r-2}(r-i)\mu_{r-i}.$$

Consequently,

$$\sum_{i=1}^{r}\mu_i = n - \sum_{i=0}^{r-2}(r-i-1)\mu_{r-i} = n - \sum_{i=2}^{r}(i-1)\mu_i.$$

The theorem is proved. □

Note that (3.2.8) is valid for μ_1 as well. Moreover, for $i = 1$ it turns into an equality.

3.2.8. The minimum of the sum (3.2.7). Leaving aside the combinatorial sense of the numbers μ_i, we find simply the minimum of the sum

$$L(\mu_r, \mu_{r-1}, \dots, \mu_2) = \sum_{i=2}^{r} (i-1)\mu_i,$$

where μ_i, $i = 2, \dots, r$, satisfy (3.2.8). Define inductively the sequences

$$\lambda'_r, \lambda'_{r-1}, \dots, \lambda'_1, \qquad \lambda_r, \lambda_{r-1}, \dots, \lambda_1,$$

where $\lambda_i = \lceil \lambda'_i \rceil$, $i = r, r-1, \dots, 1$. Let

$$\lambda'_r = (dn - (r-1)m)/(r(d-1)+1).$$

If for each integer $i \in [2, r]$ the numbers $\lambda'_r, \lambda'_{r-1}, \dots, \lambda'_i$ and, consequently, the numbers $\lambda_r, \lambda_{r-1}, \dots, \lambda_i$ are defined, then we set λ'_{i-1} equal to the ratio in the brackets in the right-hand side of (3.2.8) with $\mu_s = \lambda_s$, $s = i, i+1, \dots, r$. The integers $\lambda_1, \dots, \lambda_r$ are the parameters of a class of matrices contained in the set $\mathfrak{A}^{m,n}(\leq r, \geq d)$. Setting $\mu_s = \lambda_s$, $s = 1, \dots, r$, in (3.2.8), we find that $\sum_{s=1}^{r} \lambda_s = n$.

Let a set of numbers $\mu = (\mu_r, \mu_{r-1}, \dots, \mu_i)$, where μ_s, $s = i, i+1, \dots, r$, satisfy (3.2.8), be given. Then we denote the set $(\mu_r, \dots, \mu_i, \mu_{i-1}, \dots, \mu_2)$, where $\mu_{i-1} \dots, \mu_2$ are equal to the corresponding right-hand sides of (3.2.8), by

$$(\mu_r, \dots, \mu_i, \lambda_{i-1}^{(\mu)}, \dots, \lambda_2^{(\mu)}),$$

or briefly, by $((\mu), \lambda_{i-1}^{(\mu)}, \dots, \lambda_2^{(\mu)})$.

PROPOSITION 3.2.15. *The minimum of the sum* $L(\mu_r, \mu_{r-1}, \dots, \mu_2)$, *where the* μ_i, $i = 2, 3, \dots, r$, *satisfy* (3.2.8), *is attained for* $\mu_i = \lambda_i$, $i = 2, 3, \dots, r$.

PROOF. The proposition follows from the inequality

$$L(\mu_r, \mu_{r-1}, \dots, \mu_2) \geq L(\mu_r, \mu_{r-1}, \dots, \mu_i, \lambda_{i-1}^{(\mu)}, \dots, \lambda_2^{(\mu)}) \tag{3.2.9}$$

for any set $\mu = (\mu_r, \mu_{r-1}, \dots, \mu_2)$ of numbers satisfying (3.2.8) and any $i = 3, 4, \dots, r$. We prove (3.2.9) by induction. If $\mu = (\mu_r, \mu_{r-1}, \dots, \mu_3)$, then $L((\mu), \mu_2) \geq L((\mu), \lambda_2^{(\mu)})$ for any μ_2 satisfying (3.2.8); that is, (3.2.9) is valid for $i = 3$. Let $i > 3$, and let (3.2.9) be fulfilled for all s, $i > s \geq 3$. We show that this inequality is fulfilled for $s = i$. As is easily seen, it is sufficient to prove that

$$L((\tilde{\mu}), \lambda_{i-1}^{(\tilde{\mu})}, \dots, \lambda_2^{(\tilde{\mu})}) \geq L((\mu), \lambda_{i-1}^{(\mu)}, \dots, \lambda_2^{(\mu)}) \tag{3.2.10}$$

if $\mu = (\mu_r, \dots, \mu_{i+1}, \mu_i)$, $\tilde{\mu} = (\mu_r, \dots, \mu_{i+1}, \mu_i + 1)$, where $\mu_r, \dots, \mu_{i+1}, \mu_i$ satisfy (3.2.8). Let us prove (3.2.10). We consider the sequence $\delta_s = \lambda_s^{(\tilde{\mu})} - \lambda_s^{(\tilde{\mu})}$, $s = i-1, i-2, \dots, 2$. Since $\tilde{\mu}_i = \mu_i + 1$, we have $\delta_{i-1} = -2$ or $\delta_{i-1} = -1$. Consider these two cases.

Let $\delta_{i-1} = -2$. Then it is not difficult to see that the sequence $\delta_{i-1}, \delta_{i-2}, \dots, \delta_2$ is either of the form $-2, 1, 0, \dots$ for $i \geq 5$, or of the form $-2, 1$ for $i = 4$. Hence

$$\begin{aligned} L((\tilde{\mu})), \lambda_{i-1}^{(\tilde{\mu})}, \dots, \lambda_2^{(\tilde{\mu})}) - L((\mu), \lambda_{i-1}^{(\mu)}, \dots, \lambda_2^{(\mu)}) \\ = i - 1 + \delta_{i-1}(i-2) + \delta_{i-2}(i-3) = 0. \end{aligned}$$

Let $\delta_{i-1} = -1$. In this case, we find that the sequence $\delta_{i-1}, \delta_{i-2}, \dots, \delta_2$ is of the form either $-1, (0), -1, 1, (0)$, or $-1, (0), -1$, or $-1, (0)$, where (0) denotes a sequence of zeros, maybe empty. Hence, in the same way as in the first case, we obtain

$$L((\tilde{\mu})), \lambda_{i-2}^{(\tilde{\mu})}, \dots, \lambda_2^{(\tilde{\mu})}) - L((\mu), \lambda_{i-2}^{(\mu)}, \dots, \lambda_2^{(\mu)}) \geq 0.$$

Thus, inequality (3.2.10) and, consequently, inequality (3.2.9) are proved. □

Thus, we have established that the algorithm described in Section 3.2.5 gives a covering of the maximal cardinality if at each step the submatrix $A^{(i)}$ in the matrix A under consideration contains at most λ_i pairwise orthogonal rows with row sums i, where $i = r, r-1, \dots, 2$.

3.2.9. An explicit form of an estimate of the depth of a class of $(0,1)$-**matrices (the transversal number of a hypergraph).** Here, using Theorem 3.2.14, we derive an upper bound for the maximal depth $\bar{\varepsilon}(\mathfrak{A})$ of the class of $(0,1)$-matrices $\mathfrak{A}^{m,n}(R,S)$ depending only on the size $m\times n$ of the matrices, the maximal value r of a row sum, and the minimal value d of a column sum. In fact, this estimate is an upper estimate for the depth of an arbitrary $(0,1)$-matrix and, consequently, the transversal number of an arbitrary hypergraph.

Theorem 3.2.17 below was proved in [**105**]. In [**57**], the proof of this result was obtained independently by the same method, with the function in the right-hand side written in a more symmetric form (3.2.16).

The system of inequalities (3.2.8) can be transformed to the system

$$\sum_{s=i}^{r}(sd-i+1)\mu_s \ge dn-(i-1)m, \qquad i=2,3,\dots,r. \tag{3.2.11}$$

We introduce the following notation. For positive integers r, d, $r\ge 2$, $d\ge 1$, we set

$$\pi_{r,d} = \prod_{k=2}^{r}(kd-k+1), \tag{3.2.12}$$

$$\varphi(i) = \frac{(i-2)!\,(d-1)^{i-2}}{\pi_{i,d}}, \qquad i=2,3,\dots \tag{3.2.13}$$

In order to estimate the maximal depth, we need the following auxiliary result.

LEMMA 3.2.16. *The identities*

$$d\sum_{i=2}^{s}\varphi(i) = 1-\frac{(s-1)!\,(d-1)^{s-1}}{\pi_{s,d}}, \tag{3.2.14}$$

$$\sum_{i=2}^{s}(i-1)\varphi(i) = 1-\frac{s!\,(d-1)^{s-1}}{\pi_{s,d}} \tag{3.2.15}$$

hold.

These assertions are proved by induction.

THEOREM 3.2.17 ([**105, 40**]). *Let* $\mathfrak{A} = \mathfrak{A}^{m,n}(R,S)$ *be a nonempty class of* $(0,1)$*-matrices contained in the set* $\mathfrak{A}^{m,n}(\le r, \ge d)$. *Then the maximal depth* $\bar{\varepsilon}(\mathfrak{A})$ *of this class satisfies the inequality*

$$\bar{\varepsilon}(\mathfrak{A}) \le \left(1-\left(r-\frac{n}{m}\right)\frac{(r-1)!\,(d-1)^{r-1}}{\pi_{r,d}}\right)m, \tag{3.2.16}$$

where $\pi_{r,d}$ *is the number defined by* (3.2.12), *and* $r,d\ge 2$.

PROOF. Multiplying both sides of the ith inequality in (3.2.11) by $\varphi(i)$ defined by (3.2.13) and summing over i, $2 \le i \le r$, we obtain

$$\sum_{i=2}^{r}\left(\sum_{s=i}^{r}(sd-i+1)\mu_s\right)\varphi(i) \ge \sum_{i=2}^{r}(dn-(i-1)m)\varphi(i). \tag{3.2.17}$$

Changing the order of summation in the left-hand side of (3.2.17), we obtain

$$\sum_{s=2}^{r}\mu_s\left(sd\sum_{i=2}^{s}\varphi(i)-\sum_{i=2}^{s}(i-1)\varphi(i)\right) \ge dn\sum_{i=2}^{r}\varphi(i)-m\sum_{i=2}^{r}(i-1)\varphi(i). \tag{3.2.18}$$

Applying Lemma 3.2.16 to both sides of (3.2.18), we find from (3.2.14) and (3.2.15) that

$$\sum_{s=2}^{r}(s-1)\mu_s \ge n-m+(rm-n)\frac{(r-1)!\,(d-1)^{r-1}}{\pi_{r,d}}.$$

The required estimate follows now from inequality (3.2.6) in Theorem 3.2.14. □

COROLLARY 3.2.18. *Let A be a $(0,1)$-matrix of size $m\times n$, with $m \le n$. If r is the maximal row sum and d is the minimal column sum of A, then*

$$\varepsilon(A) \le \left(1-(r-1)\frac{(r-1)!\,(d-1)^{r-1}}{\pi_{r,d}}\right)n, \tag{3.2.19}$$

where $r, d \ge 2$. In particular, (3.2.19) *is true for any square $(0,1)$-matrix of order n.*

Indeed, for $m = n$ estimate (3.2.16) gives (3.2.19). Denote $r!\,(d-1)^{r-1}/\pi_{r,d}$ by $\psi_{r,d}$. If $m < n$, then inequality (3.2.16) is stronger than (3.2.19). This fact follows from the inequality $\psi_{r,d} < 1$, which obviously holds since

$$\psi_{r,d} = \frac{r!\,(d-1)^{r-1}}{2d\cdot 3d\cdots rd}\bigg/\prod_{k=2}^{r}\left(1-\frac{k-1}{kd}\right) = \left(1-\frac{1}{d}\right)\bigg/\prod_{k=2}^{r}\left(1-\frac{k-1}{kd}\right).$$

COROLLARY 3.2.19. *Let A be a $(0,1)$-matrix of size $m\times n$ whose row sums are at least r and whose column sums are at most d. Then the width $\varepsilon(A^T)$ of A satisfies the inequality*

$$\varepsilon(A^T) \le \left(1-\left(d-\frac{m}{n}\right)\frac{(d-1)!\,(r-1)^{d-1}}{\pi_{d,r}}\right)n, \tag{3.2.20}$$

where $r, d \ge 2$. If $n \le m$, then the inequality

$$\varepsilon(A^T) \le \left(1-(d-1)\frac{(d-1)!\,(r-1)^{d-1}}{\pi_{d,r}}\right)m \tag{3.2.21}$$

is also true.

Formulas (3.2.20) and (3.2.21) follow from Theorem 3.2.17 and Corollary 3.2.18 applied to A^T.

The following assertion concerning the transversal number $\tau(H)$ of a hypergraph H is equivalent to Theorem 3.2.17 and Corollaries 3.2.18 and 3.2.19.

THEOREM 3.2.20. *Let H be a hypergraph with n vertices and m edges, and let Δ be the maximal degree of a vertex in H. Suppose that each edge of H contains at least r vertices. Then*

$$\tau(H) \le \left(1-\left(\Delta-\frac{m}{n}\right)\frac{(\Delta-1)!\,(r-1)^{\Delta-1}}{\pi_{\Delta,r}}\right)n, \tag{3.2.22}$$

where $r, \Delta \geq 2$ *and* $\pi_{\Delta,r}$ *is the number defined by* (3.2.12). *For* $m \leq n$,

$$\tau(H) \leq \left(1 - (\Delta - 1)\frac{(\Delta - 1)!\,(r-1)^{\Delta-1}}{\pi_{\Delta,r}}\right) m. \tag{3.2.23}$$

The assertion follows directly from (3.2.20) and (3.2.21). In particular, Theorem 3.2.20 is true for r-uniform graphs.

3.2.10. An asymptotic estimate of the maximal depth. In order to look more closely at the behaviour of the maximal depth of the class $\mathfrak{A}^{m,n}(R, S)$ which is contained in the set $\mathfrak{A}^{m,n}(\leq r, \geq d)$, we obtain an asymptotic estimate of the maximal depth using inequality (3.2.16). We assume that there exist constants b, b_1, c, c_1 such that

$$0 < c_1 \leq m/n \leq c, \qquad 0 < b_1 \leq r/d \leq b \tag{3.2.24}$$

as $m, n, r, d \to \infty$. These conditions guarantee a certain proportionality in the form and structure of the matrices under consideration.

We agree that for sequences $f(n)$, $g(n)$ and a nonnegative sequence $h(n) \geq 0$ the expression $f(n) \leq g(n) + O(h(n))$ as $n \to \infty$ means that there exists a sequence $\varepsilon(n)$ such that $f(n) \leq g(n) + \varepsilon(n)$ and $\varepsilon(n) = O(h(n))$ as $n \to \infty$.

THEOREM 3.2.21 ([**105**]). *If* $m, n, r, d \to \infty$, *conditions* (3.2.24) *are fulfilled, and* $C = bc$, *then the maximal depth of a class* $\mathfrak{A} = \mathfrak{A}^{m,n}(R, S)$ *of* $(0,1)$*-matrices contained in* $\mathfrak{A}^{m,n}(\leq r, \geq d)$ *satisfies the inequality*

$$\frac{1}{n}\,\bar{\varepsilon}(\mathfrak{A}) \leq \frac{C \ln r}{r} + O\left(\frac{1}{r}\right). \tag{3.2.25}$$

PROOF. Denote by $u(r,d)$ the coefficient $1 - (r - n/m)(r-1)!\,(d-1)^{r-1}/\pi_{r,d}$ of m in the right-hand side of (3.2.16). Then

$$\begin{aligned}
1 - u(r,d) &= \left(r - \frac{n}{m}\right)\frac{(r-1)!\,(d-1)^{r-1}}{\pi_{r,d}} \\
&= \left(r - \frac{n}{m}\right)\frac{(r-1)!\,(d-1)^{r-1}}{2d \cdot 3d \cdots rd} \Big/ \prod_{k=2}^{r}\left(1 - \frac{k-1}{kd}\right) \\
&= \frac{1}{r}\left(r - \frac{n}{m}\right)\left(1 - \frac{1}{d}\right)^{r-1} \Big/ \, v(r,d),
\end{aligned}$$

where

$$v(r,d) = \prod_{k=2}^{r}\left(1 - \frac{k-1}{kd}\right).$$

Expanding $\ln v(r,d)$ into a series, we find after simple transformations that

$$\ln v(r,d) = -\frac{r}{d} + \frac{1}{d}\ln r + O\left(\frac{1}{d}\right),$$

that is,

$$\ln(1 - u(r,d)) = -\frac{\ln r}{d} + O\left(\frac{1}{d}\right).$$

Hence

$$u(r,d) = \frac{\ln r}{d} + O\left(\frac{1}{d}\right). \tag{3.2.26}$$

By Theorem 3.2.17 relation (3.2.26) implies that

$$\frac{1}{n}\,\bar{\varepsilon}(\mathfrak{A}) \leq \frac{mr\ln r}{ndr} + \frac{m}{n}\,O\left(\frac{1}{d}\right).$$

Since m, n, r, d satisfy conditions (3.2.24), it follows that (3.2.25) is valid. □

COROLLARY 3.2.22. *The maximal depth of the class* $\mathfrak{A} = \mathfrak{A}^n(r)$ *satisfies the inequality*

$$\frac{1}{n}\,\bar{\varepsilon}(\mathfrak{A}) \leq \frac{\ln r}{r} + O\left(\frac{1}{r}\right)$$

as $n, r \to \infty$.

The following result, formulated in terms of hypergraphs, is equivalent to Theorem 3.2.21.

THEOREM 3.2.23. *Let H be a hypergraph with n vertices and m edges, such that any edge of H contain at least r vertices, and let Δ be the maximal degree of a vertex in H. Assume that $m, n, r, \Delta \to \infty$ in such a way that there exist constants b, b_1, c, c_1 such that*

$$0 < c_1 \leq m/n \leq c, \qquad 0 < b_1 \leq r/\Delta \leq b.$$

Then for the transversal number $\tau(H)$ of the hypergraph H we have the estimate

$$\frac{1}{n}\,\tau(H) \leq \frac{\ln r}{r} + O\left(\frac{1}{r}\right).$$

The theorem can be proved in the same way as Theorem 3.2.21, but with the use of inequality (3.2.22) instead of (3.2.16).

The asymptotic behaviour of the α-depth of classes of $(0,1)$-matrices under various assumptions on the parameters of matrices and the values of α was studied by Leontyev [**67**] and Kuzyurin [**62**], [**63**]. In particular, a result proved in [**62**] implies that the asymptotic estimate of Theorem 3.2.21 is attainable. It should also be noted that some asymptotic relations for the maximal depth (in terms of hypergraphs), in particular, the assertion of Theorem 3.2.21, were rediscovered in [**8**] for a particular case of uniform hypergraphs.

3.2.11. An estimate of the α-depth for an arbitrary α. Let A be a $(0,1)$-matrix of size $m \times n$. Denote by t_k the number of columns of A of weight k, that is, the number of columns with the column sums equal to k, where $k = 1, 2, \ldots, m$. Let $\mathfrak{A}^{m,n}[t_1, \ldots, t_m]$ be the set of all $(0,1)$-matrices of size $m \times n$ with t_k columns of weight k, $t_1 + \cdots + t_m = n$. In studying the α-depth, we need to consider only the matrices of the set $\mathfrak{A}^{m,n}[0, \ldots, 0, t_\alpha, t_{\alpha+1}, \ldots, t_m]$. We denote by $\bar{\varepsilon}[t_\alpha, t_{\alpha+1}, \ldots, t_m; \alpha]$ the maximum of $\varepsilon(A; \alpha)$ over all

$$A \in \mathfrak{A}^{m,n}[0, \ldots, 0, t_\alpha, t_{\alpha+1}, \ldots, t_m].$$

THEOREM 3.2.24 ([**67**]). *The inequality*

$$\bar{\varepsilon}[t_\alpha, t_{\alpha+1}, \ldots, t_m; \alpha] \leq \min_{1\leq s\leq m}\left\{s + \alpha\binom{m}{s}^{-1}\sum_{v=\alpha}^{m} t_v \sum_{i=0}^{\alpha-1}\binom{v}{i}\binom{m-v}{s-i}\right\} \tag{3.2.27}$$

holds.

PROOF. Let $A = (a_{ij}) \in \mathfrak{A}^{m,n}[0,\dots,0,t_\alpha,\dots,t_m]$, and let s be a positive integer, $1 \le s \le m$. If B is a submatrix of A with n columns, then we denote by $h^{(\alpha)}(B)$ the number of columns of weight at most $\alpha - 1$ in B. Let $\mathcal{A}^{(s)}$ denote the family of submatrices of the form $A[I, N_n]$, $|I| = s$. We set

$$h^{(\alpha)}(A,s) = \binom{m}{s}^{-1} \sum_{B\in\mathcal{A}^{(s)}} h^{(\alpha)}(B). \tag{3.2.28}$$

It is clear that in $\mathcal{A}^{(s)}$ there exists a submatrix B with $h^{(\alpha)}(B) \le h^{(\alpha)}(A,s)$, since $h^{(\alpha)}(A,s)$ defined by (3.2.28) is the mean value of the function $h^{(\alpha)}$ over all submatrices in $\mathcal{A}^{(s)}$. For each column in B with column sum less than α there exists a set of at most α rows such that if we add it to B, then we obtain a submatrix with at least α ones in this column. Thus, we can obtain an α-covering of A with at most $s + \alpha h^{(\alpha)}(A,s)$ rows, that is,

$$\varepsilon(A;\alpha) \le s + \alpha h^{(\alpha)}(A,s). \tag{3.2.29}$$

Let us estimate $h^{(\alpha)}(A,s)$. For $B = A[\{i_1,\dots,i_s\}, N_n] \in \mathcal{A}^{(s)}$ and $k \in N_n$, we define the function $\chi_k^{(\alpha)}(B)$, setting $\chi_k^{(\alpha)}(B) = 1$ if $\sum_{q=1}^{s} a_{i_q k} < \alpha$ and $\chi_k^{(\alpha)}(B) = 0$ otherwise. Then

$$h^{(\alpha)}(B) = \sum_{k=1}^{n} \chi_k^{(\alpha)}(B). \tag{3.2.30}$$

Substituting (3.2.30) into (3.2.28) and changing the order of summation, we obtain

$$h^{(\alpha)}(A,s) = \binom{m}{s}^{-1} \sum_{k=1}^{n} \sum_{B\in\mathcal{A}^{(s)}} \chi_k^{(\alpha)}(B). \tag{3.2.31}$$

It is clear that the inner sum in (3.2.31) is equal to the number of ways to choose a submatrix in $\mathcal{A}^{(s)}$ such that the kth column contains less that α ones. If the kth column of A contains v ones and $m - v$ zeros, then such a choice can be realized in

$$\sum_{i=0}^{\alpha-1} \binom{v}{i}\binom{m-v}{s-i}$$

ways. Therefore

$$\sum_{B\in\mathcal{A}^{(s)}} \chi_k^{(\alpha)}(B) = \sum_{i=0}^{\alpha-1} \binom{v}{i}\binom{m-v}{s-i}. \tag{3.2.32}$$

The matrix A has exactly t_v columns of weight v. Using (3.2.32), we obtain

$$h^{(\alpha)}(A,s) = \binom{m}{s}^{-1} \sum_{v=\alpha}^{m} t_v \sum_{i=0}^{\alpha-1} \binom{v}{i}\binom{m-v}{s-i}.$$

The assertion of the theorem follows now from (3.2.29). □

Using Theorem 3.2.24, we can obtain asymptotic estimates (as $n \to \infty$) of the maximal α-depth of classes of (0, 1)-matrices with at least k ones in each column under various conditions on m, n, k, α. These questions were considered in detail in [**62**], [**63**]. In particular, it was shown that under rather general assumptions about a class $\mathfrak{A}$, most of the matrices (in the natural sense) of the class have α-depth close to the maximal value.

Problems.

1. Find the depth of the matrices

$$A = \begin{pmatrix} 0 & 1 & 1 & 1 \\ 1 & 0 & 1 & 1 \\ 1 & 1 & 0 & 1 \\ 1 & 1 & 1 & 0 \end{pmatrix}, \qquad B = \begin{pmatrix} 1 & 1 & 1 & 0 & 0 & 0 \\ 0 & 1 & 0 & 1 & 1 & 0 \\ 1 & 0 & 0 & 1 & 0 & 1 \\ 1 & 0 & 0 & 0 & 1 & 1 \\ 0 & 1 & 1 & 0 & 0 & 1 \\ 0 & 0 & 1 & 1 & 1 & 0 \end{pmatrix}.$$

   Let n be an even number, $n \geq 4$. Taking into account that in the class $\mathfrak{A}^n(3)$ for any such n there exist matrices which can be represented as a direct sum of matrices of order 6 and order 4, show that $\bar{\varepsilon}(\mathfrak{A}^n(3)) \geq n/2$.
2. Prove Theorem 3.2.9.
3. For positive integers r and a, consider the class $\mathfrak{A}^{r+a}(r)$. Show that $\bar{\varepsilon}(\mathfrak{A}^{r+a}(r)) = 2$ if $r > (a-1)^2$.
4. Let r, n_1, n_2 be positive integers such that $r \leq n_1 \leq n_2$. Suppose that the maximal depths $\bar{\varepsilon}(\mathfrak{A}^{n_2}(r))$ and $\bar{\varepsilon}(\mathfrak{A}^{n'}(r))$ for $r \leq n' \leq n_1$ are known. Let $n = qn_2 + s$, where $q = \lfloor n/n_2 \rfloor$, $0 \leq s < n_2$. Using the existence in the class $\mathfrak{A}^n(r)$ of a direct sum of matrices of less order and Theorem 3.2.8, show that
$$\bar{\varepsilon}(\mathfrak{A}^n(r)) \geq \bar{\varepsilon}(\mathfrak{A}^{n_2}(r))\lfloor n/n_2 \rfloor + \bar{\varepsilon}(\mathfrak{A}^m(r)),$$
   where $m = \min\{n_1, s\}$ (we set $\bar{\varepsilon}(\mathfrak{A}^n(r)) = 0$ for $n < r$).
5. Using the results of Problem 4,
   (a) prove that
$$\bar{\varepsilon}(\mathfrak{A}^n(r)) \geq 2\lfloor n/(r+1) \rfloor,$$
   and if $n \equiv -1 \pmod{r+1}$, then
$$\bar{\varepsilon}(\mathfrak{A}^n(r)) \geq 2\lfloor n/(r+1) \rfloor + 1;$$
   (b) prove that
$$\bar{\varepsilon}(\mathfrak{A}^n(2)) = \lfloor 2n/3 \rfloor;$$
   (c) prove that
$$\bar{\varepsilon}(\mathfrak{A}^n(3)) \geq \lfloor n/2 \rfloor;$$
   (d) prove that
$$\bar{\varepsilon}(\mathfrak{A}^n(4)) \geq \begin{cases} \lfloor 3n/7 \rfloor & \text{for } n \not\equiv 3 \pmod{7}, \\ \lfloor 3n/7 \rfloor - 1 & \text{for } n \equiv 3 \pmod{7}; \end{cases}$$
   we suggest using the results of Problem 1, and in case (d) the property that $\bar{\varepsilon}(\mathfrak{A}^n(r^2)) = 3$ if $n = r^2 + r + 1$ and r is a power of a prime number (see Chapter 8 in [**87**]).
6. Applying Theorems 3.2.17 and 3.2.14 and also the results of Problem 5, find the values of $\bar{\varepsilon}(\mathfrak{A}^n(4))$ for $n = 4, 5, \ldots, 9$. Show also that $\bar{\varepsilon}(\mathfrak{A}^{12}(4)) = 5$, $\bar{\varepsilon}(\mathfrak{A}^{14}(4)) = \bar{\varepsilon}(\mathfrak{A}^{15}(4)) = 6$.

3.3. The van der Waerden–Egorychev–Falikman Theorem

In this section we consider one of the best-known combinatorial problems, which attracted the attention of specialists for more than half a century. In substance, this is a classical problem on a conditional extremum. The problem is to find the minimum of the permanent (see Section 1.2.3) of a doubly stochastic matrix. The set of all doubly stochastic matrices of order n is denoted by Ω_n. Let J_n denote the matrix $n^{-1}\mathbf{J}_{nn}$. It is clear that $J_n \in \Omega_n$ and $\operatorname{per} J_n = n!/n^n$. In 1926 van der Waerden [**114**] put forth the following conjecture on the permanent of doubly stochastic matrices:

If $A \in \Omega_n$ *and* $A \neq J_n$, *then* $\operatorname{per} A > N!/n^n$.

Note that here we are required not only to find the minimum of the permanent, but also to prove that the matrix which provides this minimum is unique. The doubly stochastic matrix of order n whose permanent is minimal will be referred to as the minimizing matrix. For many years considerable efforts of mathematicians were devoted to the solution of this problem. Many papers related to this problem were published, and some progress was achieved. However, only in the early 1980s was the conjecture finally proved independently by two Soviet mathematicians, Egorychev [**35**] and Falikman [**38**] (Falikman proved it without uniqueness of the minimizing matrix).

In this section, we give the proof of the van der Waerden conjecture; that is, we prove the following assertion.

THEOREM 3.3.1. *If* $A \in \Omega_n$, $A \neq J_n$, *then* $\operatorname{per} A > \operatorname{per} J_n$.

First we recall some well-known facts about the permanents of doubly stochastic matrices.

3.3.1. Elementary properties of permanents. The permanent is defined for rectangular matrices of size $m \times n$, $m \leq n$, with elements in some ring. In this section, we consider only square matrices. For a matrix $A = (a_{ij})$ of order n the permanent is defined by the equality

$$\operatorname{per} A = \sum_{\sigma} a_{1\sigma(1)} a_{2\sigma(2)} \cdots a_{n\sigma(n)}, \tag{3.3.1}$$

where the summation is over all permutations σ of degree n. As mentioned in Section 1.2, the permanent is a linear function of rows and columns. For calculating the permanent of a matrix $A = (a_{ij})$ of order n, the decomposition with respect to the ith row, $i \in N_n$,

$$\operatorname{per} A = \sum_{j=1}^{n} a_{ij} \operatorname{per} A(i,j), \tag{3.3.2}$$

can be used. The decomposition with respect to the jth column, $j \in N_n$, is defined similarly as

$$\operatorname{per} A = \sum_{i=1}^{n} a_{ij} \operatorname{per} A(i,j). \tag{3.3.3}$$

Here $\operatorname{per} A(i,j)$ denotes the permanent of the minor of order $n-1$, that is, the permanent of the matrix obtained by removing the ith row and the jth column of A.

In this section, a matrix A of order n will often be represented as the sequence $a_1, \ldots, a_n$ of its column vectors. In such a case, we write $A = (a_1, \ldots, a_n)$ and $\operatorname{per} A = \operatorname{per}(a_1, \ldots, a_n)$. The permanent of the matrix $A(i,j)$ can be written as $\operatorname{per}((a_1, \ldots, a_n)(i,j))$. It is easily seen from the definition that the permanent does not change under any permutation of the columns, namely,

$$\operatorname{per}(a_{\sigma(1)}, \ldots, a_{\sigma(n)}) = \operatorname{per}(a_1, \ldots, a_n)$$

for any permutation σ of degree n. The vector $(1, 1, \ldots, 1)^T$ will be denoted by e, and the column vectors forming the standard basis of the space $\mathbf{R}^n$ by $e^{(j)}$,

$j = 1, \ldots, n$. It is obvious that

$$\operatorname{per}(a_1, \ldots, a_{n-1}, e^{(j)}) = \operatorname{per}((a_1, \ldots, a_n)(j, n))$$

for any $j \in N_n$ and any matrix $(a_1, \ldots, a_n)$ (see (3.3.3)).

3.3.2. Some properties of doubly stochastic matrices. In Section 1.1, we considered the doubly stochastic matrices as the incidence matrices of doubly stochastic configurations. They are nonnegative real-valued matrices $A = (a_{ij})$ of order n satisfying the conditions

$$\sum_{j=1}^{n} a_{ij} = \sum_{i=1}^{n} a_{ij} = 1.$$

One of the fundamental results on the doubly stochastic matrices is the Birkhoff Theorem (see Theorem 1.2.5) This theorem implies that the set Ω_n of doubly stochastic matrices is a convex polytope in $\mathbf{R}^n$ whose vertices are permutation matrices. Recall also that any doubly stochastic matrix has a positive diagonal (see Theorem 1.2.4).

It is easily established that any matrix in Ω_n is either completely indecomposable or permutationally equivalent to a direct sum of matrices B and C, where $B \in \Omega_{n_1}$, $C \in \Omega_{n_2}$, $n_1 + n_2 = n$.

A characteristic property of a doubly stochastic matrix A is the fact that the vector e is the eigenvector of the matrices A and A^T corresponding to the eigenvalue equal to 1. The following assertion is true.

THEOREM 3.3.2. *The number* 1 *is the eigenvalue of the maximal absolute value of a doubly stochastic matrix. For a completely indecomposable matrix, this eigenvalue is simple.*

PROOF. Indeed, let $A = (a_{ij}) \in \Omega_n$, and let $x = (x_1, \ldots, x_n)^T$ be the eigenvector of A corresponding to the eigenvalue λ, that is, $Ax = \lambda x$. Let the absolute value of x_i be maximal over all coordinates of x. Then

$$\lambda x_i = \sum_{j=1}^{n} a_{ij} x_j$$

and

$$|\lambda|\,|x_i| \leq |x_i| \sum_{j=1}^{n} a_{ij} = |x_i|,$$

that is, $|\lambda| \leq 1$.

In order to prove the simplicity of the eigenvalue 1 in the case of a completely indecomposable matrix A, we use the fundamental Perron–Frobenius Theorem (see Chapter 13 in [**45**]). In particular, according to this theorem a completely indecomposable nonnegative matrix has a positive eigenvalue of maximal absolute value which is a simple root of the characteristic equation. The theorem is thus proved. □

Note that the set of doubly stochastic matrices is closed with respect to multiplication, that is, Ω_n is a semigroup (see Problem 6 in Section 1.1).

3.3.3. The Lorentz spaces generated by permanents. More subtle properties of permanents are related to a representation of $\mathbf{R}^n$ as a space with symmetric inner product determined by the permanent of some matrix of order n with $n-2$ positive column vectors.

Let Q be a real-valued symmetric matrix of order n. Consider in $\mathbf{R}^n$ the inner product

$$\langle x, y\rangle = x^T Q y, \tag{3.3.4}$$

where x, y are n-dimensional real-valued column vectors. If Q has one positive and $n-1$ negative eigenvalues, then $\mathbf{R}^n$ with the inner product (3.3.4) is the Lorentz space (or the Minkowski space). The spaces

$$\mathbf{R}^2 \text{ with } Q = \begin{pmatrix} 0 & 1 \\ 1 & 0 \end{pmatrix} \quad \text{and} \quad \mathbf{R}^3 \text{ with } Q = \begin{pmatrix} 0 & 1 & 1 \\ 1 & 0 & 1 \\ 1 & 1 & 0 \end{pmatrix}$$

provide examples of Lorentz spaces. A vector $x \in \mathbf{R}^n$ is called Q-positive, Q-isotropic, or Q-negative if $\langle x, x\rangle > 0$, $\langle x, x\rangle = 0$, or $\langle x, x\rangle < 0$, respectively.

LEMMA 3.3.3. *If a is a Q-positive vector in the Lorentz space $\mathbf{R}^n$ and b is an arbitrary n-dimensional vector, then*

$$\langle a, b\rangle^2 \geq \langle a, a\rangle\langle b, b\rangle, \tag{3.3.5}$$

and the equality holds if and only if $b = \lambda a$, $\lambda \in \mathbf{R}$.

PROOF. Let a be a Q-positive vector. If a and the vector $b \neq 0$ are noncollinear, then by the Sylvester Theorem in the plane generated by the vectors a and b there exists a Q-negative vector $a + \lambda b$. The quadratic form

$$\langle a + \lambda b, a + \lambda b\rangle = \langle b, b\rangle\lambda^2 + 2\langle a, b\rangle\lambda + \langle a, a\rangle$$

has a positive discriminant, that is,

$$\langle a, b\rangle^2 > \langle a, a\rangle\langle b, b\rangle.$$

Thus, equality in (3.3.4) is possible only if $b = \lambda' a$, $\lambda' \in \mathbf{R}$. □

COROLLARY 3.3.4. *If in the Lorentz space a vector a is Q-positive and a vector $b \neq 0$, then the equality $\langle a, b\rangle = 0$ implies that b is Q-negative.*

Let $a_1, \ldots, a_{n-2}$ be vectors in $\mathbf{R}^n$ with positive coordinates. We define the matrix Q as the matrix with elements

$$q_{ij} = \operatorname{per}(a_1, \ldots, a_{n-2}, e^{(i)}, e^{(j)}), \qquad i, j = 1, \ldots, n. \tag{3.3.6}$$

It is clear that Q is a symmetric matrix. We define the inner product in $\mathbf{R}^n$, setting

$$\langle x, y\rangle = \operatorname{per}(a_1, \ldots, a_{n-2}, x, y) \tag{3.3.7}$$

for $x, y \in \mathbf{R}^n$, that is,

$$\langle x, y\rangle = x^T Q y, \tag{3.3.8}$$

where Q is defined by (3.3.6).

The following theorem leads to the inequality for permanents which has turned out to be a key point in the proof of Theorem 3.3.1.

THEOREM 3.3.5. *For a matrix $A = (a_1, \ldots, a_n)$ with $n-2$ positive column vectors $a_1, \ldots, a_{n-2}$, the space $\mathbf{R}^n$ with the inner product defined by (3.3.7) is a Lorentz space.*

PROOF. We use induction on n. For $n = 2$,

$$Q = \begin{pmatrix} 0 & 1 \\ 1 & 0 \end{pmatrix};$$

consequently, the assertion of the theorem is true (see the example above). Let $n > 2$. Suppose that the theorem is true for $\mathbf{R}^{n-1}$. First we show that 0 is not an eigenvalue of Q. Let, on the contrary, $Qc = 0$ for some $c \neq 0$; then

$$\operatorname{per}(a_1, \ldots, a_{n-2}, c, e^{(j)}) = 0 \tag{3.3.9}$$

for any $j \in N_n$. For a given $j \in N_n$, consider in $\mathbf{R}^{n-1}$ the inner product

$$\operatorname{per}((a_1, \ldots, a_{n-3}, x, y, e^{(j)})(j, n)). \tag{3.3.10}$$

It is defined for vectors

$$\hat{x}(j) = (x_1, \ldots, x_{j-1}, x_{j+1}, \ldots, x_n)^T,$$
$$\hat{y}(j) = (y_1, \ldots, y_{j-1}, y_{j+1}, \ldots, y_n)^T$$

and the corresponding column vectors $\hat{a}_1(j), \ldots, \hat{a}_{n-3}(j)$. According to the remark in Section 3.3.1 this inner product can be written in the form (3.3.10) with

$$x = (x_1, \ldots, x_j, \ldots, x_n)^T, \quad y = (y_1, \ldots, y_j, \ldots, y_n)^T \in \mathbf{R}^n,$$

where x_j, y_j are arbitrary numbers. Now apply the induction assumption and Lemma 3.3.3. Substituting $x = a_{n-2}$, $y = a_{n-2}$ into (3.3.10), we obtain a positive value, and the substitution of $x = a_{n-2}$, $y = c$ gives zero (see (3.3.9)). Applying Lemma 3.3.3, we obtain the inequalities

$$\operatorname{per}(a_1, \ldots, a_{n-3}, c, c, e^{(j)}) \leq 0, \qquad j = 1, \ldots, n. \tag{3.3.11}$$

Taking Corollary 3.3.4 into account, it is easy to see that, because $c \neq 0$, equality in (3.3.11) is possible for at most one value of j. Let

$$a_{n-2} = (a_{1,n-2}, a_{2,n-2}, \ldots, a_{n,n-2})^T.$$

By (3.3.7) and (3.3.8) we have

$$\sum_{j=1}^{n} a_{j,n-2} \operatorname{per}(a_1, \ldots, a_{n-3}, c, c, e^{(j)}) = \operatorname{per}(a_1, \ldots, a_{n-3}, c, c, a_{n-2}) = c^T Q c < 0,$$

which contradicts the assumption that $Qc = 0$. Hence $c = 0$.

For any real number θ, $0 \leq \theta \leq 1$, we define the matrix Q_θ, replacing each column a_i in (3.3.6) by $\theta a_i + (1 - \theta)e$, $i = 1, 2, \ldots, n-2$. We have shown that zero is not an eigenvalue of any matrix Q_θ. Consequently, the number of positive eigenvalues for all these matrices is the same. For $\theta = 0$, this number is equal to 1 (see Problem 2 at the end of this section). Therefore it equals 1 for $\theta = 1$ as well. The theorem is proved. □

The following result is a direct consequence of Theorem 3.3.5 and Lemma 3.3.3.

THEOREM 3.3.6. *Let $a_1, \ldots, a_{n-1}$ be vectors in $\mathbf{R}^n$ with positive coordinates, and b an arbitrary vector in $\mathbf{R}^n$. Then*

$$(\operatorname{per}(a_1, \ldots, a_{n-1}, b))^2 \geq \operatorname{per}(a_1, \ldots, a_{n-1}, a_{n-1}) \operatorname{per}(a_1, \ldots, a_{n-2}, b, b), \tag{3.3.12}$$

and the equality holds if and only if $b = \lambda a_{n-1}$, $\lambda \in \mathbf{R}$.

COROLLARY 3.3.7. *Inequality (3.3.12) is valid for any nonnegative vectors $a_1, \ldots, a_{n-1} \in \mathbf{R}^n$ and any vector $b \in \mathbf{R}^n$.*

We leave the proof of the corollary to the reader as an exercise.

Relation (3.3.12) is called the Alexandrov–Fenchel inequality for permanents because it originates from the similar inequality for mixed volumes of convex geometric bodies obtained in [**6**] and [**40**].

3.3.4. Properties of minimizing matrices. Complete indecomposability. The proof of Theorem 3.3.1 is carried out by analyzing properties of minimizing matrices. Recall that a nonnegative matrix of order n is completely indecomposable if it does not contain a zero matrix of size $r \times s$ such that $r + s = n$. By using induction on the number of multipliers, it is possible to prove that the product of a finite number of completely indecomposable matrices of order n is a completely indecomposable matrix as well (see Property 5.1.8).

LEMMA 3.3.8. *Let A be a doubly stochastic matrix of order n. If A is representable as the direct sum of a matrix B of order m and a matrix C of order $n-m$, then there exists a doubly stochastic matrix A' such that* $\operatorname{per} A' < \operatorname{per} A$.

PROOF. Let A admit a representation

$$A = \begin{pmatrix} B & 0 \\ 0 & C \end{pmatrix},$$

where $B = (b_{ij})$, $C = (c_{ij})$ are doubly stochastic matrices of orders m and $n-m$, respectively. Then $\operatorname{per} A = \operatorname{per} B \operatorname{per} C$ and, besides, $\operatorname{per} A \neq 0$ by Theorem 1.2.4. Since $\operatorname{per} B \neq 0$ and $\operatorname{per} C \neq 0$, the matrices B and C contain positive diagonals. By permutations of rows and columns it is possible to move the elements of these diagonals to the main diagonals of A, B, and C. Then

$$b = \operatorname{per} B(1,1) > 0, \qquad c = \operatorname{per} C(1,1) > 0.$$

Let $0 < \varepsilon < \min(b_{11}, c_{11})$. We pass from A to the matrix $A' \in \Omega_n$, replacing b_{11} by $b_{11} - \varepsilon$, c_{11} by $c_{11} - \varepsilon$ and zeros in the positions $(1, m+1)$ and $(m+1, 1)$ by ε. We denote by B' and C' the submatrices $A'[N_m, N_m]$ and $A'[N_n - N_m, N_n - N_m]$, respectively. It is clear that

$$\operatorname{per} B' = \operatorname{per} B - \varepsilon \operatorname{per} B(1,1) = \operatorname{per} B - \varepsilon b$$

and, similarly, $\operatorname{per} C' = \operatorname{per} C - \varepsilon c$. Then

$$\begin{aligned} \operatorname{per} A' &= \operatorname{per} B' \operatorname{per} C' + \varepsilon^2 \operatorname{per} B(1,1) \operatorname{per} C(1,1) \\ &= \operatorname{per} B \operatorname{per} C - \varepsilon c \operatorname{per} B - \varepsilon b \operatorname{per} C + 2\varepsilon^2 bc \\ &= \operatorname{per} A - \varepsilon(c \operatorname{per} B + b \operatorname{per} C - 2\varepsilon bc). \end{aligned}$$

Since ε can be chosen sufficiently small, we have $\operatorname{per} A' < \operatorname{per} A$. □

COROLLARY 3.3.9. *A minimizing matrix is completely indecomposable.*

Indeed, if a doubly stochastic matrix is not completely indecomposable, then, as mentioned in Section 3.3.1, it can be represented as a direct sum of two square matrices.

COROLLARY 3.3.10. *Any row and any column of a minimizing matrix contains at least two positive elements.*

LEMMA 3.3.11. *Let $A = (a_{ij})$ be a minimizing matrix of order n. Then for any element a_{ij} there exists a permutation σ acting on N_n such that $\sigma(i) = j$ and $a_{s\sigma(s)} > 0$ for each $s \in N_n$, $s \neq i$.*

PROOF. Suppose that such a permutation does not exist. Then $\operatorname{per} A(i,j) = 0$, and by Theorem 1.2.3 the matrix $A(i,j)$ contains a zero submatrix of size $k \times l$ such that $k + l = n$. This means that the doubly stochastic matrix A is not completely indecomposable. This contradicts Corollary 3.3.9, and the lemma is proved. □

3.3.5. Properties of minimizing matrices. Permanent minors. A milestone in the history of proving Theorem 3.3.1 was the following result obtained by Marcus and Newman in [**73**].

THEOREM 3.3.12. *If $A = (a_{ij})$ is a minimizing matrix of order n, then the inequality $a_{kl} > 0$ implies that* $\operatorname{per} A(k,l) = \operatorname{per} A$.

PROOF. In the convex polytope Ω_n, consider the face $C(A)$ of minimal dimension containing A. It is clear that if we denote by Z the set of all (i,j) such that $a_{ij} = 0$, then $C(A)$ is the set of all matrices $X = (x_{ij}) \in \Omega_n$ satisfying the conditions

$$\begin{aligned} \sum_{j=1}^{n} x_{ij} &= 1, \qquad && i \in N_n, \\ \sum_{i=1}^{n} x_{ij} &= 1, \qquad && i \in N_n, \\ x_{ij} &\geq 0, \qquad && i,j \in N_n, \\ x_{ij} &= 0, \qquad && (i,j) \in Z. \end{aligned}$$

By the condition of the theorem, $\operatorname{per} A$ is the absolute minimum of the permanents $\operatorname{per} X$ over all matrices $X \in \Omega_n$. A permutation matrix is not minimizing. Therefore A belongs to the interiority of $C(A)$. At A the permanent function $\operatorname{per} X$ has a relative minimum. To find this minimum, it is possible to introduce Lagrange multipliers λ_i, μ_j, $i,j = 1,\dots,n$, and consider the function on $C(A)$

$$F(X) = \operatorname{per} X - \sum_{i=1}^{n} \lambda_i \left(\sum_{r=1}^{n} x_{ir} - 1 \right) - \sum_{j=1}^{n} \mu_j \left(\sum_{r=1}^{n} x_{rj} - 1 \right).$$

If $(i,j) \notin Z$, then

$$\frac{\partial F(X)}{\partial x_{ij}} = \operatorname{per} X(i,j) - \lambda_i - \mu_j.$$

Since A is a point of minimum, we have

$$\operatorname{per} A(i,j) = \lambda_i + \mu_j, \tag{3.3.13}$$

and (see (3.3.2)) it follows that for $i \in N_n$

$$\operatorname{per} A = \sum_{j=1}^{n} a_{ij} \operatorname{per} A(i,j) = \lambda_i + \sum_{j=1}^{n} a_{ij} \mu_j. \tag{3.3.14}$$

Similarly, for $j \in N_n$ we have

$$\operatorname{per} A = \sum_{i=1}^{n} a_{ij} \operatorname{per} A(i,j) = \mu_j + \sum_{i=1}^{n} a_{ij} \lambda_i. \tag{3.3.15}$$

We set $\lambda = (\lambda_1, \dots, \lambda_n)^T$, $\mu = (\mu_1, \dots, \mu_n)^T$. It follows from (3.3.14) and (3.3.15) that

$$(\operatorname{per} A)e = \lambda + A\mu, \tag{3.3.16}$$

$$(\operatorname{per} A)e = A^T\lambda + \mu. \tag{3.3.17}$$

Since $A^Te = e$ (see Section 3.3.2), multiplying both sides of (3.3.16) by A^T, we obtain

$$(\operatorname{per} A)e = A^T\lambda + A^TA\mu. \tag{3.3.18}$$

Subtracting (3.3.17) from (3.3.18), we arrive at the equality

$$A^TA\mu = \mu.$$

Similarly, we obtain

$$AA^T\lambda = \lambda.$$

Since A and A^T are completely indecomposable, the matrices AA^T and A^TA are also completely indecomposable (see Section 3.3.4). These matrices are doubly stochastic (see Section 3.3.2); therefore 1 is a simple eigenvalue of each of them (by Theorem 3.3.2). Consequently, λ and μ are collinear with e, that is, $\lambda = ce$, $\mu = de$ for some c and d. It follows from (3.3.13) that

$$\operatorname{per} A(i,j) = c + d$$

for all $(i,j) \notin Z$. But then

$$\operatorname{per} A = \sum_{i=1}^{n} a_{ij} \operatorname{per} A(i,j) = \sum_{i=1}^{n} a_{ij}(c+d) = c + d = \operatorname{per} A(i,j)$$

for all $(i,j) \notin Z$. The theorem is proved. □

Later a refinement of this theorem was obtained. The first part of the proof of the following result belongs to London [**69**] and the second part to Egorychev [**36**].

THEOREM 3.3.13. *If $A = (a_{ij})$ is a minimizing matrix of order n, then we have* $\operatorname{per} A(i,j) = \operatorname{per} A$ *for all $i, j \in N_n$.*

PROOF. First we show that $\operatorname{per} A(i,j) \geq \operatorname{per} A$ for all $i, j \in N_n$. According to Lemma 3.3.11, for all $i, j \in N_n$ there exists a permutation σ such that $\sigma(i) = j$ and $a_{s\sigma(s)} > 0$ for $s \in N_n$, $s \neq i$. Let $P = (p_{ij})$ be the permutation matrix corresponding to σ. Consider the family of matrices $(1-\theta)A + \theta P$ for $0 \leq \theta \leq 1$. Since A is a minimizing matrix, for the function $f(\theta) = \operatorname{per}((1-\theta)A = \theta P)$ the inequality $f'(0) \geq 0$ is true; that is,

$$0 \leq \sum_{i=1}^{n}\sum_{j=1}^{n}(-a_{ij} + p_{ij}) \operatorname{per} A(i,j) = -n \operatorname{per} A + \sum_{s=1}^{n} \operatorname{per} A(s, \sigma(s)).$$

By Theorem 3.3.12, $\operatorname{per} A(s,\sigma(s)) = \operatorname{per} A$ for $s \neq i$; therefore $\operatorname{per} A(i,j) \geq \operatorname{per} A$.

Suppose that $\operatorname{per} A(r,s) > \operatorname{per} A$ for some $r, s \in N_n$. By Corollary 3.3.10, in the rth row there exists an element $a_{rt} > 0$, $t \neq s$. Along with $A = (a_1, \dots, a_n)$ we consider the matrix A_1 obtained from A by the replacement of the column a_s by the column a_t and the matrix A_2 obtained from A by replacing the column a_t by a_s. By Corollary 3.3.7

$$(\operatorname{per} A)^2 \geq \operatorname{per} A_1 \operatorname{per} A_2. \tag{3.3.19}$$

Further, by (3.3.3)

$$\operatorname{per} A_1 = \sum_{k=1}^{n} a_{kt} \operatorname{per} A(k,s), \qquad \operatorname{per} A_2 = \sum_{k=1}^{n} a_{ks} \operatorname{per} A(k,t).$$

Thus, it follows from (3.3.19) that

$$(\operatorname{per} A)^2 \geq \left(\sum_{k=1}^{n} a_{kt} \operatorname{per} A(k,s) \right) \left(\sum_{k=1}^{n} a_{ks} \operatorname{per} A(k,t) \right). \tag{3.3.20}$$

In the right-hand side, each permanent minor is $\geq \operatorname{per} A$, and $\operatorname{per} A(r,s) > \operatorname{per} A$. Since $\operatorname{per} A(r,s)$ is multiplied by the positive number a_{rt}, the right-hand side of (3.3.20) is greater than $(\operatorname{per} A)^2$; that is, we obtain a contradiction. □

3.3.6. The substitution rule. Completion of the proof of Theorem 3.3.1. We have shown that all permanent minors of order $n-1$ of a minimizing matrix are equal to its permanent. Note that this property holds for the matrix J_n. Based on this property, it is possible, starting with an arbitrary minimizing matrix A, to construct a sequence of minimizing matrices with the first member A and show that all members are identical and equal to J_n. For this purpose, we introduce a transformation called the substitution rule.

Let $A \in \Omega_n$ and $\operatorname{per} A = \operatorname{per} A(i,j)$ for any $i, j \in N_n$. If $x = (x_1, \dots, x_n)^T$ and $x_1 + \dots + x_n = 1$, then, replacing an arbitrary column in A by the column x, we obtain a matrix with the same value of the permanent.

LEMMA 3.3.14. *If $A = (a_{ij}) \in \Omega_n$ is such that $\operatorname{per} A = \operatorname{per} A(i,j)$ for all $i, j \in N_n$, then the replacement of arbitrary columns a and b of A by the vectors $(a+b)/2$ and $(a+b)/2$ gives a matrix with the same permanent.*

PROOF. Without loss of generality, let $a = a_{n-1}$ and $b = a_n$. Then by the substitution rule

$$\operatorname{per}(a_1, \dots, a_{n-2}, a, a) = \operatorname{per}(a_1, \dots, a_{n-2}, b, b) = \operatorname{per} A.$$

Hence

$$\operatorname{per}(a_1, \dots, a_{n-2}, (a+b)/2, (a+b)/2) = 4 \operatorname{per} A/4 = \operatorname{per} A.$$

□

Let A be a minimizing matrix and b an arbitrary, say the last, column of A. By Corollary 3.3.10, in each row of A there exist at least two positive elements. Using the substitution rule, we can, without changing the last column, perform several replacements of the form described in Lemma 3.3.14 and obtain a minimizing matrix $A' = (a'_1, \dots, a'_{n-1}, b)$ such that the first $n-1$ columns of A' are positive. Now it is possible to apply Theorem 3.3.6 $n-1$ times, moving the columns $a'_1, \dots, a'_{n-1}$ sequentially to the position $n-1$. By the substitution rule, each time we obtain equality in (3.3.12). By Theorem 3.3.6 this means that $b = \lambda_i a'_i$, $i = 1, \dots, n-1$. Hence $\lambda_1 = \dots = \lambda_{n-1}$ and $b = n^{-1}e$; that is, $A = J_n$.

Theorem 3.3.1 is proved.

3.3.7. Another proof of the minimality of $\operatorname{per} J_n$. There exists another proof of the minimality of the value $n!/n^n$ among the values of the permanents of matrices in Ω_n, different from the proof given above. The starting point of this proof is the following result, derived from Theorem 3.3.12 also by Marcus and Newman in [**73**].

THEOREM 3.3.15. *If there exists a positive minimizing matrix A, then*

$$\operatorname{per} A = \operatorname{per} J_n,$$

and, moreover, $A = J_n$.

The proof is contained in [**75**].

Denote by Ω_n^* the set of all positive matrices in Ω_n. It is easily seen from Theorem 3.3.15 that in order to show that $\operatorname{per} J_n$ is minimal over all matrices in Ω_n, it is sufficient to show that $\operatorname{per} A \geq n!/n^n$ for any matrix $A \in \Omega_n^*$ (in view of the continuity of permanents).

For a matrix A of order n, we set

$$\Pi(A) = \prod_{i=1}^{n} \prod_{j=1}^{n} a_{ij}.$$

Let $\varepsilon > 0$. On the set Ω_n^*, consider the function

$$F_\varepsilon(A) = \operatorname{per} A + \varepsilon/\Pi(A). \tag{3.3.21}$$

For small ε, $\varepsilon F_\varepsilon(A)$ is closed to $\operatorname{per} A$.

THEOREM 3.3.16. *In Ω_n^* there exists a matrix $A = (a_{ij})$ which is the point of minimum of the function $F_\varepsilon(A)$. The matrix A is such that for all $k, l \in N_n$*

$$\operatorname{per} A(k,l) = b + c/a_{kl}, \tag{3.3.22}$$

where $c = \varepsilon/\Pi(A)$ and $b = \operatorname{per} A - nc$.

PROOF. Denote by $\alpha(X)$ the minimal element of the matrix $X \in \Omega_n^*$. The geometric mean of n real numbers does not exceed the arithmetic mean; therefore

$$\Pi(X) \leq \alpha(X) \left(\frac{1}{n}\right)^{n(n-1)}.$$

Hence the infimum of the numbers $\alpha(X)$ for the set of matrices X such that $F_\varepsilon(X) \leq 2F_\varepsilon(J_n)$ is positive. Therefore the matrices X with this property form a compact subset in Ω_n^*. Consequently, there exists a matrix $A \in \Omega_n^*$ such that

$$F_\varepsilon(A) = \min_{X \in \Omega^*} F_\varepsilon(X).$$

On the set introduced above, consider the function

$$\hat{F}_\varepsilon(X) = F_\varepsilon(X) - \sum_{i=1}^{n} \left(\sum_{r=1}^{n} x_{ir} - 1\right) - \sum_{j=1}^{n} \mu_j \left(\sum_{r=1}^{n} x_{rj} - 1\right)$$

introduced in Section 3.3.5. The proof of the property (3.3.22) for the matrix A minimizing $F_\varepsilon(X)$ is carried out in the same way as the proof of Theorem 3.3.12. We leave this proof to the readers as an exercise (see Problem 4 in Section 3.3). □

Now we prove that the matrix A which provides the minimum of the function $F_\varepsilon(X)$ is in fact J_n for any $\varepsilon > 0$. Let u, v be arbitrary columns of $A(a_1, \ldots, a_n)$. Without loss of generality, we assume that $u = a_{n-1}$, $v = a_n$. Consider also the column vectors $t = (t_1, \ldots, t_n)^T = u - v$ and $s = (s_1, \ldots, s_n)^T$, where $s_i = u_i v_i$, $i = 1, \ldots, n$. Note that the vector s is positive. By Theorem 3.3.16

$$\langle t, e_i \rangle = \operatorname{per} A(i,n) - \operatorname{per} A(i, n-1) = ct_i/(u_i v_i). \tag{3.3.23}$$

Using (3.3.23), we find that

$$\langle t, e_i \rangle = c \sum_{i=1}^{n} t_i = 0.$$

Then by Corollary 3.3.4 the vector t is either zero or Q-negative. Using (3.3.23) again, we obtain

$$\langle t, t \rangle = c \sum_{i=1}^{n} t_i^2/(u_i v_i) \geq 0.$$

Thus, $t = 0$, that is, $u = v$. Therefore all columns of A are identical and, consequently, $A = J_n$. Thus, for any matrix $X \in \Omega_n^*$ the inequality

$$\operatorname{per} X + \varepsilon/\Pi(X) \geq \operatorname{per} J_n + \varepsilon n^{n^2}$$

holds for any $\varepsilon > 0$. Hence $\operatorname{per} X \geq \operatorname{per} J_n$. According to the above remark, the inequality $\operatorname{per} A \geq n!/n^n$ is proved for any $A \in \Omega_n$.

Problems.

1. Consider n urns with one ball in each of them. Suppose that the probability that the ball contained in the ith urn is moved into the jth urn is equal to a_{ij}, $i, j = 1, \ldots, n$. Let us move all balls simultaneously. What is the probability that after these replacements each urn contains exactly one ball?
2. Find the matrix Q for the inner product $\langle x, y \rangle = x^T Q y$ constructed on $\mathbf{R}^n$ by formula (3.3.6) for the matrix J_n. Find the eigenvalues of Q.
3. Which of the following bilinear forms on $\mathbf{R}^3$

$$x_1 y_1 + x_2 y_3 + x_3 y_2,$$
$$x_1 y_2 + x_2 y_1 + x_2 y_3 + x_3 y_2,$$
$$\sum_{i \neq j} x_i y_i$$

   can be generated with the use of the permanent function? Find the corresponding representation.
4. Prove the property (3.3.22) for the matrix which provides the minimum of the function $F_\varepsilon(X)$ (see (3.3.21)).
5. Using results of Chapter 1 (see Problem 6 in Section 1.2), derive the following lower bound for the number $L(r, n)$ of Latin rectangles of size $r \times n$ with elements in N_n:

$$L(r, n) \geq \frac{(n!)^r}{n^{n(r-1)}} \prod_{t=1}^{r-1} (n - t)^n.$$

6. Prove Corollary 3.3.7 using the continuity of permanents.

CHAPTER 4

Asymptotic Methods in the Study of Nonnegative Matrices

Introduction

In the first section of this chapter, we consider various elementary properties of nonnegative matrices and their supports (see Section 1.1). The majority of these properties are well known [**48, 28**] and are related to special properties of the graphs whose incidence matrices are the supports of the corresponding nonnegative matrices. This relation prompted Stanley to develop the technique of matrix functions (see [**102**]), which is also presented in the first section of this chapter. This technique is an effective tool for solving combinatorial problems, including problems on restricted permutations.

In the second section, we present the well-known results obtained by Moon and Moser [**78**], and, in particular, we give a proof of the fact that almost all $(0,1)$-matrices of order n are primitive with exponent equal to 2 as $n \to \infty$. In other words, this result means that, in the corresponding digraph, almost every pair of vertices is connected by a directed path of length 2.

In the third section, asymptotic formulas for the first two moments of the permanent of a random $(0,1)$-matrix of size $m \times n$ with N ones are derived. For the case where $m = n$, $N = n^{3/2+\varepsilon}$, $\varepsilon > 0$, O'Neil [**80**] proved that the permanent of a random matrix coincides with the mean value with probability tending to one as $n \to \infty$. Following the papers [**4**] and [**92**], we give a proof of a similar result in the case where $n = \beta n$, $\beta \geq 1$.

In Section 4.4, we consider lattices and Boolean algebras. With a Boolean algebra generated by a system of generators we associate a $(0,1)$-matrix with distinct columns. For a random family of subsets, we estimate the probability of generating a Boolean algebra and obtain the corresponding formulas repeating the results of Renyi [**83**]. Similar results for free Boolean algebras are obtained with the use of the Poisson theorem in the classical occupancy problem [**5**].

The next two sections are devoted to coverings of finite sets and their relations to $(0,1)$-matrices. The notion of a minimal covering of a finite set was introduced in [**55**], and a formula expressing the number of such coverings in terms of Stirling numbers of the second kind was obtained. Later, in the book [**93**] a relationship between minimal coverings and certain $(0,1)$-matrices was found, and a number of explicit and asymptotic formulas were derived. Finally, in [**95**] the asymptotics of the number of minimal coverings with k blocks and a nonstandard distribution of the number of blocks in a random covering were found. These results are presented in Sections 4.5 and 4.6.

In Section 4.5 we consider a relation between minimal coverings and the quasiconsistency of systems of Boolean equations.

4.1. Nonnegative matrices and graphs

4.1.1. The support and the graph of a matrix. The study of combinatorial properties of nonnegative matrices sometimes can be reduced to the study of the corresponding $(0,1)$-matrices. In such cases, the notion of the support of a nonnegative matrix, defined in Chapter 1, is used.

Recall that the support of a nonnegative matrix $\tilde{A} = (\tilde{a}_{ij})$ of size $m \times n$ is the $(0,1)$-matrix ${}^{s}A = A = (a_{ij})$ such that for $i = 1, \dots, m$, $j = 1, \dots, n$

$$a_{ij} = \begin{cases} 1, & \tilde{a}_{ij} > 0, \\ 0, & \tilde{a}_{ij} = 0. \end{cases}$$

In what follows it is convenient to assume that the $(0,1)$-matrices are viewed as defined in terms of the Boolean algebra $B_2 = \{0,1\}$, where the operations of addition and multiplication are defined by the rules

$$1 \boxplus 0 = 0 \boxplus 1 = 1 \boxplus 1 = 1, \qquad 0 \boxplus 0 = 0,$$
$$1 \cdot 0 = 0 \cdot 1 = 0 \cdot 0 = 0, \qquad 1 \cdot 1 = 1.$$

Then the $(0,1)$-matrices of order n form a semigroup with respect to the usual matrix multiplication (see Section 1.1), and are called Boolean matrices.

The notion of the support allows one to associate with each nonnegative matrix of order n the digraph G with n vertices such that the ith vertex is connected by an arc (i,j) with the jth vertex if and only if $a_{ij} = 1$, $i,j = 1, \dots, n$. In other words, the support is the adjacency matrix of the digraph G (see Section 1.1).

PROPERTY 4.1.1. *If $A = (a_{ij})$ is the adjacency matrix of a digraph G with n vertices and $A^l = (a_{ij}^{(l)})$, $l = 1, 2, \dots$, is the lth power of A, then $a_{ij}^{(l)} = 1$ if and only if the digraph G contains a directed path of length l from the vertex i to the vertex j.*

The validity of Property 4.1.1 is obvious.

If $A = (a_{ij})$ is the adjacency matrix of a digraph with n vertices and multiple arcs in the graph are allowed, then the elements a_{ij} take nonnegative integer values. In this case, we assume that the operations of addition and multiplication are defined in the ring of integers.

PROPERTY 4.1.2. *The element $a_{ij}^{(l)}$ of the matrix $A^l = (a_{ij}^{(l)})$, $i,j = 1, \dots, n$, which is the adjacency matrix of a digraph with multiple arcs, is equal to the number of directed paths of length l from the vertex i to the vertex j.*

Property 4.1.2 is similar to Property 4.1.1. The following property is also obvious.

PROPERTY 4.1.3. *If $A = (a_{ij})$ is the adjacency matrix of a digraph G and for some positive integer l_0 the condition $A^l = 0$ for all $l \geq l_0$ is fulfilled, then the digraph G has no circuits. If the digraph G has no circuits, then the element $A_{ij}^{(l)}$ of the matrix $A^l = (a_{ij}^{(l)})$ is equal to the number of simple directed paths in G from the vertex i to the vertex j.*

A nonnegative matrix $A = (a_{ij})$ is called decomposable if there exists a permutation matrix Π such that

$$\Pi A \Pi^{-1} = \begin{pmatrix} B & \mathbf{0} \\ C & D \end{pmatrix}, \tag{4.1.1}$$

where B and D are square matrices and $\mathbf{0}$ is a matrix with all elements equal to zero.

A matrix is called indecomposable if it is not decomposable.

PROPERTY 4.1.4. *An arbitrary nonnegative square matrix $\tilde{A}$ can be transformed by the same permutation of rows and columns to the canonical form*

$$\Pi\tilde{A}\Pi^{-1} = \begin{pmatrix} \tilde{A}_{11} & & & 0 \\ \tilde{A}_{21} & \tilde{A}_{22} & & \\ & & \ddots & \\ \tilde{A}_{k1} & \tilde{A}_{k2} & \dots & \tilde{A}_{kk} \end{pmatrix}, \tag{4.1.2}$$

where $\tilde{A}_{11}, \tilde{A}_{22}, \dots, \tilde{A}_{kk}$ are indecomposable square matrices and Π is a permutation matrix determining the permutation of rows and columns.

The proof is obvious.

PROPERTY 4.1.5. *The support $A = (a_{ij})$ of an indecomposable nonnegative matrix $\tilde{A}$ of order n is the adjacency matrix of a strongly connected graph G.*

Indeed, if the graph G has at least two strongly connected components (see Section 6.1), then the adjacency matrix of G can be transformed by renumbering the vertices of the graph to the form (4.1.1); that is, the matrix is decomposable. If the adjacency matrix can be transformed to the form (4.1.1), then any power of the matrix is of the same form. Consequently, there exist i, j, $1 \le i, j \le n$, such that for any l we have $a_{ij}^{(l)} = 0$. Thus, the graph is not strongly connected.

Hence, the next assertion immediately follows.

PROPERTY 4.1.6. *If a nonnegative matrix $A = (a_{ij})$ is indecomposable, then for any i, j there exists l such that $a_{ij}^{(l)} > 0$, where $A^l = (a_{ij}^{(l)})$.*

PROPERTY 4.1.7. *If a nonnegative matrix $\tilde{A}$ of order n is reduced to the form (4.1.2) so that its support is of the form*

$$^s\tilde{A} = \begin{pmatrix} A_{11} & & & 0 \\ A_{21} & A_{22} & & \\ & & \ddots & \\ A_{k1} & A_{k2} & \dots & A_{kk} \end{pmatrix},$$

and is the adjacency matrix of a directed graph with n vertices, then the matrices $A_{11}, A_{22}, \dots, A_{kk}$ determine the arcs connecting vertices inside each of the k strongly connected components, and the matrices A_{ij}, $i < j$, determine the arcs connecting vertices of the ith strongly connected component with vertices of the jth strongly connected component, $i, j = 1, \dots, k$.

An indecomposable matrix $A \ge 0$ is called primitive if there exists k such that

$$A^k > 0.$$

The smallest k satisfying this condition is called the exponent of the matrix A and is denoted by $\gamma = \gamma(A)$.

PROPERTY 4.1.8. *If A is an indecomposable nonnegative matrix of order n and $\mathbf{I}$ is the identity matrix, then the matrix $A + \mathbf{I}$ is primitive and*

$$(A + \mathbf{I})^{n-1} > 0;$$

that is, $\gamma(A + \mathbf{I}) \leq n - 1$. The bound given by the last inequality is attainable.

We denote by $\sigma(x)$ the number of zero coordinates of the column vector $x \geq 0$, $x \neq 0$. Let us show that $\sigma((A + \mathbf{I})x) < \sigma(x)$. Suppose that $\sigma((A + \mathbf{I})x) = \sigma(x)$. Then it follows from the equality $y = x + Ax$ that the column vectors x and y have the same zero coordinates. By renumbering the coordinates the vectors x and y can be brought to the form

$$x = \begin{pmatrix} u \\ \mathbf{0} \end{pmatrix}, \quad y = \begin{pmatrix} v \\ \mathbf{0} \end{pmatrix}, \quad n > 0, \quad v > 0,$$

where the vectors u and v have positive coordinates and the dimension l. We write the matrix A in the form

$$A = \begin{pmatrix} A_{11} & A_{12} \\ A_{21} & A_{22} \end{pmatrix},$$

where the matrix A_{11} is of order l. Then the equality

$$\begin{pmatrix} u \\ \mathbf{0} \end{pmatrix} + \begin{pmatrix} A_{11} & A_{12} \\ A_{21} & A_{22} \end{pmatrix} \begin{pmatrix} u \\ \mathbf{0} \end{pmatrix} = \begin{pmatrix} v \\ \mathbf{0} \end{pmatrix}$$

implies that $A_{21}u = 0$, $u > 0$; that is, A_{21} is the zero matrix. This contradicts the indecomposability of A. Thus, $\sigma((A + \mathbf{I})^{n-1}x) = 0$ and $(A + \mathbf{I})^{n-1} > 0$.

Let $A = C$, where C is the permutation matrix corresponding to the unicyclic permutation $(1\,2 \dots n)$. Then $\gamma(C + \mathbf{I}) = n - 1$, and the bound for γ given in Property 4.1.8 is attainable.

A simpler proof of Property 4.1.8 uses the digraph G corresponding to the matrix $A + \mathbf{I}$. Since the matrix A is indecomposable, the digraph G is strongly connected and, consequently, each vertex i is connected with each vertex j, $i \neq j$, by a directed path of length $l_{ij} \leq n - 1$, $i, j = 1, \dots, n$. Each vertex i has a loop, and, adding an appropriate number of passes of the loop of this vertex, we can obtain a path from i to j of length exactly $n - 1$. Consequently, $(A + \mathbf{I})^{n-1} > 0$.

PROPERTY 4.1.9. *If a matrix A is primitive and $A^\gamma > 0$, then $A^l > 0$ for all $l \geq \gamma$. A power of a primitive matrix is primitive.*

The proof is obvious.

A digraph G with the adjacency matrix which is the support of a primitive matrix is called primitive, and the exponent of the adjacency matrix is called the exponent of the graph.

PROPERTY 4.1.10. *If G is a primitive digraph with n vertices, the adjacency matrix A, and the exponent γ, then for any i, j, $1 \leq i, j \leq n$, and any integer $l \geq \gamma$ there exists a directed path of length l from i to j.*

By the primitivity of G the matrix A^γ does not contain zero rows or columns; therefore $A^\gamma A > 0$, and $A^{\gamma+k} > 0$ for all $k = 0, 1, \dots$

PROPERTY 4.1.11. *A strongly connected digraph G with $n \geq 2$ vertices is primitive if and only if the greatest common divisor of the lengths of all simple circuits is equal to one.*

Let the lengths of all simple circuits of G be $l_1, \dots, l_k$, and let $(l_1, \dots, l_k) = 1$. Consider a closed directed path going through all vertices of G. Let the length of this path be l_0. Consider also a directed path from a vertex i to a vertex j of length l. Let us go from i to j by a compound path. First let us pass along the directed path of length l_0 and return to i. Reaching the corresponding vertices, we pass also through the simple circuits. Finally we go through the path of length l from i to j. The total length of this compound path is equal to

$$L = l_0 + \sum_{\nu=1}^{k} x_{l\nu} l_\nu + l,$$

where the $x_{i\nu}$ are nonnegative integers, $1 \le \nu \le k$.

Since $(l_1, \dots, l_k) = 1$, there exist a set of positive integers $\{\alpha_1, \dots, \alpha_k\}$ and a partition $N_k = X \cup Y$, $X \cap Y = \emptyset$, such that

$$\sum_{\nu \in X} \alpha_\nu l_\nu - \sum_{\nu \in Y} \alpha_\nu l_\nu = 1.$$

For each pair of integers l, ν, $1 \le l \le n$, $1 \le \nu \le k$, and sufficiently large m, we set

$$x_{l\nu} = \begin{cases} (m + 1 - l)\alpha_\nu, & \nu \in X, \\ (l - 1)\alpha_\nu, & \nu \in Y. \end{cases}$$

It is easy to find that

$$L = l_0 + m \sum_{\nu \in X} \alpha_\nu l_\nu + l;$$

that is, for such choice of the numbers of passages of the circuits, the length of the compound directed path does not depend on i and j. Thus, if A is the adjacent matrix of the digraph G, then $A^L > 0$ and G is a primitive graph.

Now let G be a primitive graph with exponent γ. Consider a circuit of length t in G such that, starting with a vertex i and going along this circuit, we pass through each simple circuit at least once and $t \ge \gamma$. If $l_1, \dots, l_k$ are the lengths of the simple circuits, then

$$t = \sum_{\nu=1}^{k} x_\nu l_\nu,$$

where $x_\nu > 0$, $\nu = 1, \dots, k$. By Property 1.3.10 there exists a directed path from i to i of length $t + 1$. By the same property there exist similar paths of lengths $2t$ and $2t + 1$. Therefore we can write

$$2t = \sum_{\nu=1}^{k} y_\nu l_\nu, \qquad 2t + 1 = \sum_{\nu=1}^{k} z_\nu l_\nu,$$

where y_ν, z_ν are positive integers. Two consecutive integers can be represented as linear combinations with integer coefficients of a set of integers $\{l_1, \dots, l_k\}$ if and only if $(l_1, \dots, l_k) = 1$.

PROPERTY 4.1.12. *If A is the adjacency matrix of a primitive digraph G with n vertices and l is the length of the shortest simple circuit in G, then the exponent of G coinciding with the exponent of A satisfies the inequality*

$$\gamma(A) \le n + l(n - 2).$$

Consider the digraph G_l with the adjacency matrix ${}^s(A^l)$. By Property 4.1.9 the digraph G_l is primitive and has at least l vertices with loops. Let v_k be one of the vertices having loops. In G_l there exist directed paths of length $n-1$ from v_k to any vertex v_j. Therefore in G_l there exist directed paths of lengths $l(n-1)$ from v_k to any vertex v_j. There exists a path of length $n-l$ from an arbitrary vertex v_i to some vertex on the circuit of the shortest length. Thus, there exists a directed path from any vertex v_i to any vertex j of length $l(n-1)+(n-l)$,; that is, $A^{l(n-1)+(n-l)} > 0$ and, consequently,

$$\gamma(A) \leq l(n-1)+(n-l).$$

PROPERTY 4.1.13. *The exponent of a primitive digraph G with n vertices and the adjacency matrix A satisfies the inequality*

$$\gamma(A) \leq (n-1)^2+1.$$

The length l of the shortest circuit satisfies the inequality $l \leq n-1$. Therefore the estimate in Property 4.1.13 follows from Property 4.1.12.

4.1.2. A partial order for Boolean matrices. Let $A=(a_{ij})$, $B=(b_{ij})$, $i,j=1,\dots,n$, be Boolean $(0,1)$-matrices. Let us agree that $A \leq B$ if $a_{ij} \leq b_{ij}$ for all $1 \leq i,j \leq n$. If in this case there exist i and j such that $a_{ij} < b_{ij}$, then $A < B$. The binary relation $\leq$ is reflexive, antisymmetric, transitive, and so it is a partial order.

PROPERTY 4.1.14. *If $A \leq B$, then for any Boolean $(0,1)$-matrix C the relation $AC \leq BC$ is valid.*

The proof is obvious.

PROPERTY 4.1.15. *If $\mathbf{I}$ is the identity matrix and B is an arbitrary $(0,1)$-matrix of order n, then the Boolean $(0,1)$-matrix $A=(\mathbf{I}+B)$ is such that*

$$\mathbf{I} \leq A \leq A^2 \leq \cdots \leq A^{n-1} = A^n = A^{n+1} = \dots.$$

The inequalities $\mathbf{I} \leq A \leq A^2 \leq \cdots \leq A^{n-1}$ follow from Property 4.1.14. In order to prove the equalities, consider the graph G with the adjacent matrix A. If $A^r = (a_{ij}^{(r)})$, then

$$a_{ij}^{(r)} = \sum_{i_1=1}^{n} \cdots \sum_{i_{r-1}=1}^{n} a_{ii_1} a_{i_1 i_2} \dots a_{i_{r-1},j},$$

where $ii_1, i_1i_2, \dots, i_{r-1}j$ is a path from i to j in G. We set $r=n$. Then the path $ii_1, i_1i_2, \dots, i_{n-1}j$ contains $n+1$ vertices; that is, there exist k and l such that $i_k = i_l$ and the path contains a closed circuit. By removing the terms corresponding to the circuit we obtain

$$a_{ii_1} \dots a_{i_{k-1}i_k} a_{i_k i_{k+1}} \dots a_{i_{l-1}i_l} a_{i_l i_{l+1}} \dots a_{i_{n-1}j} \leq a_{ii_1} \dots a_{ik-1i_k} a_{i_k i_{k+1}} \dots a_{i_{n-1}j}.$$

If the right-hand side contains less than $n-1$ multipliers, then we complement the product by the multiplier $a_{i_k i_k} = a_{i_k i_l} = a_{i_l i_l} = 1$ which corresponds to adding the necessary number of unit circuits to the path from i to j. Thus, for each i and j we have $a_{ij}^{(n)} \leq a_{ij}^{(n-1)}$; that is, $A^n \leq A^{n-1}$. Hence, $A^{n-1} = A^n$. The remaining equalities are proved similarly.

4.1.3. Matrix generating functions. Let G be a directed graph with the set of vertices $V = \{1, 2, \dots, n\}$ and the set of arcs E (multiple arcs with the same directions are allowed). Let $w\colon E \to \mathbf{R}$, where $\mathbf{R}$ is the set of real numbers, be a weight function. If $e_1, e_2, \dots, e_l \in E$ form a path, then the weight of the path $\Gamma = e_1 e_2 \dots e_l$ is defined as $w(\Gamma) = w(e_1)w(e_2)\dots w(e_l)$. We set

$$A_{ij}(l) = \sum_{\Gamma} w(\Gamma),$$

where the summation is over all paths Γ in G from i to j, $i, j \in V$, of length l. We also set $A_{ij}(0) = \delta_{ij}$ and $A_{ij}(1) = A_{ij}$,

$$A_{ij} = \sum_{e} w(e),$$

where the summation is over all arcs e emanating from i and entering j. The matrix $A = (A_{ij})$ is called the adjacent matrix of G with respect to the weight function w.

PROPERTY 4.1.16. *For each nonnegative integer l,*

$$A^l = (A_{ij}(l)), \qquad i, j = 1, \dots, n.$$

The property follows directly from the equality representing the element in a position (i, j) of the matrix A^l in terms of elements of the matrix A:

$$(A^l)_{ij} = \sum A_{ii_1} A_{i_1 i_2} \dots A_{i_{l-1} j},$$

where the summation is over all sequences $i_1 \dots, i_{l-1}$, $1 \le i_1, \dots, i_{l-1} \le n$.

PROPERTY 4.1.17. *For the generating function*

$$F_{ij}(\lambda; G) = \sum_{l=0}^{\infty} A_{ij}(l)\lambda^l,$$

the equality

$$F_{ij}(\lambda; G) = \frac{(-1)^{i+j} \det((\mathbf{I} - \lambda A)(j, i))}{\det(\mathbf{I} - \lambda A)}$$

holds, where the A_{ij} are determined by a weight function w taking real values, and for a matrix B the matrix $B(j, i)$ is obtained by removing row j and column i.

Indeed, $F_{ij}(\lambda; G)$ is the element in the position (i, j) of the matrix $(\mathbf{I} - \lambda A)^{-1}$, and by Property 4.1.16

$$(\mathbf{I} - \lambda A)^{-1} = (F_{ij}(\lambda; G) = \sum_{l=0}^{\infty} A^l \lambda^l.$$

This matrix series converges for all λ such that $|\lambda| < 1/\lambda_0$, where λ_0 is the maximal modulus of the eigenvalues of the matrix A. For such λ, the matrix $\mathbf{I} - \lambda A$ is nonsingular and, consequently, the matrix $(\mathbf{I} - \lambda A)^{-1}$ exists and the element in the position (i, j) of this matrix is equal to the cofactor of the element (j, i) divided by the determinant of the matrix.

PROPERTY 4.1.18. *The function $F_{ij}(\lambda; G)$ is a rational function of λ whose degree is less than the multiplicity of the zero eigenvalue of the matrix A.*

Consider the characteristic polynomial of the matrix A, and write it in the form

$$\det(A - \lambda\mathbf{I}) = (-1)^n \lambda^{s_0}(\alpha_{n-s_0} + \alpha_{n-s_0-1}\lambda + \cdots + \alpha_1 \lambda^{n-s_0-1} + \lambda^{n-s_0}).$$

Then

$$\det(\mathbf{I} - \lambda A) = 1 + \alpha_1\lambda + \cdots + \alpha_{n-s_0}\lambda^{n-s_0}.$$

For the degrees, the relations

$$\deg\det(\mathbf{I} - \lambda A) = n - s_0, \qquad \deg\det((\mathbf{I} - \lambda A)(j,i)) \leq n-1$$

hold. Consequently,

$$\deg F_{ij}(\lambda; G) \leq n - 1 - (n - s_0) < s_0.$$

PROPERTY 4.1.19. *Let $B(\lambda)$ be the adjoint matrix for the matrix $\mathbf{I} - \lambda A$ satisfying the equality*

$$(\mathbf{I} - \lambda A)B(\lambda) = (\det(\mathbf{I} - \lambda A))\mathbf{I}$$

and

$$B(\lambda) = \sum_{k=0}^{n} B_k \lambda^k.$$

Then the matrices $B_0, B_1, \dots, B_n$ satisfy the relations

$$B_0 = \mathbf{I}, \quad AB_n = 0, \quad B_k = AB_{k-1} + \alpha_k \mathbf{I}, \quad k = 1, 2, \dots, n,$$

where

$$\det(\mathbf{I} - \lambda A) = 1 + \sum_{k=1}^{n} \alpha_k \lambda^k.$$

Indeed, the relations can be obtained by comparing the coefficients of λ^k in the left- and right-hand sides of the equality

$$(\mathbf{I} - \lambda A)\sum_{k=0}^{n} B_k \lambda^k = (1 + \sum_{k=1}^{n} \alpha_k \lambda^k)\mathbf{I}.$$

Note that Property 4.1.19 allows us to find the generating functions $F_{ij}(\lambda; G)$ by the use of the characteristic polynomial of the matrix A and the matrix recurrence relations without finding the inverse matrix $(\mathbf{I} - \lambda A)^{-1}$.

4.1.4. Smirnov sequences. A sequence $a_1, \dots, a_l$ with elements in the set $N_n = \{1, \dots, n\}$ is called a Smirnov sequence if $a_i \neq a_{i+1}$ for each $i = 1, \dots, l-1$. It is easy to see that the number $S(n, l)$ of Smirnov sequences of length l with elements in N_n equals $n(n-l)^{l-1}$. We prove this fact by the method of generating functions.

Consider a graph G with n vertices and the adjacent matrix $A = (1 - \delta_{ij})$, $i, j = 1, \dots, n$. In the graph G, we assign to an arc e the weight $w(e) = 1$. By Property 4.1.17

$$S(n, l) = \sum_{i=1}^{n}\sum_{j=1}^{n} A_{ij}(l-1),$$

where

$$F_{ij}(\lambda; G) = \sum_{l=0}^{\infty} A_{ij}(l)\lambda^l = \frac{(-1)^{i+j}Q_{ij}(\lambda)}{Q(\lambda)},$$

$$Q(\lambda) = \det(\mathbf{I} - \lambda A),$$

$$Q_{ij}(\lambda) = \det((\mathbf{I} - \lambda A)(j, i)).$$

Thus,

$$F_n(\lambda) = \sum_{l=0}^{\infty} S(n, l+1)\lambda^L = \frac{1}{Q(\lambda)} \sum_{i=1}^{n}\sum_{j=1}^{n} Q_{ij}(\lambda)(-1)^{i+j}.$$

We immediately obtain

$$Q(\lambda) = \det((1+\lambda)\delta_{ij} - \lambda) = (1-(n-1)\lambda)(1+\lambda)^{n-1}.$$

Observe that

$$\sum_{i=1}^{n}(-1)^{i+j}Q_{ij}(\lambda) = \det\begin{pmatrix} 1 & -\lambda & \dots & 1 & \dots & -\lambda \\ -\lambda & 1 & \dots & 1 & \dots & -\lambda \\ -\lambda & -\lambda & \dots & 1 & \dots & -\lambda \\ & & \dots & & \dots & \\ -\lambda & -\lambda & \dots & 1 & \dots & 1 \end{pmatrix}.$$

Multiplying in this matrix the jth column (which consists of ones) by λ and adding the result to each remaining column, we obtain a matrix which can be made triangular by a permutation of rows and columns. The principal diagonal of this matrix contains one element equal to 1 and $n-1$ elements equal to $1+\lambda$. Thus,

$$\sum_{i=1}^{n}(-1)^{i+j}Q_{ij}(\lambda) = (1+\lambda)^{n-1}.$$

Hence

$$F_n(\lambda) = \frac{n}{1-(n-1)\lambda},$$

and we obtain the final formula for the number of Smirnov sequences of length l:

$$S(n,l) = n(n-1)^{l-1}, \qquad n \geq 2, \qquad l = 2, 3, \dots.$$

This formula can be derived by the inclusion–exclusion principle. We say that a sequence $a_1, \dots, a_l$ of length l has the property A_i if $a_i = a_{i+1}$, $i = 1, 2, \dots, l-1$. The number of sequences having the properties $A_{i_1}, \dots, A_{i_k}$, $1 \leq i_1 < \dots < i_k \leq l-1$, is equal to n^{l-k}. Therefore the number of sequences having none of the properties $A_1, \dots, A_{l-1}$ equals

$$S(n,l) = \sum_{k=0}^{l-1}(-1)^k \binom{l-1}{k} n^{l-k} = n(n-1)^{l-1}.$$

Let any of the n^l sequences $a_1, \dots, a_l$ be chosen randomly with equal probabilities, and let $\xi(n,l)$ be the number of i's such that $a_i = a_{i+1}$, $1 \leq i \leq l-1$. The probability distribution of the random variable $\xi(n,l)$ can be obtained by the inclusion–exclusion method:

$$\mathbf{P}(\xi(n,l) = r) = \sum_{k=r}^{l-1}(-1)^{k-r}\binom{k}{r}\binom{l-1}{k}\frac{1}{n^k}, \qquad r = 0, 1, \dots, l-1.$$

Using appropriate transformations, these formulas can be reduced to the form

$$\mathbf{P}(\xi(n,l) = r) = \binom{l-1}{r}\frac{1}{n^r}\left(1 - \frac{1}{n}\right)^{l-r-1}, \qquad r = 0, 1, \dots, l-1;$$

that is, the random variable $\xi(n,l)$ has the binomial distribution with $l-1$ trials and the probability of success $1/n$.

4.1.5. Closed paths. Consider the values

$$C(l) = \sum_{\Gamma} w(\Gamma),$$

where the sum is taken over closed paths of length l in the graph G. It is clear that

$$C(l) = \sum_{i=1}^{n} A_{ii}(l) = \operatorname{tr} A^l,$$

where $\operatorname{tr} B$ is the trace of the matrix B. If $\varepsilon_1, \dots, \varepsilon_s$ are nonzero eigenvalues of the matrix A, then it is well known that

$$\operatorname{tr} A^l = \varepsilon_1^l + \dots + \varepsilon_s^l.$$

Therefore

$$\sum_{l=1}^{\infty} C(l)\lambda^l = \sum_{i=1}^{s} \frac{\varepsilon_i \lambda}{1 - \varepsilon_i \lambda}.$$

Reducing the fractions in the sum to the common denominator, we obtain

$$\sum_{l=1}^{\infty} C(l)\lambda^l = \prod_{i=1}^{s} (1 - \varepsilon_i \lambda)^{-1} \sum_{k=1}^{s} \varepsilon_k \lambda \prod_{j \neq k} (1 - \varepsilon_j \lambda).$$

Hence

$$\sum_{l=1}^{\infty} C(l)\lambda^l = -\frac{\lambda Q'(\lambda)}{Q(\lambda)},$$

where $Q(\lambda) = \det(\mathbf{I} - \lambda A)$ and $Q'(\lambda)$ is the derivative of $Q(\lambda)$ with respect to λ.

4.1.6. Cyclic Smirnov sequences. A sequence $a_1, \dots, a_l$ of elements of N_n is called a cyclic Smirnov sequence if $a_i \neq a_{i+1}$ for $i = 1, \dots, l-1$ and $a_l \neq a_1$. The number of cyclic Smirnov sequences is denoted by $\tilde{S}(n, l)$. Consider the graph G and the matrix A of the same form as for ordinary Smirnov sequences. Then

$$Q(\lambda) = \det(\mathbf{I} - \lambda A) = ((1 - (n-1)\lambda)(1 + \lambda)^{n-1}.$$

According to the results obtained above,

$$\sum_{l=1}^{\infty} \tilde{S}(n, l)\lambda^l = -\frac{\lambda Q'(\lambda)}{Q(\lambda)} = \frac{n(n-1)\lambda^2}{(1+\lambda)(1-(n-1)\lambda)}.$$

Hence

$$\tilde{S}(n, l) = (n-1)^l + (-1)^l (n-1), \quad n \geq 2, \quad l = 1, 2, \dots.$$

In particular, this formula implies the following obvious fact: for $n = 2$ there are no cyclic Smirnov sequences of odd length and there are only two such sequences of each even length.

4.1.7. Free monoids. Let A^* denote the set of all words over a finite alphabet A. For arbitrary words $u = a_1 \dots a_m$ and $v = b_1 \dots b_n$ belonging to A^*, we define the operation of concatenation of u and v, whose result is the word

$$uv = a_1 \dots a_m b_1 \dots b_m.$$

An empty word e satisfies the relations $ue = eu = u$ for all $u \in A^*$. The set A^* with the concatenation operation is called the free monoid over the alphabet A.

We define the length $l(u)$ of a word $u = a_1 \dots a_n$ to be equal to n, and we set

$$C_n = \{u \in C \colon l(u) = n\}$$

for any $C \subseteq A^*$.

Let B^* be the submonoid of A^* generated by a set of words $B \subseteq A^*$. We say that the submonoid B^* is freely generated by the set B if each word $u \in B^*$ can be uniquely represented in the form

$$u = u_1 \dots u_n, \quad u_i \in B, \quad 1 \le i \le n,$$

for some n.

For a commutative ring R, we consider a weight function $w\colon A \to R$ and set

$$w(u) = w(a_1) \dots w(a_n), \quad u = a_1 \dots a_n, \quad a_i \in A, \quad 1 \le i \le n.$$

For an arbitrary subset $C \subseteq A^*$, we consider the generating function as a formal power series

$$C(\lambda) = \sum_{u \in C} w(u) \lambda^{l(u)} = \sum_{n=0}^{\infty} \left(\sum_{u \in C_n} w(u) \right) \lambda^n.$$

It is obvious that if a subset $B \subseteq A^*$ freely generates the submonoid B^*, then

$$\sum_{u \in B_n^*} w(u) = \sum_{k=0}^{|B|} \sum_{i_1 + \dots + i_k = n} \sum_{u \in B_{i_1}} w(u) \cdots \sum_{u \in B_{i_k}} w(u).$$

Thus, for the generating functions

$$B(\lambda) = \sum_{n=0}^{\infty} \left(\sum_{u \in B_n} w(u) \right) \lambda^n,$$

$$B^*(\lambda) = \sum_{n=0}^{\infty} \left(\sum_{u \in B_n^*} w(u) \right) \lambda^n$$

the equality

$$B^*(\lambda) = (1 - B(\lambda))^{-1}$$

holds.

4.1.8. Permutations with restricted positions. Consider the set of permutations s of degree n such that

$$s(1) = 1, 2, \quad s(i) = i, i-1, \quad 2 \le i \le n-1, \quad s(n) = n-1, n.$$

The permutation graph $\Gamma(s)$ of any permutation s of this set consists of subgraphs of two types, circuits with one vertex and circuits with two vertices which determined $\Gamma(s)$ uniquely. If $A = \{a, b, c\}$ and $B = \{a, bc\}$, then a word of length n in B_n^* uniquely determines the permutation graph $\Gamma(s)$ of some permutation s of degree n. Therefore if $f(n)$ is the number of permutations of the kind under consideration, then, setting $w(u) = 1$, we obtain

$$\sum_{n=0}^{\infty} \lambda^n = B^*(\lambda) = (1 - B(\lambda))^{-1},$$

where $B(\lambda) = \lambda + \lambda^2$. Thus,

$$f(n) = \frac{1}{\sqrt{5}} \left(\left(\frac{1+\sqrt{5}}{2} \right)^{n+1} - \left(\frac{1-\sqrt{5}}{2} \right)^{n+1} \right), \qquad n = 0, 1, \dots.$$

4.1.9. Monoids with unique cyclic decompositions. Let B^* be freely generated by a set $B \subseteq A^*$, where A^* is the free monoid over an alphabet A. We say that B^* is a monoid with unique cyclic decomposition if for any word $u = u_1 \dots u_n \in B^*$, $u_i \in A$, $i = 1, \dots, n$, located on a circle in the clockwise direction without fixing the place of the first letter u_1, there exists only one way to divide the word into parts such that, reading them in the clockwise direction, we obtain words of the set B.

For the monoid B^* with unique cyclic decomposition, the word

$$u_i u_{i+1} \dots u_n u_1 \dots u_{i-1} \in A^*$$

is called A^*-conjugate to the word $u_1 \dots u_n \in B^*$, $u_i \in A$, $i = 1, \dots, n$. Consider $g(u) = \sum w(u)$, where the summation is over all distinct words u of length n which are A^*-conjugate to the words of B^*. We introduce the generating function

$$\tilde{B}(\lambda) = \sum_{n=1}^{\infty} g(n)\lambda^n.$$

Let us find the expression for $\tilde{B}(\lambda)$ in terms of the generating function

$$B^*(\lambda) = \sum_{n=0}^{\infty} f(n)\lambda^n,$$

where $f(n)$ is the total weight of words of length n in B^*.

Let $g_v(n)$ be the total weight of distinct words $u_i \dots u_n u_1 \dots, u_{i-1}$ which are A^*-conjugate to the words of B_n^* and satisfy the condition

$$u_j u_{j+1} \dots u_k = v, \quad j \le i, \quad k \ge i,$$

where v is a fixed word in B of length at least two. By the uniqueness of the cyclic decomposition of any v the indices j and k are determined uniquely. If $l(v) = m$, then

$$g_v(n) = mf(n-m), \qquad 0 \le m \le n.$$

If $b(m)$ is the number of words of length m in B, then

$$g(n) = \sum_{v \in B} g_v(n) = \sum_{m=0}^{n} mb(m)f(n-m).$$

If

$$B(\lambda) = \sum_{m=0}^{\infty} b(m)\lambda^m,$$

then the last equality implies that

$$\tilde{B}(\lambda) = B^*(\lambda)\lambda B'(\lambda),$$

where $B'(\lambda)$ is the derivative of $B(\lambda)$ with respect to λ. Since $B^*(\lambda) = (1-B(\lambda))^{-1}$, we obtain the final formula

$$\tilde{B}(\lambda) = \frac{\lambda B'(\lambda)}{1 - B(\lambda)}.$$

4.1.10. Cyclic permutations with restricted positions. Let $g(n)$ be the number of permutations $a_1a_2\dots a_n$ of degree n such that

$$a_i - i \equiv 0, \pm 1 \pmod{n}, \qquad i = 1, \dots, n.$$

Let B^* be the monoid defined in Section 4.1.8, and let $u = u_1 \dots u_n \in B^*_n$. It is clear that the words $u, u_2 \dots u_n u_1$, and $u_n u_1 \dots u_{n-1}$ give permutations of the form considered here. In addition to these permutations, in the cases $n \geq 3$ there are only two permutations, $2\ 3\ \dots\ n\ 1$ and $n\ 1\ 2\ \dots\ n-1$. For the monoid B^*, the condition of uniqueness of the cyclic decomposition is fulfilled, and therefore we can apply to B^* the argiments of Section 4.1.9.

Since $B(\lambda) = \lambda + \lambda^2$, we see that

$$\tilde{B}(\lambda) = \frac{\lambda + 2\lambda^2}{1 - \lambda - \lambda^2} = \sum_{n=1}^{\infty} (\alpha^n + \beta^n)\lambda^n,$$

where $\alpha = (1+\sqrt{5})/2$, $\beta = (1-\sqrt{5})/2$. Note that for $n = 1, 2$ the previous arguments are inapplicable, and the coefficients of λ and λ^2 do not necessarily coincide with the true values of $g(1)$ and $g(2)$, which are equal to 1 and 2, respectively. Taking into account that $\alpha + \beta = 1$ and $\alpha^2 + \beta^2 = 3$, we obtain

$$\sum_{n=1}^{\infty} g(n)\lambda^n = \sum_{n=1}^{\infty} (\alpha^n + \beta^n)\lambda^n - \lambda^2 + \sum_{n=3}^{\infty} 2\lambda^n.$$

Hence $g(1) = 1$, $g(2) = 2$, and

$$g(n) = \alpha^n + \beta^n + 2, \qquad n \geq 3,$$

where $\alpha = (1+\sqrt{5})/2$, $\beta = (1-\sqrt{5})/2$.

4.1.11. k-discordant permutations. Let $f_k(n)$ be the number of permutations $a_1 \dots a_n$ of elements of N_n such that

$$a_i - i \not\equiv 0, 1, \dots, k-1 \pmod{n}, \qquad i = 1, \dots, n.$$

Such permutations are called k-discordant. Using the inclusion–exclusion principle, it is not difficult to find that

$$f_k(n) = \sum_{i=0}^{n} (-1)^i r_i(n)(n-1)!,$$

where $r_i(n)$ is the number of i-subsets of the set

$$B_n = \{(r,s)\colon s - r \equiv 0, 1, \dots, k-1 \pmod{n},\ (r,s) \in N_n \times N_n\}$$

such that no two elements of this set have a common coordinate. If the set of integer points of B_n is interpreted as cells of a generalized chessboard, then $r_i(n)$ is the number of ways to place rooks on this chessboard so that they do not attack each other.

The generating function

$$R_n(x) = \sum_{i=1}^{n} r_i(n)x^i$$

is called the rook polynomial.

We apply the approach used above to find the generating function

$$F_k(x,\lambda) = \sum_{n=1}^{\infty} R_n(x)\lambda^n.$$

Consider the graph D_k whose vertices are all possible ways to place nonattacking rooks in the first $k-1$ horizontal lines of the board B_n. Two such placements v_1 and v_2 are connected by an edge if the last $k-2$ horizontals of the placement v_1 coincide with the first $k-2$ horizontals of v_2, and if we superpose v_1 and v_2 in the configuration consisting of k horizontals, then rooks do not attack each other as well. We assume that the weight $w(v_1, v_2)$ of the edge (v_1, v_2) is equal to $x^{\mu(v_2)}$, where $\mu(v_2)$ is the number of rooks on the last horizontal of the configuration v_2. A placement of nonattacking rooks on the chessboard B_n for $n \geq k$ corresponds to a closed path Γ of length n in the graph D_k. Consequently, if A_k is the adjacent matrix of the graph D_k with weight function w, then

$$R_n(x) = \operatorname{tr} A_k^n, \qquad n \geq k.$$

Thus, if $Q_k(\lambda) = \det(\mathbf{I} - \lambda A_k)$, then according to Section 4.1.5

$$\sum_{n=1}^{\infty} R_n(x)\lambda^n = -\frac{\lambda Q_k'(\lambda)}{Q_k(\lambda)}.$$

In the case $k = 2$, each row of the board has two cells and the ith row is shifted cyclicly to the right by one cell in comparison with the $(i-1)$th row. The vertices of the graph are of the form

$$v_1 = \boxed{\;\;|\;\;}, \quad v_2 = \boxed{\;\;|\, x}, \quad v_3 = \boxed{x\,|\;\;},$$

where the symbol x corresponds to presence of a rook.

Taking into account the step-form of contents of the rows, we find that the matrix A_k of the graph D_k is of the form

$$A_k = \begin{pmatrix} 1 & x & x \\ 1 & x & 0 \\ 1 & x & x \end{pmatrix}$$

Consequently, for $k = 2$

$$Q_k(\lambda) = \det(\mathbf{I} - \lambda A_k) = 1 - \lambda(1+2x) + \lambda^2 x^2.$$

Hence

$$\sum_{n=1}^{\infty} R_n(x)\lambda^n = \frac{(1+2x)\lambda - 2x^2\lambda^2}{1-(1+2x)\lambda + x^2\lambda^2}.$$

Thus,

$$\sum_{n=1}^{\infty} R_n(x)\lambda^n = \sum_{n=1}^{\infty} (\alpha^n + \beta^n)\lambda^n,$$

where $\alpha = (1+2x+\sqrt{1+4x})/2$, $\beta = (1+2x-\sqrt{1+4x})/2$. Hence, for $R_n(X)$ we obtain the expression

$$R_n(x) = \frac{1}{2^{n-1}} \sum_{j=0}^{\lfloor n/2 \rfloor} \binom{n}{2j} (1+2x)^{n-2j}(1+4x)^j.$$

Note that the above arguments are inapplicable for $n = 1$, and therefore the equality $R_1(x) = 1 + 2x$ does not coincide with the true expression.

Problems.

1. Derive the formula for the number of cyclic Smirnov sequences
$$\tilde{S}(n,l) = (n-1)^l + (-1)^l(n-1)$$
using the inclusion–exclusion principle.
2. Let C be some generalized chessboard and
$$R(x,C) = \sum_i r_i(C)x^i$$
the corresponding rook polynomial. Prove the equality
$$R(x,C) = xR(x,C_1) + R(x,C_2),$$
where C_1 is obtained from C by removing all cells located in the same row and the same column as some fixed cell, and C_2 is obtained from C by removing only this fixed cell.
3. For the generalized rectangular chessboard of size $m \times n$, show that the rook polynomial $R_{mn}(x)$ satisfies the equalities
$$R_{mn}(x) = R_{m-1,n}(x) + xnR_{m-1,n-1}(x),$$
$$R_{mn}(x) = R_{m,n-1}(x) + xmR_{m-1,n-1}(x).$$
4. Using the first equality of the previous problem and the initial condition $R_{1n}(x) = 1 + nx$, prove by induction on m that
$$R_{mn}(x) = \sum_{k=0}^{m} \binom{m}{k} (n)_k x^k.$$
5. Using the formula of the previous problem, show that $R_{mn}(x)$ is expressed in terms of the Laguerre polynomials
$$L_n^{\alpha}(x) = \sum_{k=0}^{n} \binom{n+\alpha}{n-k} \frac{(-1)^k}{k!}$$
by the formula
$$R_{n,n+\alpha}(x) = n!x^n L_n^{\alpha}(-x^{-1}).$$

4.2. Asymptotics of the number of primitive (0, 1)-matrices

4.2.1. Elements of the square of a random (0, 1)-matrix. Consider a family of (0, 1)-matrices A of order n such that the elements of the matrix A^2 are close to $n/4$. Below we show that the ratio of the number of (0, 1)-matrices possessing this property to the total number of (0, 1)-matrices of order n tends to one as $n \to \infty$. It is clear that such matrices are primitive and have exponent 2.

Let the uniform distribution be defined on the set of all (0, 1)-matrices which assigns the identical probability 2^{-n^2} to each matrix, and let A be a random matrix with this distribution. We denote by $M_n(\lambda)$ the expectation of the number of elements b_{ij} of the matrix $B = A^2$ such that $|b_{ij} - n/4| \geq \lambda$.

THEOREM 4.2.1. *Let* $\lambda = (\frac{3}{4}n(\ln n + \omega(n)))^{1/2}$ *and*
$$\ln n + \omega(n) \to \infty, \qquad \omega(n) = o(n^{1/3})$$
as $n \to \infty$. *Then*
$$M_n(\lambda) = \frac{1}{\sqrt{2\pi(\ln n + \omega(n))}} e^{-2\omega(n)}(1 + o(1))$$
as $n \to \infty$.

In the proof on this theorem, two auxiliary assertions are needed.

We introduce the notation

$$B_n(r) = \binom{n}{r}\left(\frac{1}{4}\right)^r\left(\frac{3}{4}\right)^{n-r}, \qquad r = 0, 1, \ldots, n, \quad n = 1, 2, \ldots.$$

LEMMA 4.2.2. *For all i, j, $1 \le i, j \le n$, and all $r = 0, 1, \ldots, n$,*

$$\frac{2}{9}B_n(r) \le \mathbf{P}(b_{ij} = r) \le 2B_n(r),$$

and if $n \to \infty$ and $\lambda n^{-1} \to 0$, then

$$\mathbf{P}(b_{ij} = r) = B_n(r)(1 + O(\lambda/n))$$

uniformly with respect to the integers r such that $|r - n/4| \le \lambda$.

PROOF. It is not difficult to check that the number of $(0,1)$-matrices A of order n such that $b_{ii} = r$ is equal to

$$\left(\binom{n-1}{r}3^{n-r-1} + \binom{n-1}{r-1}3^{n-r}\right)2^{n^2-2n+1}$$
$$= 2\binom{n}{r}\left(\frac{1}{n}\right)^r\left(\frac{3}{4}\right)^{n-r}\left(\frac{1}{3} + \frac{2r}{3n}\right)2^{n^2}.$$

For $i \ne j$, the number of $(0,1)$-matrices A of order n such that $b_{ij} = r$, $r \ge 2$, is equal to

$$\left(\binom{n-2}{r-2}3^{n-r} + 2\binom{n-2}{r-1}3^{n-r-1} + 5\binom{n-2}{r}3^{n-r-2}\right)2^{n^2-2n+1}$$
$$= \binom{n}{r}\left(\frac{1}{4}\right)^r\left(\frac{3}{4}\right)^{n-r}\frac{10}{9}\left(1 - \frac{4r}{5n}\left(1 - \frac{2(r-1)}{n-1}\right)\right)2^{n^2}.$$

The cases $r = 0$ and $r = 1$ are considered similarly. Both assertions of the lemma follows from the relations obtained above. □

LEMMA 4.2.3. *If $n \to \infty$, $\lambda^3 n^{-2} \to 0$, and $\lambda^2 n^{-1} \to \infty$, then*

$$\sum_{r\colon |r-n/4|>\lambda} B_n(r) = \left(\frac{3}{8\pi}\right)^{1/2}\frac{\sqrt{n}}{\lambda}\exp\left(-\frac{8\lambda^2}{3n}\right)(1 + o(1)).$$

PROOF. Under the conditions of the lemma it is possible to use the well-known estimate of the probabilities of the binomial distribution in the domain where $\lambda n^{-1/2} = o(n^{1/6})$ (see Section 7.6 in [**39**]). According to this estimate

$$\sum_{r\colon |r-n/4|>\lambda} B_n(r) = 2\left(1 - \Phi\left(\frac{4\lambda}{\sqrt{3n}}\right)\right)(1 + o(1)),$$

where $\Phi(x)$ is the distribution function of the standard Gaussian distribution:

$$\Phi(x) = \frac{1}{\sqrt{2\pi}}\int_{-\infty}^{x} e^{-t^2/2}dt.$$

It is not difficult to verify that

$$1 - \Phi(x) = \frac{1}{x\sqrt{2\pi}}e^{-x^2/2}(1 + o(1))$$

as $x \to \infty$. Substituting this equality into the previous relation, we obtain the assertion of the lemma. □

PROOF OF THEOREM 4.2.1. It is obvious that

$$M_n(\lambda) = \sum_{i,j=1}^{n} \sum_{|r-n/4|>\lambda} \mathbf{P}(b_{ij} = r)$$
$$= \sum_{i,j=1}^{n} \sum_{\lambda<|r-n/4|\leq 2\lambda} \mathbf{P}(b_{ij} = r) + \sum_{i,j=1}^{n} \sum_{|r-n/4|>2\lambda} \mathbf{P}(b_{ij} = r).$$

By Lemma 4.2.2 the sums in the right-hand side are estimated, respectively, as

$$n^2 \sum_{\lambda<|r-n/4|\leq 2\lambda} B_n(r)(1 + O(\lambda/n))$$

and

$$O\left(n^2 \sum_{|r-n/4|>2\lambda} B_n(r)\right),$$

so that we obtain the equality

$$M_n(\lambda) = n^2 \sum_{|r-n/4|>\lambda} B_n(r)(1 + O(\lambda/n)) + O\left(n^2 \sum_{|r-n/4|>2\lambda} B_n(r)\right).$$

It remains to use Lemma 4.2.3 and substitute the value of the parameter λ given in the statement of the theorem. □

COROLLARY 4.2.4. *Let A be a random equiprobable $(0, 1)$-matrix of order n. Then for the elements of b_{ij} of the matrix $B = A^2$ the equality*

$$\lim_{n\to\infty} \mathbf{P}\{|4b_{ij}/n - 1| \leq \varepsilon, i, j = 1, \dots, n\} = 1$$

is true for any fixed $\varepsilon > 0$.

COROLLARY 4.2.5. *A random equiprobable $(0, 1)$-matrix of order n is primitive and has exponent 2 with probability tending to 1 as $n \to \infty$.*

We leave to the reader the deduction of these assertions from Theorem 4.2.1 as exercises.

4.2.2. The number of zero elements in the square of a random $(0, 1)$-matrix. We consider the case where the matrix A contains a fixed number of zeros (or ones, which is the same). We define the uniform distribution on the set of all $(0, 1)$-matrices of order n with m ones and $n^2 - m$ zeros, assigning the probability $\binom{n^2}{m}$ to each matrix of this set. Let A be a random matrix with this distribution. We denote by $E(n, m)$ the expectation of the number of zero elements in the matrix $B = (b_{ij}) = A^2$. It is clear that

$$E(n, m) = \sum_{i=1}^{n} \mathbf{P}(b_{ij} = 0) + \sum_{i\neq j} \mathbf{P}(b_{ij} = 0).$$

Direct calculations yield

$$\mathbf{P}(b_{ii}=0)=\binom{n^2}{m}^{-1}\sum_{t=0}^{n-1}\binom{n-1}{t}2^t\binom{(n-1)^2}{m-t}, \qquad i=1,\dots,n,$$

$$\mathbf{P}(b_{ij}=0)=\binom{n^2}{m}^{-1}\sum_{t=0}^{n-2}\binom{n-2}{t}2^t\binom{(n-1)^2}{m-t}(1+O(mn^{-2})),$$

where $i\neq j$, $i,j=1,\dots,n$. Hence

$$E(n,m)=n^2\binom{n^2}{m}^{-1}\sum_{t=0}^{n-2}\binom{n-2}{t}2^t\binom{(n-1)^2}{m-t}(1+O(mn^{-2})).$$

Observe that

$$\binom{n^2}{m}\binom{(n-1)^2}{m-t}=\frac{(m)_t(n^2-m)_{2n-t-1}}{((n-1)^2+t)_t(n^2)_{2n-t-1}}.$$

It is obvious that

$$\left(\frac{m-n}{(n-1)^2+n}\right)^{-t}\le\frac{(m)_t}{((n-1)^2+t)_t}\le\left(\frac{m}{(n-1)^2}\right)^t,$$

$$\left(1-\frac{m}{(n-1)^2}\right)^{2n-t-1}\le\frac{(n^2-m)_{2n-t-1}}{(n^2)_{2n-t-1}}\le\left(1-\frac{m}{n^2}\right)^{2n-t-1}.$$

Thus,

$$n^2\left(1-\frac{m}{(n-1)^2}\right)^{n+1}\left(1-\frac{m}{(n-1)^2}+\frac{2(m-n)}{(n-1)^2+n}\right)^{n-2}$$
$$\le E(n,m)\le n^2\left(1-\frac{m}{n^2}\right)^{n+1}\left(1-\frac{m}{n^2}+\frac{2m}{(n-1)^2}\right)^{n-2}.$$

After some transformations we find that

$$n^2\left(1-\frac{m^2((n-1)^2-n)}{(n-1)^4((n-1)^2+n)}-\frac{2n}{(n-1)^2+n}\right)^n(1+O(mn^{-2}))$$
$$\le E(n,m)\le n^2\left(1-\frac{m^2}{n^4}+\frac{2m}{n^3}\right)^n(1+O(mn^{-2}))$$

as $n,m\to\infty$. Here and below, the relation $f(n)\le g(n)(1+O(h(n))$ as $n\to\infty$ means that there exists a sequence $\varepsilon(n)$ such that $f(n)\le g(n)(1+\varepsilon(n))$, where $\varepsilon(n)=O(h(n))$ as $n\to\infty$.

The estimates for $E(n,m)$ obtained above yield

$$\exp\left(-\frac{m^2}{n^3}+2\ln n-2\right)(1+O(mn^{-2}))$$
$$\le R(n,m)\le\exp\left(-\frac{m^2}{n^3}+2\ln n\right)(1+O(mn^{-2})).$$

These estimates imply the following assertion.

THEOREM 4.2.6. *If $n\to\infty$ and $m=\lfloor(2n^3(\ln n+\alpha(n)))^{1/2}\rfloor$, where $\alpha(n)$ is such that $\alpha(n)/\sqrt{n}\to 0$, then*

$$e^{-2(\alpha(n)+1)}(1+o(1))\le E(n,m)\le e^{-2\alpha(n)}(1+o(1)).$$

COROLLARY 4.2.7. *If under the conditions of Theorem 4.2.6 $\alpha(n) \to \infty$ as $n \to \infty$, then the average number of positive elements of the matrix A^2, where A is a random $(0,1)$-matrix of order n with m elements equal to one, is asymptotically equal to $n^2(1+o(1))$.*

COROLLARY 4.2.8. *If under the conditions of Theorem 4.2.6 $\alpha(n) \to \infty$ as $n \to \infty$, then a random $(0,1)$-matrix of order n with m elements equal to one is primitive and has exponent 2 with probability tending to one.*

COROLLARY 4.2.9. *Let a digraph Γ be chosen randomly (with respect to the uniform distribution) from the set of all $\binom{n(n-1)}{m}$ digraphs with n vertices and m arcs. If the conditions of Theorem 4.2.6 are fulfilled and $\alpha(n) \to \infty$, then with probability tending to one for each pair of vertices of Γ there exists a directed path of length 2 connecting these vertices.*

Problems.

1. Prove that the number C_n of connected labelled graphs with n vertices satisfies the relation
$$\frac{1}{n}\sum_{k=0}^{n} k\binom{n}{k}2^{\binom{n-k}{2}}C_k = 2^{\binom{n}{2}},$$
where $2^{\binom{n}{2}}$ is the total number of labelled graphs with n vertices.
2. Using the relation of the preceding problem, prove that
$$C_n = 2^{\binom{n}{2}}(1+o(1))$$
as $n \to \infty$.
3. Let $\mathfrak{A}_{mn}(s)$ be the class of all $m \times n$ matrices with elements taking the values $0, \dots, s-1$, and let $C_{mn}^{(s)}(r)$ be the number of such matrices without zero rows and with exactly r zero columns. Show that
$$C_{mn}^{(s)}(r) = \sum_{k=r}^{n}(-1)^{k-r}\binom{k}{r}\binom{n}{k}(s^{n-k}-1)^m.$$
4. Show that the number of matrices in the class $\mathfrak{A}_{mn}(s)$ without zero lines is equal to
$$C_{mn}^{(s)} = \sum_{k=r}^{m}(-1)^k\binom{m}{k}(s^{m-k}-1)^n.$$
5. Let $\eta^{(s)}(m,n)$ be the number of zero columns in a random equiprobable matrix without zero rows belonging to the class $\mathfrak{A}_{mn}(s)$. Show that if $m \le n$ and $m \to \infty$, then the following assertions are true.

   If $n/s^m \to \infty$, then the random variable $(\eta^{(s)}(m,n)-M)/\sigma$, where $M = n/s^m$, $\sigma^2 = m/(s^m(1-1/s^m))$, has in the limit the Gaussian distribution with parameters $(0,1)$.

   If $n/s^m \to \lambda > 0$, then $\eta^{(s)}(m,n)$ has in the limit the Poisson distribution with parameter λ.

   If $n/s^m \to 0$, then the distribution of $\eta^{(s)}(m,n)$ is asymptotically degenerate at zero.

4.3. Asymptotics of the permanent of a random $(0,1)$-matrix

4.3.1. Formulas for the first two moments. Let $\mathfrak{A}(m,n,N)$ be the set of all $(0,1)$-matrices of size $m \times n$, $m \le n$, containing N elements equal to one. We define the uniform distribution on the set $\mathfrak{A}(m,n,N)$, assigning the same probability

$\binom{mn}{N}^{-1}$ to each matrix $A \in \mathfrak{A}(m,n,N)$. Recall that $N_n = \{1,2,\ldots,n\}$. The permanent of a matrix $A = (a_{ij})$, $i = 1,\ldots,m$, $j = 1,\ldots,n$, is defined as

$$\operatorname{per} A = \sum_{\sigma\colon N_m \to N_n} a_{1\sigma(1)} a_{2\sigma(2)} \cdots a_{m\sigma(m)},$$

where the summation is over all injective mappings σ (see Section 1.2). The permanent is a random variable on the set $\mathfrak{A}(m,n,N)$.

We order in some way the $(n)_m = n(n-1)\cdots(n-m+1)$ sets of m noncollinear elements (that is, the diagonals, see Section 1.2.1) of the matrix A. Consider the random variables $\xi_1,\ldots,\xi_{(n)_m}$, where $\xi_i = 1$ if the ith diagonal of the random matrix A is positive (that is, consists of m ones) and $\xi_i = 0$ otherwise, $i = 1,\ldots,(n)_m$. Then

$$\operatorname{per} A = \xi_1 + \cdots + \xi_{(n)_m}.$$

It is obvious that

$$\mathbf{P}(\xi_1 = 1) = \binom{nm-m}{N-m} \Big/ \binom{nm}{N}, \qquad i = 1,2,\ldots,(n)_m.$$

Hence the mean value of the permanent of a random matrix defined on the set $\mathfrak{A}(m,n,N)$ is equal to

$$\mathbf{E}\operatorname{per} A = \frac{(n)_m (N)_m}{(nm)_m}.$$

Consider now the variance of a random matrix $A \in \mathfrak{A}(m,n,N)$. It is clear that

$$\begin{aligned}\mathbf{E}\xi_i\xi_j &= \mathbf{P}(\xi_i = 1, \xi_j = 1)\\ &= \binom{nm-2m+r}{N-2m+r} \Big/ \binom{nm}{N} = \frac{(N)_{2m-r}}{(nm)_{2m-r}},\end{aligned}$$

where r is the number of common ones in the ith and jth diagonals. The number of diagonals having r common ones with a fixed diagonal is equal to

$$D_{mn}(r) = \frac{(n)_m (m)_r}{r!\,(n)_r} \sum_{k=0}^{m-r} \frac{(-1)^k (m-r)_k}{k!\,(n-r)_k}, \qquad r = 0,1,\ldots,m.$$

This formula can be easily proved by the inclusion–exclusion method.

The second moment of the permanent of a random matrix is equal to

$$\mathbf{E}(\operatorname{per} A)^2 = \sum_{i,j} \mathbf{E}\xi_i\xi_j = \sum_{i=1}^{(n)_m} \sum_{r=0}^{m} \sum_{j} \mathbf{E}\xi_i\xi_j,$$

where the inner summation is over all diagonals having r common ones with a fixed diagonal.

Thus, for the second moment of the permanent of a random matrix we obtain the formula

$$\mathbf{E}(\operatorname{per} A)^2 = (n)_m^2 \sum_{r=0}^{m} \frac{(N)_{2m-r} (m)_r}{r!\,(nm)_{2m-r} (n)_r} \sum_{k=0}^{m-r} \frac{(-1)^k (m-r)_k}{k!\,(n-r)_k}.$$

4.3.2. Asymptotic formulas. The following theorem describes the asymptotic behaviour of the first two moments of the permanent of a random matrix as $m, n \to \infty$.

THEOREM 4.3.1. *If $m \to \infty$, $n = \beta m$, $\beta \geq 1$, $N = m^{3/2+\varepsilon}$, $\varepsilon > 0$, then*

$$\mathbf{E}\,\mathrm{per}\,A = (\beta m)_m \left(\frac{N}{\beta m^2}\right)^m \exp\left(-\frac{1}{2}\left(\frac{m^2}{N} - \frac{1}{\beta}\right)\right)(1 + o(1)),$$

$$\mathbf{E}(\mathrm{per}\,A)^2 = (\beta m)_m^2 \left(\frac{N}{\beta m^2}\right)^{2m} \exp\left(-\left(\frac{m^2}{N} - \frac{1}{\beta}\right)\right)(1 + o(1)).$$

PROOF. The first asymptotic formula is obtained from the exact formula by the use of the asymptotics

$$(N)_m = N^m \exp\left(-\frac{m^2}{2N} + O\left(\frac{1}{m^2 \varepsilon}\right)\right).$$

To obtain the second asymptotic formula, we represent the sum in the exact formula as a sum of two subsums S_1 and S_2 with the domains of summation $0 \leq r \leq \lfloor m^{5/8} \rfloor$ and $\lfloor m^{5/8} \rfloor + 1 \leq r \leq m$. Then

$$\mathbf{E}(\mathrm{per}\,A)^2 = (n)_m^2 (S_1 + S_2).$$

It is not difficult to see that

$$S_1 = \left(\frac{N}{\beta m^2}\right)^{2m} \exp\left(-\left(\frac{m^2}{N} - \frac{1}{\beta}\right)\right) \tilde{S}_1(1 + o(1)),$$

where

$$\tilde{S}_1 = \sum_{r=0}^{\lfloor m^{5/8} \rfloor} \frac{1}{r!}\left(\frac{m^2}{N} \exp\left(2\left(\frac{m}{N} - \frac{1}{\beta m}\right) - \frac{r}{2m}\left(1 - \frac{1}{\beta}\right)\right)\right)^r e^{-m^2/N}.$$

We set

$$\lambda_1 = \frac{m^2}{N} \exp\left(2\left(\frac{m}{N} - \frac{1}{\beta m}\right) - \frac{1}{2m^{3/8}}\left(1 - \frac{1}{\beta}\right)\right),$$

$$\lambda_2 = \frac{m^2}{N} \exp\left(2\left(\frac{m}{N} - \frac{1}{\beta m}\right)\right).$$

Using the inequalities

$$e^{\lambda_1 - m^2/N} \sum_{r=0}^{\lfloor m^{5/8} \rfloor} \frac{\lambda_1^r}{r!} e^{-\lambda_1} \leq \tilde{S}_1 \leq e^{\lambda_2 - m^2/N} \sum_{r=0}^{\lfloor m^{5/8} \lfloor} \frac{\lambda_2^r}{r!} e^{-\lambda_2},$$

we obtain the estimate

$$\tilde{S}_1 = 1 + o(1)).$$

Thus,

$$S_1 = \left(\frac{N}{\beta m^2}\right)^{2m} \exp\left(-\left(\frac{m^2}{N} - \frac{1}{\beta}\right)\right)(1 + o(1)).$$

Similarly we derive the estimate

$$S_2 = \left(\frac{N}{\beta m^2}\right)^{2m} \exp\left(-\left(\frac{m^2}{N} - \frac{1}{\beta}\right)\right) O\left(\left(\frac{1}{m^{1/8}}\right)^{m^{5/8}}\right).$$

Using these estimates for S_1 and S_2, we obtain the desired asymptotic formula for $\mathbf{E}(\mathrm{per}\,A)^2$. □

THEOREM 4.3.2. *If $m \to \infty$, $N = m^{3/2+\varepsilon}$, $\varepsilon > 0$, then for any fixed $\delta > 0$*

$$\mathbf{P}\left\{\left|\frac{\operatorname{per} A}{\mathbf{E}(\operatorname{per} A)} - 1\right| \geq \delta\right\} \to 0.$$

Indeed, under the conditions of Theorem 4.3.1 the estimate

$$\operatorname{Var}\operatorname{per} A = \mathbf{E}(\operatorname{per} A)^2 - (\mathbf{E}\operatorname{per} A)^2 = o((\mathbf{E}\operatorname{per} A)^2)$$

for the variance of the permanent of a random matrix holds. Therefore, applying the Chebyshev inequality, we find that

$$\mathbf{P}\left\{\left|\frac{\operatorname{per} A}{\mathbf{E}\operatorname{per} A} - 1\right| \geq \delta\right\} \leq \frac{\operatorname{Var}\operatorname{per} A}{(\mathbf{E}\operatorname{per} A)^2\delta^2} = o(1)$$

for any fixed $\delta > 0$.

Thus, as $m \to \infty$, the great majority of the matrices in $\mathfrak{A}(m, n, N)$ have permanents asymptotically equal to the mean value $\mathbf{E}\operatorname{per} A$ of a random permanent.

Problems.

1. Let A be a random $(0,1)$-matrix of size $m \times n$, $m \leq n$, such that each element of A takes the values 0 and 1 with equal probabilities independently of each other, and let E be the event that $\operatorname{per} A = 0$ and H the event that the matrix A contains either a zero row or $n - m + 1$ zero columns. Show that for the conditional probability $\mathbf{P}(H \mid E)$ the equality
$$\lim_{n\to\infty} \mathbf{P}(H \mid E) = 1$$
holds for any m, $m \leq n$.
2. A Latin rectangle $[s_1, \ldots, s_m]_n$ of size $m \times n$ is defined by an ordered set of m permutations $s_1, \ldots, s_m$ of degree n such that $s_i(j) \neq s_k(j)$ for any $j \in N_n$ and each pair i, k, $i \neq k$. Show that for a fixed $m \geq 1$ and $n \to \infty$ the number L_{mn} of Latin rectangles satisfies the relation
$$L_{mn} = (n!)^m e^{m(m-1)/2}(1 + o(1)).$$
3. Let C be the permutation matrix corresponding to the unicyclic permutation $c = (1\ 2 \ldots n)$ of degree n. Show that
$$\operatorname{per}(C^0 + C^i + C^j) \geq n + 3$$
for any degrees i, j, $1 \leq i < j \leq n - 1$.
4. Using the previous problem, prove that
$$\lim_{n\to\infty} \min_{A \in \mathfrak{A}^n(k)} \operatorname{per} A = \infty, \qquad k \geq 3,$$
where $\mathfrak{A}^n(k)$ is the class of $(0,1)$-matrices of order n with exactly k ones in each row and each column.

4.4. Random lattices and Boolean algebras

4.4.1. Lattices. A set S is called a lattice, or a structure, if two operations, $\vee$ and $\wedge$, are defined on S, which possess the following four properties, namely, for any $a, b, c \in S$

$$\begin{aligned}
a \vee b &= b \vee a, & a \wedge b &= b \wedge a && \text{(commutativity)},\\
a \vee (b \vee c) &= (a \vee b) \vee c, & a \wedge (b \wedge c) &= (a \wedge b) \wedge c && \text{(associativity)},\\
a \vee a &= a, & a \wedge a &= a && \text{(idempotency)},\\
a \wedge (a \vee b) &= a, & a \wedge (a \vee b) &= a && \text{(absorption)}.
\end{aligned}$$

A lattice is called distributive if for any $a, b, c \in S$

$$(a \vee b) \wedge c = (a \wedge c) \vee (b \wedge c), \qquad (a \wedge b) \vee c = (a \vee c) \wedge (b \vee c).$$

Elements a_0 and a_1 are called, respectively, the zero and the identity of a lattice S if for any $b \in S$

$$a_0 \wedge b = a_0, \qquad a_1 \wedge b = b.$$

A lattice containing the zero a_0 and the identity a_1 is called a complemented lattice if for any $a \in S$ there exists an element $\bar{a} \in S$ such that $a \wedge \bar{a} = a_0$, $a \vee \bar{a} + a_1$. A distributive complemented lattice is called a Boolean algebra.

The order $\leq$ on a lattice S is determined by the equalities

$$(a \leq b) \equiv (a \vee b = b),$$

or, what is the same (by the absorption property),

$$(a \leq b) \equiv (a \wedge b = a)$$

for any $a, b \in S$. The maximal element 0 and the minimal element 1, if they exist, are defined by the conditions

$$a \vee 0 = a, \qquad a \wedge 1 = a$$

for all $a \in S$. Therefore $0 \leq a \leq 1$ for all $a \in S$.

4.4.2. Boolean algebras. If $X = \{x_1, \ldots, x_n\}$, then the set of all subsets of X, denoted by 2^X, forms the Boolean algebra $S = S(X)$, where the role of the operations $\vee$ and $\wedge$ is played by the set-theoretic operations of union $\cup$ and intersection $\cap$, the zero element is the empty set and the identity element is the set X itself. Any finite Boolean algebra is isomorphic to the Boolean algebra corresponding to 2^X for some finite set X.

If a Boolean algebra S coincides with the minimal Boolean algebra containing a system of sets $E = \{e_1, \ldots, e_m\}$, then we say that S is generated by the system of generators E. A system of subsets $E = \{e_1, \ldots, e_m\}$ of the finite set $X = \bigcup_{\mu=1}^m e_\mu$ is called separating if for any distinct elements $x, x' \in X$ there exists a set e_k, $1 \leq k \leq m$, such that either $x \in e_k$, $x' \notin e_k$ or $x \notin e_k$, $x' \in e_k$.

THEOREM 4.4.1. *A system of subsets $E = \{e_1, \ldots, e_m\}$ of a finite set X is a system of generators of the Boolean algebra $S = S(X)$ if and only if E is a separating system.*

PROOF. Let E be a system of generators of the algebra $S(X)$ and $x, x' \in X$, $x \neq x'$. Then $\{x\} = z_1 \cup z_2 \cup \cdots \cup z_k \cup \ldots$, where z_k is expressed in terms of generators as follows:

$$z_k = e_{i_1} \cap \cdots \cap e_{i_r} \cap \bar{e}_{i_{r+1}} \cap \cdots \cap \bar{e}_{i_l}, \qquad 1 \leq i_1, \ldots, i_l \leq m,$$

for some r and l. Hence there exists an implicant z_k such that $\{x\} \subseteq z_k$, $1 \leq k \leq m$. Since $\{x\} \cap \overline{\{x\}} = \emptyset$, we have $\{x'\} \not\subseteq z_k$. Consequently, there exists either p, $1 \leq p \leq r$, such that $x \in e_{i_p}$, $x' \notin e_{i_p}$ or q, $r+1 \leq q \leq l$, such that $x \in \bar{e}_{i_q}$, $x' \notin \bar{e}_{i_q}$. Thus, E is a separating system.

Now let E be a separating system of subsets and $a = \{x_1, \ldots, x_s\} \in S(X)$. Let us show that a can be expressed in terms of elements of E in the Boolean algebra $S(X)$. Since $a = \{x\} \cup \cdots \cup \{x_s\}$, it suffices to show that any $x \in X$ can be expressed in terms of elements of E. If $x \neq x'$, then there exists an $e_\mu \in E$

such that either $x \in e_\mu$, $x' \notin e_\mu$ or $x \notin e_\mu$, $x' \in e_\mu$, $1 \le \mu \le m$. We consider consecutively all elements $x' \in X$, $x' \ne x$. The elements of E are of two types: the element x is contained either in an element itself or in its complement. Therefore $\{x\}$ is the intersection of the elements of E which contain x and the complements of the elements which do not contain x; that is, $\{x\} \in S(X)$ is expressed in terms of elements of E. □

To each element $a \in S(X)$, where $X = \{x_1, \dots, x_n\}$, we put in correspondence the $(0,1)$-vector $(\varepsilon_1(a), \dots, \varepsilon_n(a))$, where $\varepsilon_i(a) = 1$ if $x_i \in a$ and $\varepsilon_i(a) = 0$ if $x_i \notin a$. For a generating system $E = \{e_1, \dots, e_m\}$ of the Boolean algebra $S(X)$, we consider the $(0,1)$-matrix

$$\varepsilon = \begin{pmatrix} \varepsilon_1(e_1) & \varepsilon_2(e_1) & \dots & \varepsilon_n(e_1) \\ \varepsilon_1(e_2) & \varepsilon_2(e_2) & \dots & \varepsilon_n(e_2) \\ & & \dots & \\ \varepsilon_1(e_m) & \varepsilon_2(e_m) & \dots & \varepsilon_n(e_m) \end{pmatrix}.$$

This matrix will be called the matrix of the generating system $E = \{e_1, \dots, e_m\}$.

THEOREM 4.4.2. *A $(0,1)$-matrix of size $n \times n$ without zero columns is a matrix of a generating system $E = \{e_1, \dots, e_m\}$ of the Boolean algebra $S(X)$ constructed on the set 2^X of some finite set X if and only if the columns of this matrix are pairwise distinct.*

Indeed, if the column vectors are distinct, then E is a separating system, and vice versa.

Let the uniform distribution be defined on the 2^n elements of the Boolean algebra $S(X)$ with $|X| = n$, and let $E = \{e_1, \dots, e_m\}$ be a set of independent random elements of $S(X)$. We denote by B_{nm} the event that E is a generating system of $S(X)$.

THEOREM 4.4.3. *The probability that $E = \{e_1, \dots, e_m\}$ is a generating system of the Boolean algebra $S(X)$ with $|X| = n$ is equal to*

$$\mathbf{P}(B_{nm}) = \frac{(2^m - 1)_n}{2^{nm}}.$$

Consider the classical scheme of allocating n distinct particles to $2^m - 1$ distinct cells. The choice of columns of the matrix of a generating system corresponds to such an allocation with the condition that each cell contains at most one particle. This reduction implies the formula given in the theorem.

THEOREM 4.4.4. *Let (m_j) and (n_j), $j = 1, 2, \dots$, be sequences of positive integers such that there exists a limit*

$$\lim_{j \to \infty} (m_j - 2\log_2 n_j) = \alpha.$$

Then

$$\lim_{j \to \infty} \mathbf{P}(B_{n_j m_j}) = \begin{cases} 1, & \alpha = \infty, \\ \exp(-1/2^{\alpha+1}), & -\infty < \alpha < \infty, \\ 0, & \alpha = -\infty. \end{cases}$$

Indeed, since

$$\mathbf{P}(B_{nm}) = \prod_{j=1}^{n} \left(1 - \frac{j}{2^m}\right),$$

by applying the inequalities

$$e^{-t/(1-t)} < 1 - t < e^{-t}, \qquad 0 < t < 1,$$

we obtain the estimates

$$\exp\left(-\frac{n(n+1)}{2^{m+1} - n(n+1)}\right) \le \mathbf{P}(B_{nm}) \le \exp\left(-\frac{n(n+1)}{2^{m+1}}\right),$$

which yield the assertions of the theorem.

Let $S^{(m)}$ be a Boolean algebra generated by m random generators chosen from the set of all subsets of some X with $|X| = n$ independently with equal probabilities, and let ν_n be the random variable equal to the minimum number m such that $S^{(m)} = S(X)$.

Since $\mathbf{P}(\nu_n \le m) = \mathbf{P}(B_{nm})$, Theorems 4.4.3 and 4.4.4 give the following results.

COROLLARY 4.4.5. *The following equalities hold:*

$$\mathbf{P}(\nu_n \le m) = \frac{(2^m - 1)_n}{2^{nm}}, \qquad m = 1, 2, \dots,$$

$$\lim_{n\to\infty} \mathbf{P}(\nu_n \le 2\log_2 n + \alpha) = \exp(-1/2^{\alpha+1}).$$

4.4.3. Free Boolean algebras. A generating system $E = \{e_1, \dots, e_m\}$ of a Boolean algebra $S \subseteq S(X)$, where $|X| = n$, is called an independent generating system if it is nonempty and for any disjoint subsets $\Delta, \Delta' \subseteq N_m$ the condition

$$\left(\bigcap_{i\in\Delta} e_i\right) \cap \left(\bigcap_{j\in\Delta'} \bar{e}_j\right) \ne \emptyset$$

is fulfilled.

A Boolean algebra which has an independent generating system is called free. If $E = \{e_1, \dots, e_m\}$ is an independent generating system of a free Boolean algebra $S(X)$, where $|X| = n$, then the matrix of this system contains all 2^m distinct columns, and the converse assertion is also true. In the allocation scheme interpretation, this property corresponds to an allocation of n distinct particles to 2^m distinct cells such that each cell is nonempty.

THEOREM 4.4.6. *Let subsets $e_1, \dots, e_m$ be chosen independently with equal probabilities from the set of all subsets of a set X, where $|X| = n$, and let $\tilde{B}_{nm}$ be the event that the system $E = \{e_1, \dots, e_m\}$ generates a free Boolean algebra. Then*

$$\mathbf{P}(\tilde{B}_{nm}) = \sum_{k=0}^{2^m} (-1)^k \binom{2^m}{k} \left(1 - \frac{k}{2^m}\right)^n.$$

Indeed, if we allocate n distinct particles independently with equal probabilities to 2^m distinct cells, then the probability that there are no empty cells can be expressed with the use of the difference operator Δ as follows:

$$\mathbf{P}(\tilde{B}_{nm}) = \frac{\Delta^{2^m} 0^n}{2^n m} = \sum_{k=0}^{2^m} (-1)^k \binom{2^m}{k} \left(1 - \frac{k}{2^m}\right)^n.$$

It is obvious that the probability $P_r(n, m)$, that after allocating n distinct particles independently with equal probabilities to 2^m distinct cells there are exactly

r empty cells, is equal to

$$\begin{aligned} P_r(n,m) &= \binom{2^m}{r} \frac{\Delta^{2^m-r}0^n}{2^{nm}} \\ &= \binom{2^m}{r} \sum_{k=0}^{2^m-r} (-1)^k \binom{2^m-r}{k} \left(1-\frac{r+k}{2^m}\right)^n, \qquad r=0,1,\dots,2^m, \end{aligned}$$

where $P_0(nm) = \mathbf{P}(\tilde{B}_{nm})$. Let ξ_{nm} be a random variable equal to the number of empty cells. Then

$$\mathbf{P}(\xi_{nm} = r) = P_r(n,m), \qquad r = 0,1,\dots,2^m,$$

and the kth binomial moment of ξ_{nm} is

$$B_k(n,m) = \binom{2^m}{k}\left(1-\frac{k}{2^m}\right)^n, \qquad k=0,1,\dots,2^m.$$

It is clear that

$$\frac{(2^m-k^k}{k!}\left(1-\frac{k}{2^m}\right)^n \le B_k(n,m) \le \frac{2^{mk}}{k!}\left(1-\frac{k}{2^m}\right)^n.$$

These inequalities imply that under the condition

$$\lim_{n\to\infty} \exp\left(-\frac{n}{2^m} + m\ln 2\right) = \lambda,$$

the equalities

$$\lim_{n\to\infty} B_k(n,m) = \frac{\lambda^k}{k!}, \qquad k=0,1,\dots,$$

hold. Therefore

$$\lim_{n\to\infty} P_r(n,m) = \frac{\lambda^k}{k!}e^{-\lambda}, \qquad k=0,1,\dots.$$

In particular,

$$\lim_{n\to\infty} \mathbf{P}(\tilde{B}_{nm}) = e^{-e^{-\gamma}},$$

where

$$\gamma = \lim_{n\to\infty} (n/2^m - m\ln 2).$$

Thus, the following result is proved.

THEOREM 4.4.7. *If* $\gamma(n,m) = n/2^m - m\ln 2$, *then*

$$\lim_{n\to\infty} \mathbf{P}(\tilde{B}_{nm}) = \begin{cases} 0, & \gamma(n,m)\to -\infty, \\ e^{-e^{-\gamma}}, & \gamma(n,m)\to\gamma,\ -\infty<\gamma<\infty, \\ 1, & \gamma(n,m)\to\infty. \end{cases}$$

Problems.

1. Show that the minimum number m_0 of generators of the Boolean algebra with n elements is given by the equality $m_0 = \lceil \log_2 n \rceil$.
2. Prove Theorem 4.4.4 for the scheme of sampling of generators of a Boolean algebra without replacement.

4.5. Coverings of sets and $(0,1)$-matrices

4.5.1. Coverings. Let $X = \{x_1, \dots, x_n\}$, and let $\mathcal{X} = \{X_1, \dots, X_k\}$ be a family of distinct nonempty subsets of the set X. Recall that the family $\mathcal{X}$ is called a covering of the set X by blocks $X_1, \dots, X_k$ if $X = X_1 \cup \dots \cup X_k$ (see Section 3.2). If $X_i \cap X_j = \emptyset$, $i \neq j$, then the covering is a partition of X. The incidence matrix $A = (a_{ij})$, $i = 1, \dots, k$, $j = 1, \dots, n$, of the covering $\mathcal{X}$ is the matrix with elements

$$a_{ij} = \begin{cases} 1, & x_j \in X_i, \\ 0, & x_j \notin X_i. \end{cases}$$

The matrix A has no zero lines, and all rows of A are distinct. It is clear that $k!$ $(0,1)$-matrixes of size $k \times n$ which differ by $k!$ permutations of rows correspond to the same k-block covering.

Let D_{nk} be the number of k-block coverings of an n-set, $k = 1, 2, \dots, 2^n - 1$. Then

$$D_{nk} = \sum_{j=0}^{n} (-1)^j \binom{n}{j} \binom{2^{n-j} - 1}{k}, \qquad k = 1, 2, \dots, 2^n - 1. \tag{4.5.1}$$

Here $D_{n0} = 0$, and $D_{0k} = 0$ for $k \geq 1$.

Consider the set of all $(0,1)$-matrices of size $k \times n$ with distinct rows. We say that a matrix in this set possesses property A_j if the jth column of A is the zero vector. The number of matrices possessing the properties $A_{j_1}, \dots, A_{j_i}$, $1 \leq j_1 < \dots < j_i \leq n$, is equal to

$$M(A_{j_1} \dots A_{j_i}) = (2^{n-i} - 1)(2^{n-i} - 2) \dots (2^{n-i} - k) = (2^{n-i} - 1)_k.$$

Applying the inclusion–exclusion principle, we obtain

$$k! D_{nk} = \sum_{j=0}^{n} (-1)^j \binom{n}{j} (2^{n-i} - 1)_k.$$

Formula (4.5.1) follows from this equality.

Hence, using the Bonferroni inequalities, we obtain

$$1 - n \frac{(2^{n-1} - 1)_k}{(2^n - 1)_k} \leq D_{nk} \Big/ \binom{2^n - 1}{k} \leq 1.$$

Let $k = \omega(n) + \log_2 n$, where $\omega(n) \to \infty$ as $n \to \infty$. Then these inequalities imply

$$\lim_{n \to \infty} D_{nk} \Big/ \binom{2^n - 1}{k} = 1.$$

Let D_n be the total number of coverings of an n-set. Then

$$D_n = \sum_{k=1}^{2^n - 1} D_{nk},$$

and by summing up both sides of (4.5.1) we obtain

$$D_n = \sum_{j=0}^{n} (-1)^j \binom{n}{j} 2^{4^{n-j} - 1}, \qquad n = 0, 1, \dots. \tag{4.5.2}$$

Let $\tilde{D}_n$ be the number of nonempty subsets of the elements of the Boolean algebra 2^X, $|X| = n$, covering all elements of the set X. By the inclusion–exclusion principle it is easy to show that

$$\tilde{D}_n = \sum_{k=0}^{n} (-1)^k \binom{n}{k} 2^{2^{n-k}-1}$$

and, consequently, $\tilde{D}_n = D_n$. Applying the Bonferroni inequalities, we obtain the estimates

$$1 - n/2^{2^{n-1}} \leq \tilde{D}_n/2^{2^{n-1}} \leq 1. \tag{4.5.3}$$

Hence the fraction of nonempty subsets of elements of 2^X that do not cover all elements of the set X does not exceed $n/2^{2n-1}$ and tends to zero as $n \to \infty$.

4.5.2. Minimal coverings. A covering of a set $X = \{x_1, \ldots, x_n\}$ by blocks $X_1, \ldots, X_k$ is called minimal if for any i, $1 \leq i \leq k$, there exists an element $x \in X$ such that $x \in X_i$ but $x \notin X_j$ for all $j \neq i$. In this case, x is called a single-covered element. At least one single-covered element corresponds to each block of a minimal covering.

To a minimal k-block covering with j single-covered elements there corresponds an incidence matrix of size $k \times n$ with exactly j columns each of which contains exactly one element equal to 1, $k \leq j \leq n$. For the number $L_n(k)$ of minimal k-block coverings we have

$$L_n(k) = \frac{1}{k!} \sum_{s=0}^{k} (-1)^s \binom{k}{s} (2^k - s - 1)^n. \tag{4.5.4}$$

Indeed, consider the set of $(2^k - 1)^n$ $(0,1)$-matrices of size $k \times n$ without zero columns. We say that a matrix of this class has property A_i if the ith row, $1 \leq i \leq k$, has no element equal to 1 such that the removal of this element leads to the appearance of a zero column. The number of matrices each possessing properties $A_{i_1}, \ldots, A_{i_s}$, $1 \leq i_i < \cdots < i_s \leq k$, is equal to $(2^k - s - 1)^n$. Therefore, applying the inclusion–exclusion principle, we obtain a formula for the number of matrices which possess none of the properties $A_1, \ldots, A_k$:

$$\tilde{L}_n(k) = \sum_{s=0}^{k} (-1)^s \binom{k}{s} (2^k - s - 1)^n.$$

It is clear that $\tilde{L}_n(k)$ equals the number of $(0,1)$-matrices with distinct rows which have no zero lines and contain at least k columns with single ones. Since $\tilde{L}_n(k) = k! L_n(k)$, formula (4.5.4) is proved.

Let the uniform distribution be defined on the set of all $(0,1)$-matrices of size $k \times n$ without zero columns, and let $\xi(k,n)$ be the random variable equal to the number of rows of a random $(0,1)$-matrix such that the removal of any of them gives a zero column. Using the inclusion–exclusion principle, we find that

$$\mathbf{P}(\xi(k,n) = r) = \sum_{s=r}^{k} (-1)^{s-r} \binom{s}{r} \binom{k}{s} \left(1 - \frac{s}{2^k - 1}\right)^n, \qquad r = 0, 1, \ldots, k.$$

After some simplifications we obtain

$$\mathsf{P}(\xi(k,n)=r)=\binom{k}{r}\sum_{s=0}^{k-r}(-1)^s\binom{k-r}{s}\left(1-\frac{s+r}{2^k-1}\right)^n,\qquad r=0,1,\dots,k.$$

The binomial moment $B_s(k,n)$ of order s of the random variable $\xi(k,n)$ is given by the formula

$$B_s(k,n)=\binom{k}{s}\left(1-\frac{s}{2^k-1}\right)^n,\qquad s=0,1,\dots,k. \tag{4.5.5}$$

THEOREM 4.5.1. *If $n\to\infty$ and $\gamma_{nk}=n/(2^k-1)-\ln k\to\gamma<\infty$, then the random variable $\xi(k,n)$ tends to the Poisson distribution with parameter $\lambda=e^{-\gamma}$.*

The theorem immediately follows from fact that under the condition $\gamma_{nk}\to\gamma$ the equality

$$\lim_{n\to\infty}B_s(k,n)=\frac{1}{s!}(e^{-\gamma})^s$$

holds for any fixed $s=0,1,\dots$

THEOREM 4.5.2. *If $n\to\infty$, $k=k(n)$, and $\gamma_{kn}\to\gamma$, then*

$$\frac{k!L_n(k)}{(2^k-1)^n}\to e^{-e^{-\gamma}}.$$

Theorem 4.5.2 follows from Theorem 4.5.1, since

$$\frac{k!L_n(k)}{(2^k-1)^n}=\mathsf{P}(\xi(k,n)=0)$$

and

$$\lim_{n\to\infty}\mathsf{P}(\xi(k,n)=0)=e^{-e^{-\gamma}}.$$

Note that if $n\to\infty$ and $\gamma_{nk}\to\infty$, then

$$L_n(k)=\frac{(2^k-1)^n}{k!}(1+o(1)).$$

4.5.3. t-minimal coverings. A covering of an n-set $X=X_1\cup\dots\cup X_k$ by distinct nonempty subsets $X_1,\dots,X_k$ is called t-minimal if for any $X_{i_1},\dots,X_{i_t}$, $1\le i_1<\dots<i_t\le k$, there exists an element $x\in X$ such that

$$x\in X_{i_1},\dots,x\in X_{i_t},\qquad x\notin X_{j_1},\dots,x\notin X_{j_{k-t}},$$

where $\{i_1,\dots,i_t,j_1,\dots,j_{k-t}\}=\{1,2,\dots,k\}$. It is clear that in this case $n\ge\binom{k}{t}$. If $t=1$, then the notion of a t-minimal covering coincides with the notion of a minimal covering.

Let $C_k^t=\{\alpha_1,\alpha_2,\dots,\alpha_{\binom{k}{t}}\}$ be the family of all distinct t-subsets of the set N_k. We identify the sets $\alpha_1,\alpha_2,\dots,\alpha_{\binom{k}{t}}$ with $(0,1)$ column vectors with exactly t coordinates equal to 1. Let $B(C_k^t)$ be the $(0,1)$-matrix of size $k\times\binom{k}{t}$ consisting of the columns $\alpha_1,\alpha_2,\dots,\alpha_{\binom{k}{t}}$. On $B(C_k^t)$, the direct product $\Pi_k^t=\mathfrak{S}_k\times\mathfrak{S}(C_k^t)$ acts, where $\mathfrak{S}_k$ is the symmetric group acting on k elements and $B(C_k^t)$ is the symmetric group of all permutations acting on C_k^t. The family C_k^t is a separating system for $t<k$ (see Section 4.4); therefore not only all columns but also all rows of $B(C_k^t)$ are distinct. Hence the group $\operatorname{Aut}_p B(C_k^t)$ of permutational automorphisms (see Section 1.1) is isomorphic to the group of admissible permutations, say, of rows (see Theorem 1.1.3). Each permutation π of rows is admissible, since it leads to a matrix

with the same set of columns. Therefore there exists a unique permutation ρ_π on C_k^t which transfers this matrix to $B(C_k^t)$. Thus, $|\operatorname{Aut} B(C_k^t)| = k!$. Consequently, by Theorem 1.1.1 there exist exactly

$$k!\binom{k}{t}!\Big/ k! = \binom{k}{t}!$$

distinct matrices permutationally equivalent to $B(C_k^t)$. The permutations ρ_π form a subgroup in $\mathfrak{S}(C_k^t)$. Consider a complete system of representatives R of cosets of Π_k^t with respect to the subgroup formed by $k!$ pairs $(\varepsilon_1, \rho\pi)$, where ε_1 is the identical permutation. Add to R all pairs of the form (π, ε_2), where ε_2 is the identical permutation in $\mathfrak{S}(C_k^t)$. It is clear that $\binom{k}{t}!$ distinct matrices can be obtained by the pairs in R acting on $B(C_k^t)$. An element $x \in X$ is called t-covered if the corresponding column in the incidence matrix $B(\mathcal{X})$ of the k-block covering $\mathcal{X}$ contains exactly t coordinates equal to 1. We denote by $B^{(t)}(\mathcal{X})$ the submatrix of $B(\mathcal{X})$ formed by the columns corresponding to t-covered elements. We define the action of a pair $(\pi, \rho) \in \Pi_k^t$ on $B^{(t)}(\mathcal{X})$ as the permutation of the rows according to π followed by the transformation of columns by the rule

$$\alpha_i \to \rho(\alpha_i), \qquad i = 1, 2, \ldots, \binom{k}{t}.$$

Let $\mathcal{X}$ be a t-minimal covering. Then, in the same way as it was done above for the matrix $B(C_k^t)$, one can show that the described action of Π_k^t on $B^{(t)}(\mathcal{X})$ gives $\binom{k}{t}!$ distinct matrices when all pairs of R act on the matrix $B^{(t)}(\mathcal{X})$.

Let $\mathcal{X} = (X_1, \ldots, X_k)$ be a k-block t-minimal covering of an n-set X, and let $B(\mathcal{X})$ be the incidence matrix of this covering. For each possible $(\pi, \rho) \in R$, we form the matrix by the permutation of rows in $B(\mathcal{X})$ according to π followed by the transformation of columns according to ρ. As shown above, we obtain $\binom{k}{t}!$ distinct matrices. Each of these matrices has distinct rows, nonzero columns, and is the incidence matrix of a t-minimal covering. We do not distinguish these t-minimal coverings, considering them as the same covering. It is clear that for $t = 1$ this property is reduced to the usual condition that a covering does not depend on the order of blocks.

Consider the number $L_n^{(t)}(k)$ of the k-block t-minimal coverings of an n-set. For $L_n^{(t)}(k)$, the formula

$$L_n^{(t)}(k) = \frac{1}{\binom{k}{t}!} \sum_{s=0}^{\binom{k}{t}} (-1)^s \binom{\binom{k}{t}}{s} (2^k - s - 1)^n \tag{4.5.6}$$

is valid.

Let $\alpha_1, \ldots, \alpha_{\binom{k}{t}}$ be distinct t-subsets of the set $\{1, 2, \ldots, k\}$. We say that a $(0,1)$-matrix of size $k \times n$ has the property A_{α_i}, $1 \le i \le \binom{k}{t}$, if for the rows with numbers in α_i there is no column such that all ones of this column lie in the rows with numbers in α_i. The number of matrices possessing the properties $A_{\alpha_{i_1}}, \ldots, A_{\alpha_{i_s}}$ is equal to

$$M(A_{\alpha_{i_1}}, \ldots, A_{\alpha_{i_s}}) = (2^k - s - 1)^n.$$

Applying the inclusion–exclusion principle, we obtain the following formula for the number of $(0,1)$-matrices $\tilde{L}_n^{(t)}(k)$ without zero columns which possess none of the properties $A_{\alpha_1}, \ldots, A_{\alpha_{\binom{k}{t}}}$:

$$\tilde{L}_n^{(t)}(k) = \sum_{s=0}^{\binom{k}{t}} (-1)^s \binom{\binom{k}{t}}{s} (2^k - s - 1)^n.$$

Since

$$\tilde{L}_n^{(t)}(k) = \binom{k}{t}! L_n^{(t)}(k),$$

the last formula implies (4.5.6).

We set $\gamma_{kn}(t) = n/(2^k - 1) - \ln \binom{k}{t}$. The following assertion is true.

THEOREM 4.5.3. *Let $n \to \infty$, $k = k(n)$, $t = t(n)$ and $\gamma_{kn}(t) \to \gamma$. Then*

$$\lim_{n\to\infty} \frac{\binom{k}{t}! L_n^{(t)}(k)}{(2^k - 1)^n} = \begin{cases} e^{-e^{-\gamma}}, & -\infty < \gamma < \infty, \\ 1, & \gamma = \infty, \\ 0, & \gamma = -\infty. \end{cases}$$

The proof of Theorem 4.5.3 is similar to the proof of Theorems 4.5.1 and 4.5.2.

Consider the number $L_n^{(t)}(k, j)$ of k-block t-minimal coverings of an n-set with the number of t-covered element equal to j. For $L_n^{(t)}(k, j)$, the formula

$$L_n^{(t)}(k, j) = \binom{n}{j} \sigma\left(j, \binom{k}{t}\right) \left(2^k - \binom{k}{t} - 1\right)^{n-j}, \qquad \binom{k}{t} \le j \le n, \tag{4.5.7}$$

holds, where $\sigma(j, \binom{k}{t})$ is the Stirling number of the second kind.

The right-hand side of the formula is the product of the number of ways to choose j elements which are t-covered, the number of partitions of a j-set into $\binom{k}{t}$ blocks, and the number of ways to add to these blocks the remaining $n - j$ not t-covered elements.

It follows from the obvious equality

$$L_n^{(t)}(k) = \sum_{j=\binom{k}{t}}^{n} L_n^{(t)}(k, j) \tag{4.5.8}$$

that

$$L_n^{(t)}(k) = \sum_{j=\binom{k}{t}}^{n} \binom{n}{j} \sigma\left(j, \binom{k}{t}\right) \left(2^k - \binom{k}{t} - 1\right)^{n-j}. \tag{4.5.9}$$

Using the well-known formula for the Stirling numbers of the second kind, we arrive at the formula

$$L_n^{(t)}(k) = \frac{1}{\binom{k}{t}!} \sum_{s=0}^{\binom{k}{t}} (-1)^s \binom{\binom{k}{t}}{s} (2^k - s - 1)^n \tag{4.5.10}$$

obtained above.

In terms of the finite difference function this function can be written in the form

$$L_n^{(t)}(k) = \frac{1}{\binom{k}{t}!} \Delta^{\binom{k}{t}} \left(2^k - \binom{k}{t} - 1\right)^n. \tag{4.5.11}$$

Taking into account the relations defining the Stirling numbers of the second kind, we can rewrite this formula in the form

$$L_n^{(t)}(k) = \sum_{r=\binom{k}{t}}^{n} \binom{r}{\binom{r}{t}} \sigma(n,r) \left(2^k -] \binom{k}{t} - 1\right)_{\binom{r-k}{t}}. \tag{4.5.12}$$

The generating function

$$F_n^{(t)}(x) = \sum_{k=1}^{n} L_n^{(t)} x^k \tag{4.5.13}$$

can be written in the form

$$F_n^{(t)}(x) = \sum_{r=1}^{n} \sigma(n,r) \varphi_r^{(t)}(x), \tag{4.5.14}$$

where

$$\varphi_r^{(t)}(x) = \sum_{k=\lfloor \log_2(r+1) \rfloor}^{\bar{k}} \binom{r}{\binom{k}{t}} \left(2^k - \binom{k}{t} - 1\right)_{r-\binom{k}{t}} x^k \tag{4.5.15}$$

and $\bar{k}$ is the maximal k such that

$$\binom{k}{t} \le r.$$

The polynomials $\varphi_r^{(t)}(x)$ will be referred to as the polynomials of t-minimal coverings. For asymptotic analysis, as $r \to \infty$, we shall use the formula

$$\varphi_r^{(t)}(x) = \sum_{k=\lfloor \log_2(r+1) \rfloor}^{(rt!)^{1/t}} \binom{r}{\binom{k}{t}} \left(2^k - \binom{k}{t} - 1\right)_{r-\binom{k}{t}} x^k. \tag{4.5.16}$$

For $t = 1$, we set $\varphi_r(x) = \varphi_r^{(1)}(x)$ and obtain the polynomials of minimal coverings

$$\varphi_r(x) = \sum_{k=\lfloor \log_2(r+1) \rfloor}^{r} \binom{r}{k} (2^k - k - 1)_{r-k} x^k, \qquad r = 1, 2, \ldots. \tag{4.5.17}$$

Setting $t = 1$ in (4.5.12), we obtain the following formula for the number of k-block coverings of an n-set:

$$L_n(k) = \sum_{r=k}^{n} \binom{r}{k} \sigma(n,r) (2^k - k - 1)_{r-k}. \tag{4.5.18}$$

4.5.4. Partitions and minimal coverings. Let $1 \le k \le r \le n$. Consider a partition of an n-set into r blocks

$$X = X_1 \cup \cdots \cup X_k \cup X_{k+1} \cup \cdots \cup X_r.$$

Using this partition, we can construct a minimal covering of the n-set by k blocks. Without loss of generality we choose $X_1, \ldots, X_k$ as these k blocks. We include the remaining blocks $X_{k+1}, \ldots, X_r$ with repetitions into the blocks $X_1, \ldots, X_k$. The block X_{k+1} can be included into blocks $X_1, \ldots, X_k$ in $2^k - k - 1$ ways. In turn, we include the block X_{k+2} into the blocks $X_1, \ldots, X_k$ in the ways which differ from the way chosen for the block X_{k+1}. The number of such ways is equal to $2^k - k - 2$. The blocks $X_{k+3} \ldots, X_r$ are included in a similar way. Thus, the total number of ways to include the blocks $X_{k+1}, \dots, X_r$ into the blocks $X_1, \dots, X_k$, provided that the ways of their inclusions are distinct, is equal to $(2^k - k - 1)_{r-k}$. As a result we obtain a covering of the n-set X by k blocks. This covering is minimal, since the elements of the blocks $X_1, \dots, X_k$ (and only they) are single-covered.

Consider now a k-block covering of an n-set $X = Y_1 \cup \dots \cup Y_k$ such that the single-covered elements form a partition of a part of the set X with blocks $X_1, \dots, X_k$. Take the maximal set X_{k+1} of elements of X which do not belong to $X_1 \cup \dots \cup X_k$ and are contained simultaneously only in blocks $Y_{i_1}, \dots, Y_{i_\mu}$. We take X_{k+1} as a block of the constructed partition of X. Then we choose the maximal set X_{k+2} of elements of X which do not belong to $X_1 \cup \dots \cup X_k \cup X_{k+1}$ and are contained simultaneously in blocks $Y_{j_1}, \dots, Y_{j_\nu}$. It follows from the maximality of the sets that $\{i_1, \dots, i_\mu\} \neq \{j_1, \dots, j_\nu\}$ and $X_{k+1} \cap X_{k+2} = \emptyset$. Consequently, we can take X_{k+2} as a block of the constructed partition. Continuing the process in the same way, we uniquely determine the blocks of the partition

$$X = X_1 \cup \dots \cup X_k \cup X_{k+1} \cup \dots \cup X_r.$$

The algorithm described above allows us to give a combinatorial interpretation of equality (4.5.18). Indeed, $\sigma(n, r)$ is the number of partitions of an n-set into r blocks, $\binom{r}{k}$ is the number of ways of choosing single-covered elements of a k-block minimal covering, and $(2^k - k - 1)_{r-k}$ is the number of way of adding the remaining $r - k$ blocks according to the procedure described above. Multiplying these numbers and summing over r, we obtain the total number of k-block coverings of an n-set.

4.5.5. Coverings of sets and systems of Boolean equations. A Boolean function of n variables is a function of binary vectors of length n taking the values 0 and 1.

For Boolean functions $f_1, f_2 \dots, f_k$ of n variables and an arbitrary binary vector $(\delta_1, \delta_2, \dots, \delta_k)$, we consider the system of equations

$$\begin{aligned} f_1(x_1, \dots, x_n) &= \delta_1, \\ f_2(x_1, \dots, x_n) &= \delta_2, \\ &\dots \\ f_k(x_1, \dots, x_n) &= \delta_k. \end{aligned} \tag{4.5.19}$$

We assume that none of the equations $f_i(x_1, \dots, x_n) = \delta_i$, $1 \le i \le k$, is an identity and all of them are distinct. We denote by Y_i the set of solutions of the ith equation and determine the family $X_1, \dots, X_k$ by the relations

$$X_i \cup Y_i = \mathfrak{A}_n, \quad X_i \cap Y_i = \emptyset, \quad 1 \le i \le k, \tag{4.5.20}$$

where $\mathfrak{A}_n$ is the set of all binary n-dimensional vectors. The set

$$Y = Y_1 \cap Y_2 \cap \dots \cap Y_k \tag{4.5.21}$$

is the set of all solutions of the system (4.5.19). The following simple lemma is true.

LEMMA 4.5.4. *The system* (4.5.19) *is inconsistent if and only if the family of the sets* $X_1, X_2, \dots, X_k$ *defined by* (4.5.20) *forms a covering of* $\mathfrak{A}_n$, *that is,*

$$\mathfrak{A}_n = X_1 \cup X_2 \cup \dots \cup X_k. \tag{4.5.22}$$

Indeed, if the system (4.5.19) is inconsistent, then $Y = \emptyset$ and

$$\bar{Y} = \bar{Y}_1 \cup \bar{Y}_2 \cup \dots \cup \bar{Y}_k = \mathfrak{A}_n.$$

Since $X_i = \bar{Y}_i$, $1 \le i \le k$, we have

$$X_1 \cup X_2 \cup \cdots \cup X_k = \mathfrak{A}_n.$$

Now let the condition (4.5.22) be fulfilled. Then

$$\bar{X}_1 \cap \bar{X}_2 \cap \cdots \cap \bar{X}_k = \emptyset,$$

and it follows from the equalities $X_i = \bar{Y}_i$, $1 \le i \le k$, that

$$Y = Y_1 \cap Y_2 \cap \cdots \cap Y_k = \emptyset;$$

that is, the system (4.5.19) is inconsistent.

An inconsistent system of the form (4.5.19) is called t-quasiconsistent if for any t equations of the system there exists a binary vector $(x_1^0, x_2^0, \dots, x_n^0) \in \mathfrak{A}_n$ such that it is not a solution of each of these t equations and is a solution of the subsystem of the remaining $k - t$ equations.

For $t = 1$ such a system is called quasiconsistent. In this case, for any equation there exists a vector $(x_1^0, x_2^0, \dots, x_n^0) \in \mathfrak{A}_n$ which is not a solution of this equation but is a solution of the system consisting of the remaining $k - 1$ equations.

It is clear that for $k = 2$ any inconsistent system of the form (4.5.19) is quasiconsistent. For $k > 2$, this is obviously not true. As an example of a quasiconsistent system for $k = 3$, $n = 3$, we consider the system of linear equations over the field $GF(2)$

$$\begin{aligned} x_1 + x_2 &= 1, \\ x_2 + x_3 &= 1, \\ x_1 + x_3 &= 1. \end{aligned}$$

The rank of the matrix of this system is 2, and the rank of the extended matrix is 3; therefore the system is inconsistent. At the same time, the subsystem consisting of the first and second equations has the solutions (1 0 1) and (0 1 0), the subsystem consisting of the first and third equations has the solutions (0 1 1) and (1 0 0), and the subsystem consisting of the second and third equations has the solutions (1 1 0) and (0 0 1).

THEOREM 4.5.5. *A system of Boolean equations of the form* (4.5.19) *is t-quasiconsistent if and only if the family of sets $X_1, \dots, X_k$ defined by* (4.5.20) *is a t-minimal covering of the set $\mathfrak{A}_n$ of all binary n-dimensional vectors.*

PROOF. Let the system (4.5.19) be t-quasiconsistent, and let $(x_1^0, x_2^0, \dots, x_n^0)$ be a solution of the subsystem with numbers $j_1, \dots, j_{k-t}$ but not a solution of any other equation with the numbers $i_1, \dots, i_t$, $\{i_1, \dots, i_t\} \cup \{j_1, \dots, j_{k-t}\} = N_k$. Then

$$(x_1^0, \dots, x_n^0) \in X_{i_1} \cap \cdots \cap X_{i_t}, \qquad (x_1^0, \dots, x_n^0) \notin X_{j_1} \cup \cdots \cup X_{j_{k-t}}.$$

Consequently, the element $(x_1^0, \dots, x_n^0) \in \mathfrak{A}_n$ is t-covered. Since such an element exists for any t equations of the system (4.5.19), any t sets of the family $X_1, X_2, \dots, X_k$ contain a t-covered element. Consequently, $X_1, X_2, \dots, X_k$ is a t-minimal covering of $\mathfrak{A}_n$.

Now let the family $X_1, X_2, \dots, X_n$ be a t-minimal covering. Then for any t indices $1 \le i_1 < \cdots < i_t \le k$ there exists a t-covered element $(x_1^0, \dots, x_n^0) \in \mathfrak{A}_n$ corresponding to the sets $X_{i_1}, \dots, X_{i_t}$. Since $(x_1^0, \dots, x_n^0) \in X_{i_1} \cap \cdots \cap X_{i_t}$, the element $(x_1^0, \dots, x_n^0)$ is not a solution of equations with numbers $i_1, \dots, i_t$. In turn, $(x_1^0, \dots, x_n^0) \notin X_{j_1} \cup \dots \cup X_{j_{k-t}}$, where $\{j_1, \dots, j_{k-t}\} = N_k - \{i_1, \dots, i_t\}$. Thus,

$$(x_1^0, \dots, x_n^0) \in Y_{j_1} \cap \dots \cap Y_{j_{k-1}},$$

that is, $(x_1^0, \dots, x_n^0)$ satisfies the subsystem of equations with numbers $j_1, \dots, j_{k-t}$. □

COROLLARY 4.5.6. *A system of Boolean equations of the form* (4.5.19) *is quasi-consistent if and only if the family of sets defined by* (4.5.20) *is a minimal covering of the set* $\mathfrak{A}_n$ *of all binary* n*-dimensional vectors.*

Let functions $f_1, \dots, f_n$ be randomly chosen from the set of all Boolean functions of n variables with equal probabilities, and independently of each other. Then the corresponding sets $X_1, \dots, X_n$ are random equiprobable subsets of the set $\mathfrak{A}_n$. If $\xi_n^{(t)}$ is the random variable equal to the number of equations in a t-quasiconsistent system, then it coincides with the number of blocks in the corresponding t-minimal covering of $\mathfrak{A}_n$. In the following section, it is shown that if $n \to \infty$ and $t = 1$, then the random variable $\xi_n^{(1)}$ tends to a nonstandard discrete distribution, and for $t > 1$ the random variable $\xi_n^{(t)}$ is asymptotically degenerate at zero. For comparison, note that the distribution of the number of blocks in a random covering of $\mathfrak{A}_n$ is asymptotically Gaussian. The number of equations when for the first time the system (4.5.19) becomes inconsistent in the process of consecutive addition of equations has also the same limit distribution.

Problems.

1. Prove Theorem 4.5.3.
2. Prove that the number of k-block t-minimal coverings of an n-set is equal to the number of k-block $(k-t)$-minimal coverings of an n-set.
3. Let $L_n^{(k)}$ be the number of k-block k-minimal coverings, and $n/2^k \to \alpha < \infty$ as $n \to \infty$. Show that
$$\lim_{n\to\infty} L_n^{(k)} 2^{-kn} = e^{-\alpha}(1 - e^{-\alpha}).$$

4.6. Random coverings of sets

Using results of the previous section, we consider probabilistic properties of random coverings of an n-set X as $n \to \infty$. First we analyze random coverings without any restrictions. Then we investigate some limit characteristics of minimal and t-minimal random coverings.

4.6.1. Random coverings. Take the uniform distribution on the set of all coverings of an n-set X, and let ξ_n be the random variable equal to the number of blocks in a random covering. It is obvious that

$$\mathsf{P}(\xi_n = k) = \frac{D_{nk}}{D_n}, \qquad k = 1, 2, \dots, 2^n - 1,$$

where D_{nk} and D_n are given by formulas (4.5.1) and (4.5.2) respectively. The following theorem describes the limit distribution of ξ_n as $n \to \infty$.

THEOREM 4.6.1. *Let* $x_{nk} = (k - (2^{n-1} - 1/2))/2^{n/2-1}$. *If* $n \to \infty$, *then for all integer* k *such that* $x_{nk} 2^{-n/6} \to 0$

$$\mathsf{P}\left\{ \frac{\xi_n - (2^{n-1} - 1/2)}{2^{n/2-1}} = x_{nk} \right\} = \frac{1}{\sqrt{\pi 2^{n-1}}} e^{-x_{nk}^2/2}(1 + o(1));$$

that is, for ξ_n the local limit theorem is valid with centering and normalizing parameters

$$(2^{n-1} - 1/2, 2^{n/2-1}).$$

The remainder tends to zero uniformly with respect to the integers k such that $|x_{nk}|$ is bounded.

PROOF. It follows from (4.5.1), (4.5.2) and the Bonferroni inequalities that

$$\left|\mathbf{P}(\xi_n = k) - \frac{1}{D_n}\binom{2^n - 1}{k}\right| \leq \frac{n}{D_n}\binom{2^{n-1} - 1}{2^{n-2}},$$

$$1 - \frac{n}{2^{2^n-1}} \leq \frac{D_n}{2^{2^n-1}} \leq 1.$$

Hence, applying the Stirling formula, we obtain

$$\left|\mathbf{P}(\xi_n = k) - \frac{1}{2^{2^n-1}}\binom{2^n - 1}{k}\right| \leq \frac{n}{2^{2^{n-1}-1+(n-1)/2}}.$$

Using the Gaussian approximation for the binomial distribution, we find that

$$\frac{1}{2^{2^n-1}}\binom{2^n - 1}{2^{n-1} - 1/2 + x2^{n/2-1}} = \frac{1}{\sqrt{2\pi 2^{n-2}}}e^{-x^2/2}\left(1 + O\left(\frac{x}{2^{n/6}}\right)\right)$$

as $n \to \infty$ and $2^{-n/6}x \to \infty$. This estimate proves the theorem. □

The following integral theorem is a direct corollary of the local theorem proved above.

THEOREM 4.6.2. *If $n \to \infty$, then for any constants α, β, $\alpha < \beta$,*

$$\mathbf{P}\left\{\alpha \leq \frac{\xi_n - (2^{n-1} - 1/2)}{2^{n/2-1}} \leq \beta\right\} \sim \Phi(\beta) - \Phi(\alpha),$$

where

$$\Phi(x) = \frac{1}{\sqrt{2\pi}}\int_{-\infty}^{x} e^{-u^2/2}du.$$

4.6.2. Random minimal coverings. Here we show that random minimal coverings differ significantly from the random coverings without any restrictions. In particular, the limit distribution of the number of blocks of a minimal covering as $n \to \infty$ is of a nonstandard form, and thus it is not Gaussian.

We begin by finding estimates for the polynomials of minimal coverings defined by (4.5.17).

THEOREM 4.6.3. *For all z, $-\delta \leq z \leq \delta$, and all $r = 1, 2, \ldots$ the following estimates are true. If r is even, then*

$$|\varphi_r(e^{iz})| \leq \frac{2^{(r/2+1)^2}e}{\sqrt{2\pi r}}(1 + 2C_0), \tag{4.6.1}$$

and if r is odd, then

$$|\varphi_r(e^{iz})| \leq \frac{2^{(r/2+1)^2-1/4}e}{\sqrt{2\pi r}}(2C_1), \tag{4.6.2}$$

where

$$C_0 = \sum_{j=1}^{\infty}\frac{1}{2^{j^2}}, \qquad C_1 = \sum_{j=0}^{\infty}\frac{1}{2^{j(j+1)}}, \tag{4.6.3}$$

and $i = \sqrt{-1}$.

Indeed, it follows from (4.5.17) that

$$|\varphi_r(e^{iz})| \le \sum_{k=1}^{r} \binom{r}{k} 2^{k(r-k)}.$$

Applying the Stirling inequalities

$$\sqrt{2\pi} r^{r+1/2} e^{-r} < r! < \sqrt{2\pi} r^{r+1/2} e^{-r+1}, \qquad r = 1, 2, \dots,$$

we see that for even r

$$|\varphi_r(e^{iz})| \le \frac{2^{(r/2+1)^2} e}{\sqrt{2\pi r}} \left(1 + \sum_{j=1}^{r/2-1} \frac{1}{2^{j^2}} + \sum_{j=1}^{r/2} \frac{1}{2^{j^2}} \right).$$

Similarly, for odd r,

$$|\varphi_r(e^{iz})| \le \frac{2^{(r/2+1)^2-1/4} e}{\sqrt{2\pi r}} \left(\sum_{j=0}^{(r-1)/2-1} \frac{1}{2^{j(j+1)}} + \sum_{j=0}^{(r-1)/2} \frac{1}{2^{j(j+1)}} \right).$$

The assertions of the theorem follow from the two last estimates.

The following theorem provides an asymptotic estimate of $\varphi(e^{iz})$ as $r \to \infty$.

THEOREM 4.6.4. *If* $r \to \infty$, *taking even values, then*

$$\varphi_r(e^{iz}) = \frac{2^{(r/2+1)^2} e^{izr/2}}{\sqrt{2\pi r}} (\Psi_0(e^{iz}) = o(1)). \tag{4.6.4}$$

and if $r \to \infty$, *taking odd values, then*

$$\varphi_r(e^{iz}) = \frac{2^{(r/2+1)^2-1/4} e^{iz(r-1)/2}}{\sqrt{2\pi r}} (\Psi_1(e^{iz}) = o(1)), \tag{4.6.5}$$

where Ψ_0 *and* Ψ_1 *are analytic functions defined for* $-\delta \le z \le \delta$, $\delta > 0$, *by absolutely and uniformly converging series*

$$\Psi_0(e^{iz}) = 1 + \sum_{j=1}^{\infty} \frac{e^{-izj}}{2^{j^2}} + \sum_{j=1}^{\infty} \frac{e^{izj}}{2^{j^2}}, \tag{4.6.6}$$

$$\Psi_0(e^{iz}) = \sum_{j=1}^{\infty} \frac{e^{-izj}}{2^{j(j+1)}} + \sum_{j=0}^{\infty} \frac{e^{iz(j+1)}}{2^{j(j+1)}}, \tag{4.6.7}$$

and the remainders in (4.6.4) *and* (4.6.5) *tend to zero uniformly in* z *such that* $-\delta \le z \le \delta$, $\delta > 0$.

PROOF. We set $\gamma = \lfloor 2 \log_2 r + \omega(r) \rfloor$, where $\omega(r) \to \infty$ as $r \to \infty$ in such a way that $\omega(r)/\log_2 r \to 0$. In (4.5.17), we put $x = e^{iz}$ and split the sum in the right-hand side into two sums so that

$$\varphi_r(e^{iz}) = \varphi_r^{(1)}(e^{iz}) + \varphi_r^{(2)}(e^{iz}),$$

where

$$\varphi_r^{(1)}(e^{iz}) = \sum_{k=\lfloor \log_2(r+1)\rfloor}^{\gamma-1} \binom{r}{k}(2^k-k-1)_{r-k}e^{ikz},$$

$$\varphi_r^{(2)}(e^{iz}) = \sum_{k=\gamma}^{r} \binom{r}{k}(2^k-k-1)_{r-k}e^{ikz}.$$

It is obvious that

$$|\varphi_r^{(1)}(e^{iz})| \le 2^{\gamma(r+1)}. \tag{4.6.8}$$

If $r \to \infty$ and $k \ge 2\log_2 r + \omega(r)$, then

$$(2^k-k-1)_{r-k} = 2^{k(r-k)}(1+O(2^{-\omega(r)})).$$

Therefore

$$\varphi_r^{(2)}(e^{iz}) = \sum_{k=\gamma}^{r} \binom{r}{k} 2^{k(r-k)}e^{ikz}(1+O(2^{-\omega(r)})).$$

If $r \to \infty$ taking even values, then we can represent the sum in the right-hand side of the previous relation in the form

$$\sum_{k=\gamma}^{r} \binom{r}{k} 2^{k(r-k)}e^{ikz} = \frac{2^{(r/2+1)^2}e^{irz/2}}{\sqrt{2\pi r}} = (1+S_1+S_2+o(1)), \tag{4.6.9}$$

where

$$S_1 = S_1(r,z) = \sum_{j=1}^{r/2-\gamma} \frac{e^{-izj}}{2^{j^2}} H_{rj},$$

$$S_2 = S_2(r,z) = \sum_{j=1}^{r/2} \frac{e^{izj}}{2^{j^2}} H_{rj},$$

and

$$H_{rj} = \prod_{\nu=1}^{j} \frac{1-2(\nu-1)/r}{1+2\nu/r}.$$

If $n \to \infty$, $0 < \varepsilon < 1/2$, $1 \le j \le r^{1/2-\varepsilon}$, then

$$H_{rj} = 1 + O(r^{-2\varepsilon}).$$

Using this estimate, we find that

$$S_1 = \sum_{j=1}^{r^{1/2-\varepsilon}} \frac{e^{-izj}}{2^{j^2}}(1+O(r^{-2\varepsilon})) + O(r2^{-r^{1-2\varepsilon}}),$$

$$S_2 = \sum_{j=1}^{r^{1/2-\varepsilon}} \frac{e^{izj}}{2^{j^2}}(1+O(r^{-2\varepsilon})) + O(r2^{-r^{1-2\varepsilon}}).$$

Extending the summation in the sums in the right-hand sides of S_1 and S_2 to infinity, we make an error of the order of the last remainder term. Therefore substituting the estimates for S_1 and S_2 into (4.6.9), we obtain

$$\varphi_r^{(2)}(e^{iz}) = \frac{2^{(r/2+1)^2}e^{irz/2}}{\sqrt{2\pi r}}(\Psi_0(e^{iz})+o(1)). \tag{4.6.10}$$

In addition, estimates (4.6.8) and (4.6.10) imply that for $-\delta \le z \le \delta$ we have

$$\varphi_r^{(1)}(e^{iz})/\varphi_r^{(2)}(e^{iz}) \to 0$$

as $r \to \infty$. Thus, the assertion of the theorem for even n is proved. The asymptotic expression of $\varphi_r(e^{iz})$ for odd n is derived similarly. □

We put $t = 1$ in (4.5.14) and consider the function

$$F_n(x) = \sum_{r=1}^{n} \sigma(n,r)\varphi_r(x). \tag{4.6.11}$$

Let us find the asymptotic representation of $F_n(e^{iz})$ for real z such that $-\delta \le z \le \delta$, $\delta > 0$, as $n \to \infty$.

THEOREM 4.6.5. *If $n \to \infty$ taking even values, then*

$$F_n(e^{iz}) = \frac{2^{(n/2+1)^2} e^{inz/2}}{\sqrt{2\pi n}}(\Psi_0(e^{iz}) + o(1)), \tag{4.6.12}$$

and if $n \to \infty$ taking odd values, then

$$F_n(e^{iz}) = \frac{2^{(n/2+1)^2-1/4} e^{i(n-1)z/2}}{\sqrt{2\pi n}}(\Psi_1(e^{iz}) + o(1)), \tag{4.6.13}$$

where the remainders tend to zero uniformly in z such that $-\delta \le z \le \delta$, $\delta > 0$, and Ψ_0 and Ψ_1 are defined by (4.6.6) *and* (4.6.7), *respectively.*

PROOF. By (4.6.11),

$$\frac{F_n(e^{iz})}{\varphi_n(1)} = \frac{\varphi_n(e^{iz})}{\varphi_n(1)} R_n(z), \tag{4.6.14}$$

where

$$R_n(z) = \sum_{r=1}^{n} \sigma(n,r)\frac{\varphi_r(e^{iz})}{\varphi_n(1)}. \tag{4.6.15}$$

In order to estimate the terms of the last sum, we must consider four cases, depending on the parity of r and n. These cases are treated similarly, and the estimates have the same order; therefore we consider only the case where r and n are even.

For the Stirling numbers of the second kind, the estimate

$$\sigma(n,r) \le \frac{r^n}{r!}, \qquad 1 \le r \le n,$$

is obviously valid. Using this estimate and Theorems 4.6.3 and 4.6.4, we find that, for sufficiently large n and all r such that $1 \le r \le n-4$,

$$\sigma(n,r)\left|\frac{\varphi_r(e^{iz})}{\varphi_n(1)}\right| \le \frac{1}{r}\sqrt{\frac{n}{2\pi}}\left(\frac{re^{(r+1)/(n-r)}}{2^{(n+r)/4+1}}\right)^{n-r}.$$

The right-hand side of this inequality increases as r increases; therefore, the value for $r = n-4$ gives an estimate which is valid for all r, $1 \le r \le n-4$, and is uniform, as $n \to \infty$, with respect to z, $-\delta \le z \le \delta$, $\delta > 0$:

$$\sigma(n,r)\left|\frac{\varphi_r(e^{iz})}{\varphi_n(1)}\right| = O(n^4(e/4)^n). \tag{4.6.16}$$

To estimate the Stirling numbers of the second kind for r, $n-3 \le r \le n-1$, as $n \to \infty$, we use the asymptotic formula (see [**19**])

$$\sigma(n,r) = \frac{(r^2/2)^{n-r}}{(n-r)!}(1+o(1)).$$

Then it follows from Theorem 4.6.4 that for r, $n-3 \le r \le n-1$,

$$\sigma(n,r)\left|\frac{\varphi_r(e^{iz})}{\varphi_n(1)}\right| = O(n^2 2^{-n/2}) \tag{4.6.17}$$

as $n \to \infty$. Estimates (4.6.16), (4.6.17) and equalities (4.6.14), (4.6.15) imply the representation (4.6.12). The representation (4.6.13) is derived similarly. □

COROLLARY 4.6.6. *For the number L_n of minimal coverings of an n-set, the following asymptotic formulas are true. If $n \to \infty$ taking even values, then*

$$L_n = \frac{2^{(n/2+1)^2}}{\sqrt{2\pi n}}(1+2C_0+o(1)), \tag{4.6.18}$$

and if $n \to \infty$ taking odd values, then

$$L_n = \frac{2^{(n/2+1)^2-1/4}}{\sqrt{2\pi n}}(2C_1+o(1)), \tag{4.6.19}$$

where C_0 and C_1 are the constants defined in Theorem 4.6.3.

Corollary 4.6.6 follows from Theorem 4.6.5 and the obvious equality

$$L_n = F_n(1), \qquad n = 1, 2, \ldots.$$

Let $\xi_n^{(1)}$ be the random variable equal to the number of blocks in a random minimal covering of an n-set. It is clear that

$$\mathbf{P}(\xi_n^{(1)} = k) = \frac{L_n(k)}{L_n}, \qquad k = 1, 2, \ldots, n,$$

where $L_n(k)$ is defined by (4.5.4) and

$$L_n = \sum_{k=1}^{n} L_n(k).$$

It follows from (4.5.13) with $t = 1$ that the characteristic function of $\xi_n^{(1)}$ is equal to

$$f_n(z) = \frac{1}{L_n}F_n(e^{iz}), \qquad n = 1, 2, \ldots.$$

COROLLARY 4.6.7. *If $n \to \infty$ taking even values, then*

$$f_n(z) = e^{inz/2}\frac{\Psi_0(e^{iz})}{1+2C_0} + o(1),$$

and if $n \to \infty$ taking odd values, then

$$f_n(z) = e^{i(n-1)z/2}\frac{\Psi_1(e^{iz})}{2C_1} + o(1),$$

where the remainders tend to zero uniformly in z such that $-\delta \le z \le \delta$, $\delta > 0$.

The following theorem gives the limit distribution of $\xi_n^{(1)}$ as $n \to \infty$.

THEOREM 4.6.8. *The random variable $\xi_n^{(1)}$ has the following limit distributions as $n \to \infty$. If $n \to \infty$ taking even values, then for any fixed $j = 0, 1, 2, \ldots$*

$$\lim_{n\to\infty} \mathsf{P}(\xi_n^{(1)} = n/2 - j) = \lim_{n\to\infty} \mathsf{P}(\xi_n^{(1)} = n/2 + j) = \frac{2^{-j^2}}{1 + 2C_0},$$

and if $n \to \infty$ taking odd values, then for any fixed $j = 0, 1, 2, \ldots$

$$\lim_{n\to\infty} \mathsf{P}(\xi_n^{(1)} = (n-1)/2 - j) = \lim_{n\to\infty} \mathsf{P}(\xi_n^{(1)} = (n+1)/2 + j) = \frac{2^{-j(j+1)}}{2C_1}.$$

The theorem is a straightforward consequence of Corollary 4.6.7 and the fact that a characteristic function uniquely determines the corresponding probability distribution.

Thus, as $n \to \infty$, for the random variable $\xi_n^{(1)}$ there are two limit discrete distributions with characteristic functions $\Psi_0(e^{iz})/(1 + 2C_0)$ or $\Psi_1(e^{iz})/(2C_1)$ depending on whether the parameter n tends to infinity assuming even or odd values. By taking the derivative at zero of these characteristic functions, we obtain the expressions for the kth moments M_k^0 and M_k^1, $k = 1, 2, \ldots$, in the first and the second cases respectively:

$$M_k^0 = \frac{(-1)^k + 1}{1 + 2C_0} \sum_{j=1}^{\infty} \frac{j^k}{2^{j^2}}, \qquad M_k^1 = \frac{1}{2C_1} \sum_{j=0}^{\infty} \frac{(j+1)^k + (-j)^k}{2^{j(j+1)}}.$$

Hence the mean values of the limit distributions of $\xi_n^{(1)}$ are equal to 0 and 1/2 and the variances are of the form

$$D^0 = \frac{2}{1 + 2C_0} \sum_{j=1}^{\infty} \frac{j^2}{2^{j^2}}, \qquad D^1 = \frac{1}{C_1} \sum_{j=1}^{\infty} \frac{j(j+1)}{2^{j(j+1)}} + \frac{1}{2}$$

in the first and the second cases respectively.

4.6.3. Single-covered elements. Let η_n be the number of single-covered elements in a random equiprobable minimal covering of an n-set.

THEOREM 4.6.9. *The limit distribution of the random variable η_n as $n \to \infty$ coincides with the limit distribution of the random variable $\xi_n^{(1)}$ given in Theorem 4.6.8.*

PROOF. Indeed,

$$\mathsf{P}(\eta_n = j) = L_{nj}/L_n, \qquad j = 1, 2, \ldots, n,$$

where, according to (4.5.7) with $t = 1$,

$$L_{nj} = \binom{n}{j} \sum_{k=1}^{j} (2^k - k - 1)^{n-j} \sigma(j, k).$$

It follows from this formula and Corollary 4.6.6 that if $n \to \infty$ taking even values, then for any fixed integer ν

$$\mathsf{P}(\eta_n = n/2 + \nu) = \frac{2^{-\nu^2}}{1 + 2C_0}(1 + R_n),$$

where

$$|R_n| \le \sum_{k=1}^{n/2+\nu-1} \frac{\sigma(n/2 + \nu, k)}{2^{n/2 - \nu(n/2+\nu-k)}}.$$

Using the estimates

$$\sigma(n/2+\nu,k) \le \frac{k^{n/2+\nu}}{k!}, \qquad 1 \le k \le \frac{n}{2}+\nu-2,$$
$$\sigma(n/2+\nu,n/2+\nu-1) = \frac{1}{2}(n/2+\nu-1)^2(1+o(1)),$$

we find that

$$R_n = O(n2^{-n/2}) + O(n^3(\sqrt{e}/4)^n).$$

Consequently, for any fixed $\nu = 0, \pm 1, \dots,$

$$\lim_{n\to\infty} \mathbf{P}(\eta_n = n/2+\nu) = \frac{2^{-\nu^2}}{1+2C_0}.$$

The case of odd n is treated similarly. □

Theorem 4.6.9 means that for large n with probability close to one each block of a minimal covering contains exactly one single-covered element of an n-set. A more precise formulation of this property is given in the following theorem.

THEOREM 4.6.10. *For the conditional probabilities, the following equalities hold. If $n \to \infty$ taking even values, then for any fixed $\mu = 0, \pm 1, \dots$*

$$\mathbf{P}(\eta_n = n/2+\mu \mid \xi_n^{(1)} = n/2+\mu) \to 1,$$

and if $n \to \infty$ taking odd values, then for any fixed $\mu = 0, 1, 2, \dots$

$$\mathbf{P}(\eta_n = (n+1)/2-\mu \mid \xi_n^{(1)} = (n+1)/2-\mu) \to 1,$$
$$\mathbf{P}(\eta_n = (n+1)/2+\mu \mid \xi_n^{(1)} = (n+1)/2+\mu) \to 1.$$

PROOF. Formula (4.5.7) and Corollary 4.6.6 imply that if $n \to \infty$ taking even values and ν and μ are fixed, then

$$\mathbf{P}(\xi_n^{(1)} = n/2+\mu,\ \eta_n = n/2+\mu+\nu) = \frac{1}{\nu!}\left(\frac{(n/2+\mu)^2}{2^{n/2+\mu+1}}\right)^{\nu} \frac{2^{-\mu^2}}{1+C_0}(1+o(1)).$$

Hence these probabilities in limit as $n \to \infty$ are not zero only if $\nu = 0$. Consequently, for any fixed $\mu = 0, \pm 1, \dots,$

$$\mathbf{P}(\xi_n^{(1)} = n/2+\mu,\ \eta_n = n/2+\mu) = \frac{2^{-\mu^2}}{1+2C_0} + o(1).$$

The first assertion of the theorem follows from these equalities.

For $n \to \infty$ taking odd values, we similarly obtain

$$\mathbf{P}(\xi_n^{(1)} = (n-1)/2+\mu, \eta_n = (n-1)/2+\mu+\nu)$$
$$= \frac{1}{\nu!}\left(\frac{((n-1)/2+\mu)^2}{2^{(n-1)/2+\mu+1}}\right)^{\nu} \frac{2^{-\mu(\mu-1)}}{2C_1}(1+o(1)).$$

Hence

$$\mathbf{P}(\xi_n^{(1)} = (n-1)/2-\mu, \eta_n = (n-1)/2-\mu)$$
$$= \mathbf{P}(\xi_n^{(1)} = (n+1)/2+\mu, \eta_n = (n+1)/2+\mu) = \frac{2^{-\mu(mu-1)}}{2C_1} = o(1).$$

The assertion of the theorem in the case of odd n follows from these equalities. □

CHAPTER 5

Totally Indecomposable, Chainable, and Prime Matrices

Introduction

This chapter is devoted to the study of integral properties of nonnegative matrices related to such concepts as total indecomposability, r-indecomposability, chainability, etc. All these properties depend on the supports of a matrix, which in many cases allows us to consider $(0, 1)$-matrices. Moreover, the support of a matrix of order n can be interpreted as the adjacency matrix of some directed graph with n vertices. This interpretation allows us to reduce the analysis of properties of various types of indecomposability to problems on connectivity and strong connectivity of the corresponding graphs.

In Section 5.1, we give basic properties of totally indecomposable square matrices. In particular, we show that the multiplication of a totally indecomposable matrix by a vector which is neither a zero vector nor a strictly positive vector yields a vector with fewer zero components. Hence we immediately obtain that the exponent of a totally indecomposable matrix of order n does not exceed $n-1$. The product of a finite number of totally indecomposable matrices is a totally indecomposable matrix as well.

Further, we study properties of chainable matrices which have chains of positive elements. It turns out that a nonnegative matrix is totally indecomposable if and only if it is chainable and possesses a doubly stochastic model. We give the canonical form of a chainable matrix. Then we consider nearly decomposable and nearly partially decomposable matrices. It turns out that a nearly decomposable matrix corresponds to a minimal connected graph. We give the canonical form of a nearly decomposable matrix. The presentation of the well-known results in Section 5.1 follows the books [**45, 75**]. The presentation of those results which have not been covered adequately in books is based on the papers [**51, 68, 101**].

In Section 5.2, we consider rectangular nonnegative $m \times n$ matrices, for which we introduce the notions of total indecomposability, semi-total indecomposability, and weak total indecomposability. For $m = n$, these three notions coincide with the notion of the total indecomposability of a square matrix. We show that an $m \times n$ matrix, $m \leq n$, is totally indecomposable if and only if any submatrix of order m is totally indecomposable. In addition, if a matrix is totally indecomposable, then it is semi-totally indecomposable. In turn, if a matrix is semi-totally indecomposable, then it is weakly totally indecomposable. These classes of matrices are characterized using the notion of the nullity of a matrix. The nullity of a matrix is the maximum of the sum $m+n$ for all zero $m \times n$ submatrices of this matrix. In its turn, as in the classical König Theorem, the nullity is related to the proper cover rank, which takes into account both rows and columns. The study of total indecomposability leads us to the necessity of considering bipartite graphs and of reducing the question of whether a matrix belongs to one of the above classes to the determination of decomposability properties of the corresponding graph. The positivity condition for the permanent of an $m \times n$ matrix A, $m \leq n$,and the condition of positivity of the permanent of every order-m submatrix of A result in two more classes of matrices, called essentially nonsingular and strictly essentially nonsingular. All five classes are characterized according to the corresponding nullity value. For matrices of these classes, as well as for products of such matrices, the proper rank is estimated. We give a lower bound for the term rank and for the nullity of a product of matrices. In conclusion, we estimate the class exponents of totally indecomposable matrices. Section 5.2 is based for the most part on the papers [**41, 33, 90**].

In Section 5.3, we consider the special class of rectangular nonnegative matrices called chainable. This class of nonnegative matrices is characterized by the property that we can reach any positive element of an arbitrary matrix of the class by starting from any other positive element of the matrix and moving from one positive element of the matrix to another in the same way as a rook moves on a chessboard. If a $(0, 1)$-matrix of order $m \times n$ is viewed as the adjacency matrix of a bipartite graph, then the fact that the matrix is chainable is equivalent to the connectivity of the graph. A chainable matrix that stops being chainable as soon as a single positive element is removed is called minimal chainable. An $m \times n$ matrix is minimal chainable if and only if it contains exactly $m + n - 1$ positive elements. For chainable matrices we introduce a transformation related to permutations of zeros and ones in chains. We show that two chainable $(0, 1)$-matrices with the same number of ones can be transformed one into another by the use of this transformation. Finally, we show that the product of square chainable matrices is chainable and, moreover, the set of chainable matrices is a two-sided ideal in the semigroup of matrices without zero rows or columns. The main part of Section 5.3 is based on [**52**].

By the König–Frobenius Theorem, a nonnegative $m \times n$ matrix, $m \leq n$, possesses a partial positive diagonal of length $t \leq m$ if and only if it does not contain a zero $p \times q$ submatrix, $p + q = m + n - t + 1$, $p, q \geq 1$. For $t = m$ the diagonal is said to be a row diagonal. In [**25**], for totally indecomposable matrices it is proved that any partial diagonal of length t, $1 \leq t \leq m$, can be extended to a row diagonal if and only if the matrix has no zero submatrix of order $p \times q$, $p + q = n - t + 1$, $1 \leq p \leq m - t$, $1 \leq q \leq n - t$. This theorem solves the problem of whether partial diagonals of totally indecomposable matrices can be extended. The question on uniqueness of such extension in the class of totally indecomposable matrices leads us to a quite narrow class of matrices with permanent equal to two. These matrices are also considered in [**25**]. Along with the results from that paper, in Section 5.4 we propose new necessary and sufficient conditions for an element to be uniquely extended to the diagonal of a matrix. At the end of the section, we introduce a new notion of d-variant extendibility of a partial diagonal for $(0, 1)$-matrices, which coincides with unique extendibility for $d = 1$. We give necessary and sufficient conditions for d-variant extendibility of a partial diagonal for $(0, 1)$-matrices with two ones in each row and each column.

Sections 5.5 and 5.6 are devoted to the factorization of square nonnegative matrices. In Section 5.5, for the Boolean operations we consider conditions under which a $(0, 1)$-matrix A is algebraically prime, that is, the factorization $A = BC$ means that either B or C is a permutation matrix, whereas A itself is not a permutation matrix. For r-indecomposable matrices, these conditions are related to the number and arrangement of order two submatrices consisting of ones. For totally indecomposable matrices we establish a connection between the primality of a matrix and the value of the multiplicative rank of some submatrix; the multiplicative rank of an $m \times n$ matrix is defined as the minimal k such that the matrix admits the representation as a product of $m \times k$ and $k \times n$ matrices. We demonstrate that the primality of a matrix A is related to its scope, which is the set of the Boolean sums of its columns together with the zero column.

In Section 5.6, we study the property of a matrix to be algebraically prime in the semigroup of nonnegative matrices. It turns out that in this case the primality of a matrix A depends on the properties of its support ${}^sA$. For totally indecomposable matrices, we formulate and prove a theorem which gives sufficient conditions for a matrix A to be prime. These conditions are related to the fact that the inner products of columns of the support do not exceed one. It is worthwhile to notice that any prime matrix can be represented as a direct sum of totally indecomposable matrices. The main results on algebraically prime matrices are contained in [**32, 84**] and in the book [**15**].

5.1. Totally indecomposable and chainable matrices

5.1.1. Totally indecomposable matrices. We consider square nonnegative matrices. A matrix of order n is said to be partially decomposable if it possesses an $r \times s$ zero submatrix, $r+s=n$. In other words, a matrix A is partially decomposable if there exist permutation matrices Π and Π_1 such that

$$\Pi A \Pi_1 = \begin{pmatrix} A_{11} & A_{12} \\ \mathbf{0} & A_{22} \end{pmatrix}, \tag{5.1.1}$$

where A_{11}, A_{22} are square matrices. A matrix A is totally indecomposable if it is not partially decomposable. We consider some properties of the totally indecomposable matrices.

PROPERTY 5.1.1. *A matrix A of order n is totally indecomposable if and only if every permanent minor of order $n-1$ is positive.*

This property follows immediately from the Frobenius–König Theorem (Theorem 1.2.3).

A matrix $\mathbf{I} = (d_{ij})$, $i, j = 1, \ldots, n$, is said to be diagonal if $d_{ii} > 0$, $1 \le i \le n$, and $d_{ij} = 0$, $i \neq j$.

PROPERTY 5.1.2. *Let A and B be nonnegative matrices, let $\mathbf{I}$ be a diagonal matrix, and let*

$$A = \mathbf{I} + B. \tag{5.1.2}$$

Then the matrix A is totally indecomposable if and only if B is indecomposable.

Let the matrix A be totally indecomposable, and let there exist a permutation matrix Π such that

$$\Pi B \Pi^T = \begin{pmatrix} B_{11} & B_{12} \\ \mathbf{0} & B_{22} \end{pmatrix}, \tag{5.1.3}$$

where B_{11}, B_{12} are square matrices and $\mathbf{0}$ is the zero matrix. Then

$$\Pi A \Pi^T = \begin{pmatrix} B_{11} + \mathbf{I}_1 & B_{12} \\ \mathbf{0} & B_{22} + \mathbf{I}_2 \end{pmatrix},$$

where $\mathbf{I}_1$ and $\mathbf{I}_2$ are diagonal matrices of the same orders as B_{11} and B_{22}. Hence A is partially decomposable, which is impossible.

Let the matrix B be indecomposable and the matrix A be partially decomposable, that is, let there exist permutation matrices Π and Π_1 such that

$$\Pi A \Pi_1 = (\mathbf{I} + \Pi B \Pi^T)\Pi\Pi_1 \tag{5.1.4}$$

is of the form (5.1.1). Let A_{11} be a matrix of order r and A_{22} a matrix of order $n-r$. The permutation matrix $\Pi\Pi_1$ permutes columns of the matrix $\mathbf{I} + \Pi B \Pi^T$. Since $\Pi A \Pi_1$ is of the form (5.1.1), the permutation matrix $\Pi\Pi_1$ interchanges the first r columns and the last $n-r$ columns of the matrix $\mathbf{I} + \Pi B \Pi^T$; otherwise the positive elements of the diagonal matrix $\mathbf{I}$ fall into the zero submatrix $\mathbf{0}$. Hence

$$\Pi A \Pi^T = (\mathbf{I} + \Pi B \Pi^T) = \begin{pmatrix} B_{11} + \mathbf{I}_1 & B_{12} \\ \mathbf{0} & B_{22} + \mathbf{I}_2 \end{pmatrix},$$

where B_{11}, B_{22} are square matrices and $\mathbf{0}$ is the zero matrix; that is, the matrix B is partially decomposable, which is impossible.

If T_n is the set of totally indecomposable matrices, D_n is the set of indecomposable matrices, and M_n is the set of nonnegative matrices of order n, then

$$T_n \subset D_n \subset M_n. \tag{5.1.5}$$

Let P and Q be nonempty subsets of the set N_n, and let $A[P, Q]$ denote the submatrix of order $|P| \times |Q|$ of the matrix $A = (a_{ij})$, $i, j = 1, \dots, n$ consisting of the a_{ij} with $i \in P$, $j \in Q$; let $A(P, Q)$ denote the submatrix consisting of the a_{ij} with $i \notin P$, $j \notin Q$. Similarly, the matrices $A[P, Q)$ and $A(P, Q]$ consist of the a_{ij} with $i \in P$, $j \notin Q$ and the a_{ij} with $i \notin P$, $j \in Q$, respectively.

PROPERTY 5.1.3. *If for any nonempty subset $P \subset \mathbf{N}$ we have*

$$A[P, N_n \setminus P] \neq \mathbf{0}$$

(that is, the submatrix is not zero), then $A \in D_n$.

If $A[P, Q] \neq \mathbf{0}$ for any nonempty subsets P and Q of the set $\mathbf{N}$ such that $|P| + |Q| = n$, then $A \in T_n$.

The validity of this property follows immediately from the definitions of indecomposability and total indecomposability.

PROPERTY 5.1.4. *If $A \in D_n \setminus T_n$, then $A = (a_{ij})$, $i, j = 1, \dots, n$, has a zero element on the main diagonal.*

Indeed, the condition $A \notin T_n$ implies the existence of a zero submatrix $A[P, Q]$, $|P| + |Q| = n$. From the condition $A \in D_n$ it follows that $Q \neq N_n \setminus P$; that is, $P \cap Q \neq \emptyset$. Therefore, there exists an element i, $i \notin P$, $i \in Q$, such that $a_{ii} = 0$. Property 5.1.4 implies the sufficiency in Property 5.1.2.

COROLLARY 5.1.1. *If $B \in D_n$, then $A = (\mathbf{I} + B) \in T_n$.*

Indeed, from Property 5.1.4 it follows that $A \in T_n$, because all elements on the main diagonal of A are positive and $B \in D_n$.

Let $A = (a_{ij}) \in M_n$ and

$$I_k = \{i \colon a_{ik} = 0\}, \qquad J_k = \{j \colon a_{kj} = 0\}.$$

For the vector $v = (a_1(v), \dots, a_n(v))$ we set

$$I_0(v) = \{i \colon a_i(v) = 0\}, \qquad I(v) = \{i \colon a_i(v) > 0\}.$$

Let R_n denote the set of n-dimensional nonnegative vectors whose components may take both positive and zero values.

PROPERTY 5.1.5. *A matrix* $A = (a_{ij}) \geq 0$, $i, j = 1, \ldots, n$, *is totally indecomposable if and only if*

$$|I_0(Ax)| < |I_0(x)| \tag{5.1.6}$$

for any $x \in R_n$.

Let A be a totally indecomposable matrix but

$$|I_0(Ay)| \geq |I_0(y)|, \qquad y \in R_n.$$

If $|I_0(y)| = s$, then there exist permutation matrices Π and Π_1 such that

$$\Pi A y = \begin{pmatrix} \mathbf{0} \\ u \end{pmatrix}, \qquad \Pi_1^T y = \begin{pmatrix} \mathbf{0} \\ v \end{pmatrix}, \tag{5.1.7}$$

where $\mathbf{0}$ is the s-dimensional zero column vector, u is an $(n-s)$-dimensional nonnegative column vector, and v is an $(n-s)$-dimensional positive column vector. Thus, from (5.1.7) it follows that

$$\Pi A \Pi_1 \begin{pmatrix} \mathbf{0} \\ v \end{pmatrix} = \begin{pmatrix} \mathbf{0} \\ u \end{pmatrix}. \tag{5.1.8}$$

Using the notation

$$\Pi A \Pi_1 = \begin{pmatrix} A_{11} & A_{12} \\ A_{21} & A_{22} \end{pmatrix}, \tag{5.1.9}$$

where A_{11}, A_{12}, A_{21}, A_{22} are matrices of orders $s \times s$, $s \times (n-s)$, $(n-s) \times s$, $(n-s) \times (n-s)$, respectively, from (5.1.8) we obtain

$$A_{12} v = \mathbf{0}, \tag{5.1.10}$$

where $\mathbf{0}$ is the s-dimensional zero vector. Since $v > 0$ and $A_{12} \geq 0$, equality (5.1.10) implies $A_{12} = \mathbf{0}$; that is, A_{12} is the zero matrix. Hence the matrix A is partially decomposable, which is impossible.

Now let $A \notin T_n$; that is, let the matrix A be not totally indecomposable and hence contain a zero submatrix $A[P, Q]$, where $P, Q \neq \emptyset$, $|P|+|Q| = n$. We choose a vector $x \in R_n$ such that $I(x) = Q$. Then $I_0(x) = N_n \setminus I(x) = \mathbf{N} \setminus Q$, and therefore, $|I_0(x)| = |P|$. Since $a_{kj} = 0$ for $k \in P$, $j \in Q$, and $J_k = \{j \colon a_{kj} = 0\}$, we see that $Q \subseteq J_k$ and therefore $I(x) \subseteq J_k$ for $k \in P$.

A necessary and sufficient condition for

$$a_k(Ax) = \sum_{j=1}^{n} a_{kj} x_j = 0,$$

where $x = (x_1, \ldots, x_n)^T$, is that $a_k(Ax) = 0$, $k \in P$. Hence

$$P \subseteq I_0(Ax) = \{k \colon a_k(Ax) = 0\}.$$

Therefore,

$$|I_0(Ax)| \geq |P| = |I_0(x)|.$$

PROPERTY 5.1.6. *If a matrix A is totally indecomposable, then for any k, $1 \leq k \leq n$,*

$$|I_k| \leq n-2, \qquad |J_k| \leq n-2. \tag{5.1.11}$$

The number of positive elements in A is at least $2n$.

Indeed, if $|I_k| = |\{i\colon a_{ik} = 0\}| \geq n-1$, then the kth column has at most one positive element, which contradicts the condition $A \in T_n$. The second inequality in (5.1.11) is proved similarly. Thus, the total number of zeros in the matrix $A \in T_n$ does not exceed $n(n-2)$; that is, the number of positive elements is at least $2n$.

PROPERTY 5.1.7. *If A is a totally indecomposable matrix of order n, then*

$$A^{n-1} > 0; \tag{5.1.12}$$

that is, for the exponent of the matrix the bound

$$\gamma(A) \leq n-1 \tag{5.1.13}$$

holds.

Since $A \in T_n$, the inequality per $A > 0$ holds and there exist a permutation s of degree n and the corresponding permutation matrix Π such that $A = \alpha\Pi(B + \mathbf{I})$, where

$$0 < \alpha < \min_{1\leq i\leq n} a_{i,s(i)},$$

$\mathbf{I}$ is the identity matrix, and the matrix $B \geq 0$ is indecomposable in view of Property 5.1.2. For $x \in R_n$ with $I_0(Ax) \neq \emptyset$, the relation

$$I_0(Ax) = I_0[\Pi(B+\mathbf{I})x] = I_0[(B+\mathbf{I})x] \subset I_0(x) \tag{5.1.14}$$

is true. Let us prove this. We have

$$I_0[(B+\mathbf{I})x] = I_0(x) \cap I_0(Bx) \subseteq I_0(x).$$

Assume that

$$I_0(x) \cap I_0(Bx) = I_0(x). \tag{5.1.15}$$

This means that

$$I_0(B(x)) \supseteq I_0(x), \qquad x \in R_n. \tag{5.1.16}$$

Let a column vector $x \in R_n$ have s positive elements, $1 \leq s \leq n-1$, and let Π_1 be a permutation matrix such that $\Pi_1(x) = (u, \mathbf{0})^T$, where u is a positive s-dimensional column vector. The column vector

$$Bx = B\Pi_1^T\Pi_1 x = B\Pi_1^T(u, \mathbf{0})^T$$

has the same zero elements as the column vector x has, and, possibly, some additional zeros. Therefore, there exists a nonnegative s-dimensional vector v such that $\Pi_1(Bx) = (v, \mathbf{0})^T$, and the equality

$$\Pi_1 B\Pi_1^T \begin{pmatrix} u \\ \mathbf{0} \end{pmatrix} = \begin{pmatrix} v \\ \mathbf{0} \end{pmatrix} \tag{5.1.17}$$

holds, where $u > 0$, $v \geq 0$ are s-dimensional column vectors. We represent the matrix $\Pi_1 B\Pi_1^T$ in the form

$$\Pi_1 B\Pi_1^T = \begin{pmatrix} B_{11} & B_{12} \\ B_{21} & B_{22} \end{pmatrix}, \tag{5.1.18}$$

where B_{11}, B_{12}, B_{21}, B_{22} are matrices of orders $s \times s$, $s \times (n-s)$, $(n-s) \times s$, $(n-s) \times (n-s)$, respectively. From (5.1.17) and (5.1.18) it follows that $B_{21}u = \mathbf{0}$, where $\mathbf{0}$ is the $(n-s)$-dimensional zero vector. Since $u > 0$ and $B_{21} \geq 0$, we find that B_{21} is the zero matrix, which contradicts the indecomposability of B. Thus, inclusion (5.1.14) is proved. Note that the inclusion remains true for $I_0(Ax) = \emptyset$, that is, in the case $Ax > 0$.

Consider the chain of inclusions

$$I_0(x) \supset I_0(Ax) \supset I_0(A^2x) \supset \cdots \supset I_0(A^{n-1}x). \tag{5.1.19}$$

This chain terminates at the kth step if $A^k x > 0$ and therefore $I_0(A^k x) = \emptyset$. Since the inclusions in (5.1.19) are strict, any such chain for a totally indecomposable matrix A necessarily terminates at $k \leq n-1$, that is, $I_0(A^{n-1}x) = \emptyset$, and hence $A^{n-1}x > 0$. Since the vector $x \in R_n$ is arbitrary, we conclude that $A^{n-1} > 0$.

PROPERTY 5.1.8. *The product of a finite number of totally indecomposable matrices of order n is totally indecomposable.*

It is sufficient to prove this property for two matrices. Let $A, B \in T_n$. Let us prove that

$$|I_0(ABx)| \leq |I_0(Bx)| < |I_0(x)| \tag{5.1.20}$$

for any $x \in R_n$. From the inclusions (5.1.19) for $Bx \in R_n$ we obtain

$$|I_0(ABx)| < |I_0(Bx)| < |I_0(x)|. \tag{5.1.21}$$

If $Bx > 0$, then $I_0(ABx) = \emptyset$, because the totally indecomposable matrix A has no zero rows. Thus, in this case

$$0 = |I_0(ABx)| = |I_0(Bx)| < |I_0(x)|. \tag{5.1.22}$$

From (5.1.21) and (5.1.22) it follows that the inequalities (5.1.20) are true, which yields

$$|I_0(ABx)| < |I_0(x)|$$

for $x \in R_n$. By Property 5.1.5, it now follows that $AB \in T_n$.

PROPERTY 5.1.9. *If $A \geq 0$ is a totally indecomposable matrix and AA^T contains a nonempty zero $s \times t$ submatrix, then $s + t \leq n - 2$.*

Let

$$AA^T = \begin{pmatrix} A_{11} & \mathbf{0} \\ A_{21} & A_{22} \end{pmatrix},$$

where $\mathbf{0}$ is the zero $s \times t$ matrix.

Let the $s \times n$ submatrix of the first s rows of A contain k zero columns. Since A is totally indecomposable, $s + k \leq n-1$. Furthermore, each of the first s rows of A is orthogonal to each of the last t rows, due to the presence of the $s \times t$ zero submatrix in AA^T. Therefore, all columns of A whose first s elements are nonzero have their t last elements equal to zero. Thus, the last t rows of A contain a $t \times (n-k)$ zero submatrix. Since A is totally indecomposable, $t + n - k \leq n - 1$; that is, $t \leq k - 1$. Since we have proved that $s \leq n - k - 1$, we conclude that $s + t \leq n - 2$.

5.1.2. Chainable matrices. Let s be a permutation of degree n. The elements $a_{1,s(1)}, \ldots, a_{n,s(n)}$ of a nonnegative matrix $A = (a_{ij})$, $i, j = 1, \ldots, n$, form the diagonal corresponding to the permutation s. If $a_{i,s(i)} > 0$, $1 \le i \le n$, then the diagonal is called positive (see Section 1.2). The product $a_{1,s(1)} \cdots a_{n,s(n)}$ is called the diagonal product.

We say that a matrix $A = (a_{ij}) \ge 0$, $i, j = 1, \ldots, n$, possesses a doubly stochastic model if there exists a doubly stochastic matrix

$$B = (b_{ij}), \quad i, j = 1, \ldots, n, \quad \sum_{i=1}^{n} b_{ij} = \sum_{j=1}^{n} b_{ij} = 1,$$

such that $a_{ij} = 0$ if and only if $b_{ij} = 0$.

By the Birkhoff Theorem (Theorem 1.2.5) the doubly stochastic matrices form a convex polytope whose vertices are permutation matrices. Hence a matrix A possesses a doubly stochastic model if and only if each positive element of it lies on some positive diagonal.

A matrix $A = (a_{ij})$, $i, j = 1, \ldots, n$, is said to be chainable if for any pair of nonzero elements $a_{i_1 j_1}$ and $a_{i_k j_k}$ there exists a sequence of nonzero elements

$$a_{i_1 j_1}, a_{i_2 j_2}, \ldots, a_{i_{k-1} j_{k-1}}, a_{i_k j_k} \tag{5.1.23}$$

such that either $i_r = i_{r+1}$ or $j_r = j_{r+1}$, $1 \le r \le k-1$. A sequence (5.1.23) is called a chain with end points $a_{i_1 j_1}$ and $a_{i_k j_k}$.

In other words, a matrix is chainable if for an arbitrary pair of distinct positive elements, moving from one positive element to another in the same way as a chess rook moves, one can pass from one element of the pair to the other.

PROPERTY 5.1.10. *A nonnegative matrix A is totally indecomposable if and only if it is chainable and possesses a doubly stochastic model.*

Let a matrix A be chainable and possess a doubly stochastic model. Let A be partially decomposable; then the corresponding doubly stochastic matrix B is partially decomposable as well. Then there exist permutation matrices Π and Π_1 such that

$$\Pi B \Pi_1 = \begin{pmatrix} B_{11} & \mathbf{0} \\ B_{21} & B_{22} \end{pmatrix},$$

where B_{11}, B_{22} are square matrices and $\mathbf{0}$ is the zero matrix. Since $\Pi B \Pi_1$ is doubly stochastic, B_{21} is the zero matrix, that is,

$$\Pi B \Pi_1 = \begin{pmatrix} B_{11} & \mathbf{0} \\ \mathbf{0} & B_{22} \end{pmatrix};$$

thus the matrix $\Pi B \Pi_1$, and hence $\Pi A \Pi_1$, are not chainable. Therefore the matrix A is not chainable, which is impossible.

Now let a matrix $A = (a_{ij})$ be totally indecomposable. Let some element $a_{ij} > 0$ not lie on any positive diagonal. Then the matrix $A(i, j)$, which is the matrix A with the ith row and the jth column removed, satisfies the equality

$$\operatorname{per} A(i, j) = 0.$$

By Property 5.1.1 the matrix A is partially decomposable. This contradiction implies that any element $a_{ij} > 0$ lies on some positive diagonal, and therefore, A possesses a doubly stochastic model.

Let a totally indecomposable matrix A be not chainable. Then there exist elements $a_{i_0j_0} > 0$ and $a_{i_1j_1} > 0$ which are not terminal elements of any chain. We put $H_1 = \{i_0\}$, $K_1 = \{j : a_{i_0j} > 0\}$, and introduce the sets

$$H_s = \left\{ i : i \notin \bigcup_{k=1}^{s-1} H_k,\ a_{ij} > 0 \text{ for some } j \in K_{s-1} \right\},$$

$$K_s = \left\{ j : j \notin \bigcup_{k=1}^{s-1} K_k,\ a_{ij} > 0 \text{ for some } i \in H_{s-1} \right\}, \qquad s = 2, 3, \ldots$$

Let

$$H = \bigcup_s H_s, \qquad K = \bigcup_s K_s.$$

From the construction it follows that $a_{i_0j_0} \in A[H,K]$ and $a_{i_1j_1} \in A(H,K)$; that is, the submatrices $A[H,K]$ and $A(H,K)$ are nonempty. Since the set of positive elements of A is divided into the elements which are joined by chains with the positive elements of the ith row and the elements which do not have this property, $A[H,K)$ and $A(H,K]$ are zero matrices. Thus, there exist permutation matrices Π and Π_1 such that

$$\Pi A \Pi_1 = \begin{pmatrix} A[H,K] & A[H,K) \\ A(H,K] & A(H,K) \end{pmatrix};$$

that is, the matrix A is partially decomposable, which is impossible.

We define $E_{ij} \geq 0$ as the matrix whose elements are all zero except for the element at the ith row and the jth column, which is 1.

PROPERTY 5.1.11. *Let a matrix A of order n be of the form*

$$A = \begin{pmatrix} A_1 & \mathbf{0} & \mathbf{0} & \ldots & \mathbf{0} & \mathbf{0} & E_1 \\ E_2 & A_2 & \mathbf{0} & \ldots & \mathbf{0} & \mathbf{0} & \mathbf{0} \\ \mathbf{0} & E_3 & A_3 & \ldots & \mathbf{0} & \mathbf{0} & \mathbf{0} \\ & & & \ldots & & & \\ \mathbf{0} & \mathbf{0} & \mathbf{0} & \ldots & A_{s-2} & \mathbf{0} & \mathbf{0} \\ \mathbf{0} & \mathbf{0} & \mathbf{0} & \ldots & E_{s-1} & A_{s-1} & \mathbf{0} \\ \mathbf{0} & \mathbf{0} & \mathbf{0} & \ldots & \mathbf{0} & E_s & A_s \end{pmatrix}, \tag{5.1.24}$$

where the square submatrices A_k are totally indecomposable and $E_k \geq 0$ has a single positive element, $k = 1, \ldots, s$, $s > 1$. Then the matrix A

(a) *is chainable,*
(b) *possesses a doubly stochastic model,*
(c) *is totally indecomposable, and*
(d) *possesses the following property: if $a_{ij} > 0$ is an element of some matrix A_k, $1 \leq k \leq s$, then partial decomposability of $A - a_{ij}E_{ij}$ implies partial decomposability of $A_k - a_{ij}E_{ij}$.*

Any matrix A_k, $1 \leq k \leq s$, is chainable, because it is totally indecomposable. There exists a chain with ends at A_i and A_{i+1} passing through a positive element of E_{i+1}. Hence, for any pair of positive elements of A there exists a chain whose end elements coincide with this pair. Thus, (a) is proved.

Any positive element of A_k, $1 \leq k \leq s$, lies on some positive diagonal, because A_k is totally indecomposable. We consider the element of A_k placed in the same row with the positive element of E_k and the same column with the positive element of

$E_{k+1 \,(\mathrm{mod}\ s)}$. If we remove this row and this column from the matrix A_k, then, since A_k is totally indecomposable, for the resulting matrix A'_k the relation per $A'_k > 0$ holds. Positive diagonals of the matrices $A'_1, \ldots, A'_s$ together with positive elements of the matrices $E_1, \ldots, E_s$ form a positive diagonal of A. Thus, (b) is proved.

The assertion (c) on the total indecomposability of the matrix A follows from (a) and (b) and Property 5.1.10.

Finally, we prove (d). Let the matrix A_k remain totally indecomposable after replacing some positive element by zero. Then, according to (c), the matrix A is totally indecomposable. Therefore, if after such a replacement A becomes partially decomposable, A_k possesses the same property.

5.1.3. Nearly decomposable and nearly partially decomposable matrices. Let E_{ij} be a $(0,1)$-matrix of order n with a single element one at the ith row and the jth column. An indecomposable matrix $A \geq 0$ of order n is said to be nearly decomposable if for any element $a_{ij} > 0$ the matrix $A - a_{ij}E_{ij}$ is decomposable. A totally indecomposable matrix $A \geq 0$ of order n is said to be nearly partially decomposable if for any element $a_{ij} > 0$ the matrix $A - a_{ij}E_{ij}$ is partially decomposable.

We consider the directed graph $\Gamma(A)$ whose adjacency matrix is the support of the matrix $A \geq 0$. It is clear that $\Gamma(A)$ is strongly connected if and only if A is indecomposable. A strongly connected graph is said to be minimal connected if after removal of any edge of it the graph ceases to be strongly connected. In a strongly connected graph, each vertex lies on an elementary, or simple, circuit. An elementary circuit containing all vertices of the graph is called complete.

For a connected subgraph Γ' of a graph Γ, we introduce the operation of contraction of Γ', which consists of removing all edges joining the vertices of Γ' and identifying all vertices of Γ' with one of its vertices. If Γ is a minimal connected graph, then each elementary circuit is a subgraph of Γ and the contraction operation can be applied to it.

If a matrix $A \geq 0$ is nearly decomposable, then the corresponding directed graph $\Gamma(A)$ is a minimal connected graph. The following lemma is obvious.

LEMMA 5.1.2. *Let Γ' be a strongly connected subgraph of a minimal connected graph Γ. The operation of contraction of Γ' transforms Γ to a minimal connected graph.*

PROPERTY 5.1.12. *Let A be a nearly decomposable $(0,1)$-matrix of order $n \geq 2$. Then there exist a permutation matrix Π and a positive integer m such that*

$$\Pi A \Pi^T = \begin{pmatrix} A_1 & \mathbf{0} & \ldots & \mathbf{0} & E_1 \\ E_2 & A_2 & \ldots & \mathbf{0} & \mathbf{0} \\ & & \ldots & & \\ \mathbf{0} & \mathbf{0} & \ldots & A_{m-1} & \mathbf{0} \\ \mathbf{0} & \mathbf{0} & \ldots & E_m & A_m \end{pmatrix},$$

where each E_i has exactly one element 1, the A_i, $1 \leq i \leq m$, are nearly decomposable matrices, and the matrices $A_1, A_2, \ldots, A_{m-1}$ are of order 1.

Let V be the set of vertices of the directed graph $\Gamma = \Gamma(A)$ corresponding to the matrix A. We choose an arbitrary vertex $v_1 \in V$. It lies on an elementary circuit in Γ. If this circuit is complete, then we leave v_1 and pass to another vertex. If v_1 lies on a circuit that is not complete, we contract this circuit to some other vertex v_1'. As a result, from the graph Γ we come to a minimal connected graph Γ_1. Depending on whether v_1' does or does not lie on a complete elementary circuit, we either leave this vertex or contract the corresponding elementary circuit to some other vertex. After m contractions we arrive at a graph Γ_m, which is a complete circuit. By the construction, if l vertices of Γ are contracted to some vertex of Γ_m, then the remaining $n-l$ vertices do not participate in the contraction procedure. It is not difficult to see that the subgraph of Γ corresponding to the set of vertices contracted to one vertex is a minimal connected graph. Thus, the digraph $\Gamma(A)$ can be represented as a complete circuit with m vertices and a subgraph which is appended to one vertex of this circuit and contracted to this vertex. Hence there exists a permutation matrix Π such that the matrix $\Pi A \Pi^T$ is of the above form.

PROPERTY 5.1.13. *Let A be a nearly partially decomposable $(0,1)$-matrix of order $n \geq 2$. Then there exist permutation matrices Π and Π' and a positive integer m such that*

$$\Pi A \Pi' = \begin{pmatrix} A_1 & \mathbf{0} & \dots & \mathbf{0} & E_1 \\ E_2 & A_2 & \dots & \mathbf{0} & \mathbf{0} \\ & & \dots & & \\ \mathbf{0} & \mathbf{0} & \dots & A_{m-1} & \mathbf{0} \\ \mathbf{0} & \mathbf{0} & \dots & E_m & A_m \end{pmatrix},$$

where each E_i has exactly one 1, the A_i, $1 \leq i \leq m$, are nearly partially decomposable matrices, and the matrices $A_1, A_2, \dots, A_{m-1}$ are of order 1.

Since A is totally indecomposable, there exist permutation matrices Π_1 and Π_1' such that the main diagonal of the matrix $B = \Pi_1 A \Pi_1'$ is positive. If $\mathbf{I}$ is the identity matrix, then the matrix $B - \mathbf{I}$ is nearly decomposable, and hence there exists a permutation matrix Π_2 such that $\Pi_2(B - \mathbf{I})\Pi_2^T$ is of the form given in Property 5.1.13. Hence the matrix $\Pi A \Pi'$ is of the form given in Property 5.1.13, where $\Pi = \Pi_2 \Pi_1$, $\Pi' = \Pi_1' \Pi_2^T$. Since the matrix A is totally indecomposable, the matrices $A_1, \dots, A_m$ are also totally indecomposable, and therefore, they are nearly partially decomposable.

PROPERTY 5.1.14. *If a doubly stochastic matrix A is nearly decomposable, then it is the permutation matrix corresponding to a unicyclic permutation.*

By Property 5.1.12 the matrix A admits the representation

$$A = \begin{pmatrix} \mathbf{0} & \mathbf{0} & \dots & \mathbf{0} & \mathbf{0} & E_1 \\ 1 & \mathbf{0} & \dots & \mathbf{0} & \mathbf{0} & \mathbf{0} \\ \mathbf{0} & 1 & \dots & \mathbf{0} & \mathbf{0} & \mathbf{0} \\ & & \dots & & & \\ \mathbf{0} & \mathbf{0} & \dots & 1 & \mathbf{0} & \mathbf{0} \\ \mathbf{0} & \mathbf{0} & \dots & \mathbf{0} & E_m & A_m \end{pmatrix},$$

where $m > 1$. Since E_1 has a single element one, A_m contains a zero column. Therefore, $A_m = (0)$; that is, A_m is the zero submatrix of order 1. Hence $E_m = (1)$, and therefore A is a permutation matrix.

PROPERTY 5.1.15. *If a nearly partially decomposable $(0,1)$-matrix A of order n has t rows with three ones and $n-t$ rows with two ones, then the permanent of the matrix A satisfies the estimate*

$$\operatorname{per} A \geq t + 2.$$

For $n = 2$, the only nearly partially decomposable matrix A has all elements equal to one. In this case, $t = 0$ and $\operatorname{per} A = 2$. We use induction on n. Assume that the inequality holds for all nearly partially decomposable matrices of order k, $2 \le k \le n-1$. By Property 5.1.12, a nearly partially decomposable matrix A admits the representation

$$A = \begin{pmatrix} 1 & \mathbf{0} & \dots & \mathbf{0} & E_1 \\ 1 & 1 & \dots & \mathbf{0} & \mathbf{0} \\ & & \dots & & \\ \mathbf{0} & \mathbf{0} & \dots & 1 & \mathbf{0} \\ \mathbf{0} & \mathbf{0} & \dots & E_m & A_m \end{pmatrix},$$

where $m > 1$. If A_m is a 1×1 matrix, then $E_1 = E_m = (1)$, $t = 0$, and $\operatorname{per} A = 2$. If A_m is of order exceeding one, it is nearly partially decomposable, and if A has t rows with three ones, then A_m contains $t-1$ such rows. By the induction assumption, $\operatorname{per} A_m \ge (t-1) + 2 = t+1$. Expanding the permanent of A with respect to the element of E_1 equal to 1, we see that $\operatorname{per} A = \operatorname{per} A_m + 1$, which proves the bound.

Problems.

1. Prove that if a matrix $A \ge 0$ is totally indecomposable, then the matrix AA^T is totally indecomposable as well.
2. Prove that if $A \ge 0$ is a totally indecomposable matrix of order n, then any row of the matrix AA^T contains at most $n-3$ zeros.
3. Let s_i be the number of zeros in the ith row of a totally indecomposable matrix A, and let t_i be the number of zeros in the ith row of AA^T. Prove that $t_i < s_i$.
4. Let $A \ge 0$ be a matrix of order n. Prove that A has k, $k > 0$, zeros on each diagonal if and only if A contains the $s \times t$ zero submatrix, $s + t = n + k$.
5. Let $A = (a_{ij})$ be a totally indecomposable matrix of order n, and let E_{ij} be the matrix of order n with one in the position (i,j) and zeros in all other positions. Prove that if $a_{ij} > 0$, then
$$\operatorname{per} A > \operatorname{per}(A - E_{ij}).$$
6. Let $A = (a_{ij})$ be a totally indecomposable matrix of order n. Prove that if $a_{ij} = 0$, then
$$\operatorname{per}(A + E_{ij}) \ge \operatorname{per} A + 1,$$
where the matrix E_{ij} is defined in the preceding problem.

5.2. Rectangular nonnegative matrices

5.2.1. Main definitions. We consider a nonnegative matrix $A = (a_{ij})$, $i = 1, \dots, m$, $j = 1, \dots, n$, $2 \le m \le n$.

Recall that a matrix A is said to be totally indecomposable if for any s, $1 \le s \le m-1$, it does not contain an $(m-s) \times s$ zero submatrix. From the definition it follows that a matrix A is totally indecomposable if and only if there exist no permutation matrices Π and Π_1 of orders m and n, respectively, such that

$$\Pi A \Pi_1 = \begin{pmatrix} A_{11} & A_{12} \\ \mathbf{0} & A_{22} \end{pmatrix}, \tag{5.2.1}$$

where A_{11} is a nonempty square submatrix and $\mathbf{0}$ is a nonempty zero submatrix. A matrix A which is not totally indecomposable is called partially decomposable. For a partially decomposable matrix A there exist permutation matrices Π and Π_1 such that $\Pi A \Pi_1$ is of the form (5.2.1).

A matrix A is called semi-totally indecomposable if for any s such that $1 \leq s \leq \min\{m, n-1\}$ it does not contain an $s \times (n-s)$ zero submatrix. A matrix A is semi-totally indecomposable if and only if there exist no permutation matrices Π and Π_1 of orders m and n, respectively, such that

$$\Pi A \Pi_1 = \begin{pmatrix} A_{11} & \mathbf{0} \\ A_{21} & A_{22} \end{pmatrix}, \tag{5.2.2}$$

where A_{11} is a nonempty square submatrix and $\mathbf{0}$ is a nonempty zero submatrix.

An $s \times t$ submatrix of a matrix A is called proper if $1 \leq s < m$, $1 \leq t < n$.

A matrix A is called weakly totally indecomposable if for any s, $1 \leq s \leq m-1$, it does not contain an $s \times (n-s)$ proper zero submatrix. A matrix A is weakly totally indecomposable if and only if there exist no permutation matrices Π and Π_1 of orders m and n, respectively, such that

$$\Pi A \Pi_1 = \begin{pmatrix} A_{11} & \mathbf{0} \\ A_{21} & A_{22} \end{pmatrix}, \tag{5.2.3}$$

where A_{11} is a square nonempty submatrix and A_{22} is an arbitrary nonempty submatrix.

For $m = n$, all three definitions reduce to the definition of a totally indecomposable matrix of order n. For $m > n$ we say that a matrix A is totally indecomposable, semi-totally indecomposable, and weakly totally indecomposable if the corresponding property holds for the transposed matrix A^T.

Let $\mathcal{F}_1(n,m)$, $\mathcal{F}_2(n,m)$, $\mathcal{F}_3(n,m)$ denote, respectively, the sets of all totally indecomposable, semi-totally indecomposable, and weakly totally indecomposable $m \times n$ matrices. Then

$$\mathcal{F}_1(n,m) \subseteq \mathcal{F}_2(n,m) \subseteq \mathcal{F}_3(n,m). \tag{5.2.4}$$

We see that if an $m \times n$ matrix A is totally indecomposable, $m \leq n$, then it does not contain an $s \times t$ zero submatrix, $s + t = m$, $1 \leq s, t \leq m-1$. A semi-totally indecomposable $m \times n$ matrix cannot contain an $s \times t$ zero submatrix, $1 \leq s \leq \min\{m, n-1\}$, $\max\{n-m,\} \leq t \leq n-1$, $s+t = n$. A weakly totally indecomposable $m \times n$ matrix cannot contain an $s \times t$ zero submatrix, $1 \leq s \leq m-1$, $n - m + 1 \leq t \leq n-1$, $s + t = n$. It is clear that a totally indecomposable matrix is semi-totally indecomposable; in its turn, a semi-totally indecomposable matrix is weakly totally indecomposable.

A set of k rows and l columns of a matrix A, $0 \leq k \leq m$, $0 \leq l \leq n$, forms a (k,l)-covering if each positive element of A belongs to either a row or a column of this set.

A (k,l)-covering with $0 \leq k < m$, $0 \leq l < n$ is called a proper (k,l)-covering.

A proper (k,l)-covering is said to be strictly proper if $0 < k < m$, $0 < l < n$.

The nullity of an $m \times n$ matrix A is

$$\beta(A) = \max(s+t), \tag{5.2.5}$$

where the maximum is over all $s \times t$ zero submatrices of A, $1 \leq s \leq m$, $1 \leq t \leq n$.

The proper nullity of an $m \times n$ matrix A is

$$\beta'(A) = \max(s+t), \tag{5.2.6}$$

where the maximum is over all proper nonempty $s \times t$ zero submatrices of A, $1 \leq s < m$, $1 \leq t < n$. For the zero matrix $\mathbf{0}$, we set

$$\beta(\mathbf{0}) = m + n, \qquad \beta'(\mathbf{0}) = m + n - 2. \tag{5.2.7}$$

If $A > 0$, we set

$$\beta(A) = \beta'(A) = 0. \tag{5.2.8}$$

Let $\rho(A)$ be the term rank of a nonnegative $m \times n$ matrix A (see Section 1.2), and let $\tilde{\sigma}(A)$ be its cover rank, that is,

$$\tilde{\sigma}(A) = \min(k + l), \tag{5.2.9}$$

where the minimum is over all (k, l)-coverings of the matrix A, $0 \le k \le m$, $0 \le l \le n$. By the König Theorem (Theorem 1.2.6) the equality

$$\tilde{\sigma}(A) = \rho(A) \tag{5.2.10}$$

is true. The characteristic

$$\sigma(A) = \min(k + l),$$

where the minimum is over all proper (k, l)-coverings of the matrix A, $0 \le k < m$, $0 \le l < n$, is called the proper cover rank of the matrix.

The strictly proper cover rank is

$$\sigma'(A) = \min(k + l),$$

where the minimum is over all strictly proper (k, l)-coverings of the matrix A, $0 < k < m$, $0 < l < n$. From the definitions it follows that $\sigma(\mathbf{0}) = 0$, $\sigma'(\mathbf{0}) = 2$. For a matrix A, its nullities and cover ranks are related as follows:

$$\sigma(A) + \beta(A) = m + n, \tag{5.2.11}$$

$$\sigma'(A) + \beta'(A) = m + n, \tag{5.2.12}$$

because $\max(s + t) + \min(k + l) = m + n$, where the maximum is over all (proper) zero submatrices and the minimum is over all (proper) coverings.

If A has no zero rows or columns, then

$$\beta'(A) = \beta(A), \qquad \sigma'(A) = \sigma(A). \tag{5.2.13}$$

If $A \neq \mathbf{0}$, then

$$\beta(A) - 1 \le \beta'(A) \le \beta(A), \qquad \sigma(A) \le \sigma'(A) \le \sigma(A) + 1. \tag{5.2.14}$$

Moreover, since

$$\rho(A) = \tilde{\sigma}(A) \le \min(m, n),$$

we conclude that

$$\rho(A) = \min(\sigma(A), m, n). \tag{5.2.15}$$

For an $m \times n$ matrix A, we define $c_A(k)$, $1 \le k \le m$, as the minimum number of nonzero columns over all $k \times n$ submatrices of A. Similarly, $r_A(s)$, $1 \le s \le n$, is defined as the minimum number of nonzero rows over all $m \times s$ submatrices of A. We set $c_A(0) = r_A(0) = 0$, $c_A(m + 1) = n + 1$, $r_A(n + 1) = m + 1$. From the definition it is clear that $c_A(k) = r_{A^T}(k)$ for all k. These characteristics are monotone nondecreasing functions; namely,

$$0 = c_A(0) \le c_A(1) \le c_A(2) \le \cdots \le c_A(m - 1) \le c_A(m) \le n, \tag{5.2.16}$$

$$0 = r_A(0) \le r_A(1) \le r_A(2) \le \cdots \le r_A(n - 1) \le r_A(n) \le m. \tag{5.2.17}$$

For the zero matrix, we set $c_{\mathbf{0}}(k) = r_{\mathbf{0}}(k) = 0$ for all k. From the definitions of $\beta(A)$, $c_A(k)$, $r_A(k)$ for $A \not> 0$ we obtain

$$\beta(A) = \max_{1 \le k \le m} \{k + n - c_A(k) \mid 0 \le c_A(k) < n\}, \tag{5.2.18}$$

$$\beta(A) = \max_{1 \le s \le n} \{s + m - r_A(s) \mid 0 \le r_A(s) < m\}. \tag{5.2.19}$$

Similarly, from the definition of $\beta'(A)$ for a matrix A without zero rows we obtain

$$\beta'(A) = \max_{1 \le k < m} \{k + n - c_A(k) \mid 0 < c_A(k) < n\}, \tag{5.2.20}$$

and for a matrix A without zero columns,

$$\beta'(A) = \max_{1 \le s < n} \{s + m - r_A(s) \mid 0 < r_A(s) < m\}. \tag{5.2.21}$$

From (5.2.11), (5.2.12), and (5.2.18)–(5.2.21) the corresponding expressions for $\sigma(A)$ and $\sigma'(A)$ can be derived.

The properties below follow immediately from the definitions.

PROPERTY 5.2.1. *An $m \times n$ matrix A, $m \le n$, is totally indecomposable if and only if $\beta(A) \le m - 1$ or, which is the same, $\beta'(A) \le m - 1$.*

PROPERTY 5.2.2. *An $m \times n$ matrix A, $m \le n$, is semi-totally indecomposable if and only if $\beta(A) \le n - 1$.*

PROPERTY 5.2.3. *An $m \times n$ matrix A, $m \le n$, is weakly totally indecomposable if and only if $\beta'(A) \le n - 1$.*

Using (5.2.11), (5.2.12) and (5.2.18)–(5.2.21), we can express the corresponding properties in terms of $\sigma(A)$ and $\sigma'(A)$.

From Properties 5.2.1, 5.2.2, 5.2.3, and formulas (5.2.18) and (5.2.20) we derive the following properties.

PROPERTY 5.2.4. *An $m \times n$ matrix A, $m \le n$, is totally indecomposable if and only if*

$$c_A(k) \ge k + 1 + (n - m), \qquad 1 \le k \le m - 1.$$

An $m \times n$ matrix A, $m \le n$, is semi-totally indecomposable if and only if

$$c_A(k) \ge k + 1, \qquad 1 \le k \le \min(m, n - 1).$$

An $m \times n$ matrix A, $m \le n$, is weakly totally indecomposable if and only if

$$c_A(k) \ge k + 1, \qquad 1 \le k \le m - 1.$$

Similar properties can be formulated in terms of $r_A(s)$.

5.2.2. Bipartite graphs. With each nonnegative matrix $A = (a_{ij})$, $i = 1, \dots, m$, $j = 1, \dots, n$, we associate the bipartite graph $K(A)$ whose set of vertices is

$$X \cup Y, \quad X = \{x_1, \dots, x_m\}, \quad Y = \{y_1, \dots, y_n\},$$

and whose set of edges is

$$E = \{(x_i, y_j) \colon x_i \in X,\ y_j \in Y,\ a_{ij} > 0\}.$$

We consider a matrix A without zero rows or columns, that is, a matrix such that the corresponding graph $K(A)$ has no isolated vertices. The graph $K(A)$ is said to be irreducible if $m = n$ and A is totally indecomposable. The graph $K(A)$ is semi-irreducible if $m \neq n$ and the matrix A is semi-totally indecomposable. The graph $K(A)$ is called reducible if A is not semi-totally indecomposable, which, when there are no zero rows or columns, is equivalent to the fact that the matrix A is not weakly totally indecomposable.

The neighbourhood $K_A(x)$ of a vertex $x \in X$ is the set of vertices of $K(A)$ adjacent to the vertex x; that is,

$$K_A(x) = \{y\colon (x, y) \in E\}. \tag{5.2.22}$$

We introduce the neighbourhood of a set of vertices $S \subseteq X$ as

$$K_A(S) = \bigcup_{x \in S} K_A(x). \tag{5.2.23}$$

Now we define $c_A(k)$ as

$$c_A(k) = \min_{1 \leq k \leq m} \{|K_A(S)|\colon |S| = k,\ S \subseteq X\}. \tag{5.2.24}$$

From (5.2.22) and (5.2.23) it follows that $K_A(S)$ is the set of vertices in Y which are adjacent to the vertices of $S \subseteq X$ in the graph $K(A)$, and moreover, $K_A(S)$ is the set of columns which completely covers the positive elements of the rows of A with indices in S. Now (5.2.24) follows from the definition of $c_A(K)$.

Similarly we can give a definition of $r_A(s)$ from the viewpoint of graph theory.

PROPERTY 5.2.5. *An $m \times n$ matrix A, $m \leq n$, is totally indecomposable if and only if*

$$|K_A(S)| > |S| \tag{5.2.25}$$

for any $S \subset X$, $S \neq \emptyset$.

PROPERTY 5.2.6. *An $m \times n$ matrix A, $m \leq n$, is semi-totally indecomposable if and only if*

$$|K_A(S)| > |S| \tag{5.2.26}$$

for any $S \subseteq X$, $S \neq \emptyset$, $|S| < |Y| = n$.

PROPERTY 5.2.7. *An $m \times n$ matrix A, $m \leq n$, is weakly totally indecomposable if and only if*

$$|K_A(S)| > |S| + (n - m) \tag{5.2.27}$$

for any $S \subset X$, $S \neq \emptyset$.

Properties 5.2.5, 5.2.6, 5.2.7 follow, respectively, from the first, second, and third parts of Property 5.2.4, and equality (5.2.24). For $m = n$ all these properties are equivalent.

PROPERTY 5.2.8. *A matrix A of order m is totally indecomposable if and only if*

$$|K_A(S)| > |S| \tag{5.2.28}$$

for any $S \subset X$, $S \neq \emptyset$.

Let $\nu_+(z)$ denote the number of positive elements of a nonnegative n-dimensional vector z. Then

$$r_A(s) = \min_{z \geq 0} \{\nu_+(Az) \mid \nu_+(z) = s\}, \qquad 0 \leq s \leq n. \tag{5.2.29}$$

First we assume that

$$z = (z_1, \dots, z_s, z_{s+1}, \dots, z_n)^T, \quad z_1 > 0, \dots, z_s > 0, \quad z_{s+1} = \dots = z_n = 0.$$

Then the ith element of the vector Az is of the form

$$(Az)_i = \sum_{j=1}^{s} a_{ij} z_j.$$

The condition $(Az)_i > 0$ means that at least one of the elements $a_{i1}, \dots, a_{is}$ is positive. Hence the number of positive elements of the vector Az is equal to the number of rows of A with positive elements in the first s columns. If we consider the vector z with $\nu_+(z) = s$, where s positive elements are placed arbitrarily, then the minimum number of positive elements of the vector Az equals the minimum number of rows with nonzero elements in an $m \times s$ submatrix, that is, equals $r_A(s)$, $1 \le s \le n$.

PROPERTY 5.2.9. *An $m \times n$ matrix A, $m \le n$, is totally indecomposable if and only if the inequality*

$$\nu_+(Az) > \nu_+(z) \tag{5.2.30}$$

holds for any vector $z \ge 0$ whose number of positive elements $\nu_+(z)$ satisfies the condition $0 < \nu_+(z) < m$.

From (5.2.19) and Property 5.2.1 it follows that the matrix A is totally indecomposable if and only if

$$r_A(s) \ge s + 1, \qquad 1 \le s \le m - 1. \tag{5.2.31}$$

In view of (5.2.29), inequality (5.2.31) is equivalent to (5.2.30).

PROPERTY 5.2.10. *An $m \times n$ matrix A, $m \le n$, is totally indecomposable if and only if any submatrix of A of order m is totally indecomposable.*

Assume that A is totally indecomposable. By Property 5.2.1, $\beta(A) \le m - 1$. Therefore, any submatrix of order m is of nullity not exceeding $m - 1$; that is, it contains no $(m - s) \times s$ zero submatrices, $1 \le s \le m - 1$, and hence it is totally indecomposable.

Assume that any submatrix of order m of A is totally indecomposable. Then A is not partially decomposable, because otherwise there would exist an $(m - s) \times s$ zero submatrix, $1 \le s \le m - 1$, contained in some submatrix of A of order m, and this submatrix would be not totally indecomposable.

PROPERTY 5.2.11. *Let an $m \times n$ matrix A, $m < n$, be weakly totally indecomposable but not semi-totally indecomposable; that is, $A \in \mathcal{F}_3 \setminus \mathcal{F}_2$. Then there exist permutation matrices Π and Π_1 of orders m and n, respectively, such that*

$$\Pi A \Pi_1 = \begin{pmatrix} A_{11} & \mathbf{0} \end{pmatrix}, \tag{5.2.32}$$

where A_{11} is a totally indecomposable submatrix of order m and the zero submatrix $\mathbf{0}$ is nonempty.

Let $A \in \mathcal{F}_3 \setminus \mathcal{F}_2$. By Property 5.2.6 there exists $S \subseteq X$, $S \ne \emptyset$, $|S| < |Y|$, such that $|K_A(S)| \le |S|$. But by Property 5.2.7 $|K_A(S)| > |S|$ for all $S \subset X$. Therefore, the inequality $|K_A(S)| \le |S|$ can hold only for $S = X$. Thus, $|X| < |Y|$; that is, $m < n$, and therefore, $|K_A(X)| \le |X| = m < n$. Since $K_A(X)$ is the set of columns covering all rows of A, the matrix A has $n-m$ zero columns and is reducible to (5.2.32).

Now we prove that the submatrix A_{11} is totally indecomposable. The graph $K(A_{11})$ has the vertex sets X and $K_A(X)$, and its edge set coincides with the edge set E of the graph $K(A)$, so $K_{A_{11}}(S) = K_A(S)$ for any $S \subseteq X$. By Property 5.2.7, $|K_A(S)| > |S|$ for any $S \subset X$, $S \neq \emptyset$. This means that $|K_{A_{11}}(S)| > |S|$ for any $S \subset X$, $S \neq \emptyset$; that is, the matrix A_{11} is totally indecomposable in view of Property 5.2.8.

Assume that the representation (5.2.32) holds with some totally indecomposable matrix A_{11} of order m and some nonempty matrix $\mathbf{0}$. Then $m < n$ and $|K_A(X)| \leq |X|$, so the matrix is not semi-totally indecomposable in view of Property 5.2.6. The fact that A_{11} is totally indecomposable and Property 5.2.8 imply that $|K_{A_{11}}(S)| > |S|$ for any $S \subset X$, $S \neq \emptyset$. Since $K_{A_{11}}(S) = K_A(S)$ for any $S \subseteq X$, we find that $|K_A(S)| > |S|$ for any $S \subset X$, $S \neq \emptyset$; that is, the matrix A is weakly totally indecomposable by Property 5.2.7.

5.2.3. Essentially nonsingular matrices. A matrix of order n is called essentially nonsingular if it has a positive permanent, that is, if its term rank is equal to n.

PROPERTY 5.2.12. *For any $m \times n$ matrix A, $m \leq n$, and any v, $0 \leq v \leq m-1$,*

$$\beta(A) \leq m - v \tag{5.2.33}$$

if and only if any submatrix of A of order $m - v$ is essentially nonsingular.

Let $\beta(A) \leq m-v$, and let some submatrix of A of order $m-v$ be not essentially nonsingular. Then this submatrix, and hence A itself, have the proper nullity exceeding $m - v$; that is, $\beta(A) > m - v$, which is impossible.

Now let $\beta(A) \geq m - v + 1$; that is, for some s, $1 \leq s \leq m - v$, the matrix A contains an $s \times (m - v + 1 - s)$ zero submatrix. Then no submatrix of A of order $m - v$ which contains this zero submatrix can be essentially nonsingular.

PROPERTY 5.2.13. *An $m \times n$ matrix A, $m \leq n$, is totally indecomposable if and only if each submatrix of A of order $m - 1$ is essentially nonsingular.*

In view of Property 5.2.1, the validity of Property 5.2.13 follows from Property 5.2.12 with $v = 1$.

We now give the definition of essential nonsingularity for a rectangular matrix. We say that an $m \times n$ matrix A, $m \leq n$, is essentially nonsingular if $\rho(A) = m$, that is, if A has a positive diagonal and therefore $\operatorname{per} A > 0$.

From equalities (5.2.11) and (5.2.15) it follows that a matrix A is essentially nonsingular if and only if either

$$\sigma(A) \geq m \tag{5.2.34}$$

or

$$\beta(A) \leq n. \tag{5.2.35}$$

PROPERTY 5.2.14. *An $m \times n$ matrix A, $m \leq n$, is essentially nonsingular if and only if it has an essentially nonsingular submatrix of order m.*

The condition $\rho(A) = m$ is equivalent to the existence of a submatrix A' of A of order m such that $\operatorname{per} A' = m$; that is, the submatrix A' is essentially nonsingular.

An $m \times n$ matrix A, $m \geq n$, is called strictly essentially nonsingular if any submatrix of A of order m is essentially nonsingular.

PROPERTY 5.2.15. *An $m \times n$ matrix A, $m \leq n$, is strictly essentially nonsingular if and only if either*

$$\sigma(A) \geq n \tag{5.2.36}$$

or

$$\beta(A) \leq m. \tag{5.2.37}$$

From Property 5.2.12 it follows that $\beta(A) \leq m$ if and only if each submatrix of A of order m is essentially nonsingular; that is, the matrix A itself is strictly essentially nonsingular. Inequality (5.2.37) is proved. Inequality (5.2.36) follows from (5.2.11) and (5.2.37).

PROPERTY 5.2.16. *Let A be an $m \times n$ matrix, $m \leq n$, and A_{ij} the $(m-1) \times (n-1)$ matrix obtained from A by removing the ith row and the jth column. Then*

$$\beta'(A) = \max\{\beta(A_{ij}) \mid i \in N_m,\ j \in N_n\}. \tag{5.2.38}$$

Let $\beta'(A) = k$, and let B be a proper $a \times b$ zero submatrix of A, $a + b = k$. Then there exist s and t such that the sth row and the tth column of A do not belong to B, so that $\beta(A_{st}) \geq k$. Let $l = \max\{\beta(A_{ij}) \mid i \in N_m,\ j \in N_n\}$, and let the maximum be attained at A_{pq}; that is, A_{pq} has a $c \times d$ zero submatrix such that $c + d = l$. Since A_{pq} is a proper submatrix of A, $\beta'(A) \geq l$. Thus,

$$\beta'(A) \geq \max\{\beta(A_{ij}) \mid i \in N_m,\ j \in N_n\}.$$

On the other hand, it has been proved that there exist s and t such that $\beta'(A) \leq \beta(A_{st})$, that is,

$$\beta'(A) \leq \max\{\beta(A_{ij}) \mid i \in N_m,\ j \in N_n\}.$$

Property 5.2.16 is thus proved.

PROPERTY 5.2.17. *For an $m \times n$ matrix A, $m \leq n$, the inequality*

$$\beta'(A) \leq k, \tag{5.2.39}$$

where k is an arbitrary integer, $0 \leq k \leq m+n-2$, holds if and only if the inequality

$$\beta(A_{ij}) \leq k \tag{5.2.40}$$

holds for any $i \in N_m$, $j \in N_n$.

If $\beta(A_{ij}) \leq k$, then $\max\{\beta(A_{ij}) \mid i \in N_m,\ j \in N_n\} \leq k$, and (5.2.38) implies (5.2.39). Conversely, if $\beta'(A) \leq k$, then by (5.2.38) $\beta(A_{ij}) \leq k$ for any $i \in N_m$, $j \in N_n$.

PROPERTY 5.2.18. *An $m \times n$ matrix A, $m \leq n$, is totally indecomposable if and only if each $(m-1) \times (n-1)$ submatrix is strictly essentially nonsingular.*

From Property 5.2.1 it follows that the matrix A is totally indecomposable if and only if $\beta'(A) \leq m - 1$. Hence, in view of Property 5.2.17, $\beta(A_{ij}) \leq m - 1$ for any $i \in N_m$, $j \in N_n$. Therefore, by Property 5.2.15, any submatrix A_{ij} is strictly essentially nonsingular.

To prove the sufficiency, the above reasoning has to be performed in the reverse order.

PROPERTY 5.2.19. *An $m \times n$ matrix A, $m \leq n$, is weakly totally indecomposable if and only if each $(m-1) \times (n-1)$ submatrix is essentially nonsingular.*

From Property 5.2.3 it follows that $\beta'(A) \leq n-1$ for a weakly totally indecomposable matrix A. From Property 5.2.17 it follows that $\beta(A_{ij}) \leq n-1$ for any $i \in N_m$, $j \in N_n$. Now Property 5.2.13 (see (5.2.35)) implies that the matrix A_{ij} is essentially nonsingular. To prove the sufficiency, the above reasoning has to be performed in the reverse order.

5.2.4. Products of matrices. We turn to the study of characteristics of products of nonnegative matrices.

LEMMA 5.2.1. *If $A = (a_{ij})$, $i = 1, \ldots, m$, $j = 1, \ldots, n$, and $B = (b_{uv})$, $u = 1, \ldots, n$, $v = 1, \ldots, p$, then*

$$c_{AB}(k) \geq c_B(c_A(k)), \qquad 0 \leq k \leq m, \tag{5.2.41}$$

$$r_{AB}(s) \geq r_A(c_B(s)), \qquad 0 \leq s \leq p. \tag{5.2.42}$$

PROOF. Let $I \subseteq N_m = \{1, \ldots, m\}$, $|I| = k$. The set of rows of the matrix AB whose indices belong to I is obtained by multiplying the rows of A with indices in I by all columns of B. For subsets $I \subset N_m$, $J \subset N_n$, let $R(I)$, $R(J)$ denote the minimal sets of columns which cover the rows of A with indices in I or, respectively, the rows of A with indices in J. While multiplying the rows of A with indices in I by columns of B, it suffices to consider the submatrix of B whose row indices lie in the set $R(I)$. Let $\tilde{R}(R(I))$ be the minimal set of columns which cover this $|R(I)| \times p$ submatrix of B. Then the set of columns which cover the rows with indices in I of the matrix AB is the set $\tilde{R}(R(I))$. Let $c_A(k) = l$. If $|I| = k$, $|J| = l$, then

$$\min_I |R(I)| = c_A(k), \qquad \min_J |\tilde{R}(J)| = c_B(l);$$

therefore, the number of columns of the matrix AB which cover the rows with indices in the k-set I is at least $c_B(c_A(k))$. Hence (5.2.41) follows.

Inequality (5.2.42) follows from the equality $c_A(k) = r_{A^T}(k)$ and inequality (5.2.41). □

LEMMA 5.2.2. *Let A be an $m \times n$ matrix without zero rows, $m \leq n$, and let B be an $n \times p$ matrix without zero columns such that AB is the zero $m \times p$ matrix. Then A contains an $m \times r_B(p)$ zero submatrix and B contains an $(n - r_B(p)) \times p$ zero submatrix.*

PROOF. The matrix B has no zero columns, and therefore $B \neq \mathbf{0}$. Hence $r_B(p) > 0$. From the hypothesis $AB = \mathbf{0}$ it follows that $c_{AB}(1) = 0$. The matrix A has no zero rows; hence $c_A(1) \geq 1$. Now Lemma 5.2.1 yields

$$0 = c_{AB}(1) \geq c_B(c_A(1)) \geq c_B(1) \geq 0.$$

Hence $c_B(1) = 0$; that is, the matrix B has at least one zero row. Since the matrix B has $n - r_B(p)$ zero rows, we see that $n - r_B(p) > 0$, and therefore B contains an $(n - r_B(p)) \times p$ zero submatrix, $0 < r_B(p) < n$.

The rows of B have $n - r_B(p)$ zero elements, and the remaining $r_B(P)$ elements can be nonzero. From the condition $AB = \mathbf{0}$ it follows that the rows of A must have at least $r_B(p)$ zero elements which form an $m \times r_B(p)$ submatrix. □

THEOREM 5.2.3. *Let $A \neq \mathbf{0}$ be an $m \times n$ matrix, and let $B \neq \mathbf{0}$ be an $n \times p$ matrix. Then the nullity of the matrix AB satisfies the following inequalities.*

(i) *If A has no zero rows and B has no zero columns, then*

$$\beta(AB) \leq \max\{\beta(A) + \beta(B) - n, 0\}. \tag{5.2.43}$$

(ii) *If A has no zero rows, then*

$$\beta(AB) \leq \max\{\beta(A) + \beta(B) - n, m + p - c_B(n)\}. \tag{5.2.44}$$

(iii) *If B has no zero columns, then*

$$\beta(AB) \leq \max\{\beta(A) + \beta(B) - n, m + p - r_A(n)\}. \tag{5.2.45}$$

(iv) *In the general case,*

$$\beta(AB) \leq \max\{\beta(A) + \beta(B) - n, m + p - c_B(n), m + p - r_A(n)\}. \tag{5.2.46}$$

PROOF. We begin by proving assertion (i). If $AB > 0$, then $\beta(AB) > 0$, and the assertion is true. Let $AB \not> 0$. Then the matrix AB contains an $s \times t$ zero submatrix, and $\beta(AB) = s + t \geq 2$. By interchanging rows and columns we reduce the matrix AB to the form

$$AB = \begin{pmatrix} \mathbf{0} & C_{12} \\ C_{21} & C_{22} \end{pmatrix}, \tag{5.2.47}$$

where the zero submatrix $\mathbf{0}$ is of size $s \times t$.

It is clear that

$$AB = \begin{pmatrix} A_{11} \\ A_{21} \end{pmatrix} \begin{pmatrix} B_{11} & B_{12} \end{pmatrix} = \begin{pmatrix} A_{11}B_{11} & A_{11}B_{12} \\ A_{21}B_{11} & A_{21}B_{12} \end{pmatrix}, \tag{5.2.48}$$

where A_{11} is an $s \times n$ matrix and B_{11} is an $n \times t$ matrix. Then $A_{11}B_{11} = \mathbf{0}$, and by Lemma 5.2.2, A_{11} contains an $s \times r_{B_{11}}(t)$ zero submatrix and B_{11} contains an $(n - r_{B_{11}}(t)) \times t$ zero submatrix. Therefore,

$$\beta(A) \geq \beta(A_{11}) \geq s + r_{B_{11}}(t),$$
$$\beta(B) \geq \beta(B_{11}) \geq n - r_{B_{11}}(t) + t.$$

Hence

$$\beta(A) + \beta(B) - n \geq s + t = \beta(AB).$$

Let us prove assertion (ii). It is clear that if B has zero columns with indices $j_1, \ldots, j_t$, then the columns of AB with these indices are zero as well; that is, AB has at least t zero columns. Without loss of generality we assume that the indices of zero columns are $1, 2, \ldots, t$. Then the matrix B can be represented as $B = \begin{pmatrix} \mathbf{0} & B_{12} \end{pmatrix}$, where B_{12} is an $n \times (p - t)$ matrix. Hence $AB = \begin{pmatrix} \mathbf{0} & AB_{12} \end{pmatrix}$. For $t < p$, we see that $B \neq \mathbf{0}$; hence

$$\beta(B) = \max\{\beta(B_{12}) + t, n + t\}, \tag{5.2.49}$$

because if B_{12} has a maximal $\mu \times \nu$ zero submatrix, then in B the size of this submatrix grows up to $\mu \times (\nu + t)$. Similarly we obtain

$$\beta(AB) = \max\{\beta(AB_{12}) + t, m + t\}.$$

Since A and B_{12} have no zero rows and zero columns, respectively, assertion (i) yields

$$\beta(AB) = \max\{\beta(A) + \beta(B_{12}) - n + t, m + t\}. \tag{5.2.50}$$

From (5.2.49) we obtain

$$\beta(B) \geq \beta(B_{12}) + t. \tag{5.2.51}$$

Substituting the bound for $\beta(B_{12})$ from (5.2.51) into (5.2.50), we obtain

$$\beta(AB) \leq \max\{\beta(A) + \beta(B) - n, m + t\}. \tag{5.2.52}$$

From the definition it follows that $p - t = c_B(n)$. Substituting $t = p - c_B(n)$ into (5.2.51), we arrive at (5.2.44). If $t = p$ we find that $c_B(n) = 0$, and (5.2.44) is again true.

Assertion (iii) is proved similarly to (ii); we assume that the first $s > 0$ rows of A are zero rows, so that A takes the form

$$A = \begin{pmatrix} \mathbf{0} \\ A_{21} \end{pmatrix}.$$

Finally, let us prove (iv). Without loss of generality we assume that the first s rows of A and the first t columns of B are zero, $s, t > 0$. Then from (5.2.48) it follows that

$$AB = \begin{pmatrix} \mathbf{0} & \mathbf{0} \\ \mathbf{0} & A_{21}B_{12} \end{pmatrix},$$

where A_{21} has no zero rows, B_{12} has no zero columns, $A_{21}B_{12}$ is an $(m-s) \times (p-t)$ matrix, and moreover, this matrix is nonempty, since otherwise either A or B would be a zero matrix. It is clear that

$$\beta(AB) = \max\{\beta(A_{21}B_{12}) + s + t, p + s, m + t\}.$$

Using assertion (i), we obtain

$$\beta(AB) \leq \max\{\beta(A_{21}) + \beta(B_{12}) - n + s + t, p + s, m + t\}. \tag{5.2.53}$$

Since

$$\beta(A) \geq \beta(A_{21}) + s, \qquad \beta(B) \geq \beta(B_{12}) + t,$$

inequality (5.2.53) implies that

$$\beta(AB) \leq \max\{\beta(A) + \beta(B) - n, m + p - c_B(n), m + p - r_A(n)\}. \tag{5.2.54}$$

The theorem is thus proved. □

COROLLARY 5.2.4. *The following bounds are true for the proper cover rank of the matrix AB.*

(i) *If A has no zero rows and B has no zero columns, then*

$$\sigma(AB) \geq \min\{\sigma(A) + \sigma(B) - n, m + p\}. \tag{5.2.55}$$

(ii) *If A has no zero rows, then*

$$\sigma(AB) \geq \min\{\sigma(A) + \sigma(B) - n, c_B(n)\}. \tag{5.2.56}$$

(iii) *If B has no zero columns, then*

$$\sigma(AB) \geq \min\{\sigma(A) + \sigma(B) - n, r_A(n)\}. \tag{5.2.57}$$

(iv) *In the general case,*

$$\sigma(AB) \geq \min\{\sigma(A) + \sigma(B) - n, c_B(n), r_A(n)\}. \tag{5.2.58}$$

These inequalities follow from (5.2.11) and Theorem 5.2.3.

COROLLARY 5.2.5. *The term ranks of an $m \times n$ matrix A, an $n \times p$ matrix B, and the matrix AB, $m \le n \le p$, satisfy the inequality*

$$\rho(AB) \ge \rho(A) + \rho(B) - n. \tag{5.2.59}$$

From (5.2.15) and (5.2.58) we obtain

$$\begin{aligned}\rho(AB) &= \min\{\sigma(AB), m, p\}\\ &\ge \min\{\rho(A) + \rho(B) - n, c_B(n), r_A(n), m, p\}.\end{aligned} \tag{5.2.60}$$

If $r_A(n) < m$, then the $(r_A(n), 0)$-covering of the positive elements of A consisting only of nonzero rows is proper, and therefore $\sigma(A) \le r_A(n) < m$. Hence $\rho(A) = \min\{\sigma(A), m, n\} \le r_A(n)$. Similarly, $\rho(B) \le c_B(n)$ for $c_B(n) < p$. Thus,

$$\min\{\rho(A) + \rho(B) - n, c_B(n), r_A(n), m, p\} = \rho(A) + \rho(B) - n \tag{5.2.61}$$

for $r_A(n) < m$, $c_B(n) < p$, because $\rho(A) \le n$, $\rho(B) \le n$. For $r_A(n) = m$ by the inequality $\rho(A) \le m$ and for $c_B(n) = p$ by the inequality $\rho(B) \le n$, equality (5.2.61) is true as well. Now (5.2.61) and (5.2.60) imply (5.2.59).

We consider a matrix $X = (x_{ij})$, $i = 1, \dots, m$, $j = 1, \dots, n$, where the x_{ij} are independent variables. The Hadamard product of a matrix $A = (a_{ij})$, $i = 1, \dots, m$, $j = 1, \dots, n$, by the matrix X is defined by the equality

$$M = A * X = (a_{ij}x_{ij}), \quad i = 1, \dots, m, \quad j = 1, \dots, n. \tag{5.2.62}$$

The matrix M will be called the formal incidence matrix corresponding to A.

LEMMA 5.2.6. *The term rank $\rho(A)$ of a matrix A is equal to the usual rank $r(M)$ of the formal incidence matrix M corresponding to A.*

PROOF. If A' is a square submatrix of A, then from the condition $\rho(A') = 0$ it follows that $r(M') = 0$, where M' is the submatrix of the matrix M corresponding to A'; indeed, from the condition $\operatorname{per} A' = 0$ it follows that $\det M' = 0$ for any values of the variables x_{ij}, $1 \le i \le m$, $1 \le j \le n$. Conversely, let $\det M' = 0$ for any values of the variables. Then, by the continuity of the determinant, each diagonal of A' has to contain a zero element. □

By Corollary 5.2.5 and Lemma 5.2.6, for the usual rank of formal incidence matrices we have the bound

$$r[(A * X)(B * Y)] \ge r(A * X) + r(B * Y) - n. \tag{5.2.63}$$

5.2.5. Characterization of matrices and estimation of the exponent. In what follows, the symbol $\Longleftrightarrow$ stands for equivalence and $\Longrightarrow$ for implication.

Recall that $\mathcal{F}_1(m,n)$, $\mathcal{F}_2(m,n)$, $\mathcal{F}_3(m,n)$ denote, respectively, the classes of totally indecomposable, semi-totally indecomposable, and weakly totally indecomposable $m \times n$ matrices. Let $\mathcal{G}_1(m,n)$ and $\mathcal{G}_2(m,n)$ be the classes of essentially nonsingular and strictly essentially nonsingular $m \times n$ matrices, respectively.

THEOREM 5.2.7. *If A is an $m \times n$ matrix, then*

$$A \in \mathcal{F}_1(n,m) \Longleftrightarrow \beta(A) \le \min(m,n) - 1; \tag{5.2.64}$$

$$A \in \mathcal{G}_2(m,n) \Longleftrightarrow \beta(A) \le \min(m,n); \tag{5.2.65}$$

$$A \in \mathcal{F}_2(m,n) \Longleftrightarrow \beta(A) \le \max(m,n) - 1; \tag{5.2.66}$$

$$A \in \mathcal{F}_3(m,n) \Longleftrightarrow \beta'(A) \le \max(m,n) - 1; \tag{5.2.67}$$

$$A \in \mathcal{G}_1(m,n) \Longleftrightarrow \beta(A) \le \max(m,n). \tag{5.2.68}$$

PROOF. If $A \in \mathcal{F}_1(n,m)$, $m \le n$, then A does not contain an $(m-s)\times s$ zero submatrix, $1 \le s \le m-1$; that is, $\beta(A) \le m-1$. In the general case, $\beta(A) \le \min(m,n)-1$.

If $A \in \mathcal{G}_2(m,n)$, $m \le n$, then $\beta(A) \le m$, by (5.2.37). In the general case, $\beta(A) \le \min(m,n)$.

If $A \in \mathcal{F}_2(m,n)$, $m \le n$, then A does not contain an $s \times (n-s)$ zero submatrix, $1 \le s \le \min(m, n-1)$; that is, $\beta(A) \le n-1$. In the general case, $\beta(A) \le \max(m,n)-1$.

If $A \in \mathcal{F}_3(m,n)$, $m \le n$, then A does not contain an $s \times (n-s)$ zero submatrix, $1 \le s \le m-1$; that is, $\beta'(A) \le n-1$. In the general case, $\beta'(A) \le \max(m,n)-1$.

If $A \in \mathcal{G}_1(m,n)$, $m \le n$, then $\beta(A) \le n$, by (5.2.35). In the general case, $\beta(A) \le \max(m,n)$. □

COROLLARY 5.2.8. *(1) $A \in \mathcal{F}_1(m,n) \Longrightarrow A \in \mathcal{F}_2(m,n) \Longrightarrow A \in \mathcal{F}_3(m,n) \Longrightarrow A \in \mathcal{G}_1(m,n)$;*

(2) $A \in \mathcal{F}_1(m,n) \Longrightarrow A \in \mathcal{G}_2(m,n) \Longrightarrow A \in \mathcal{G}_1(m,n)$; that is, if A is totally indecomposable, then it is semi-totally indecomposable, etc.

Since $(AB)^T = B^T A^T$, in what follows, while dealing with products of an $m\times n$ matrix A and an $n \times p$ matrix B, it suffices to assume that $m \le p$ and to restrict the study to the three cases

$$m \le p \le n, \quad m \le n \le p, \quad n \le m \le p. \tag{5.2.69}$$

COROLLARY 5.2.9. *In any of the cases*

(a) $A \in \mathcal{F}_1(m,n)$, $B \in \mathcal{F}_1(n,p)$,
(b) $A \in \mathcal{F}_1(m,n)$, $B \in \mathcal{G}_2(n,p)$,
(c) $A \in \mathcal{G}_2(m,n)$, $B \in \mathcal{F}_1(n,p)$,

the inclusion $AB \in \mathcal{F}_1(m,p)$ holds. If

(d) $A \in \mathcal{G}_2(m,n)$, $B \in \mathcal{G}_2(n,p)$,

then $AB \in \mathcal{G}_2(m,p)$.

According to Theorem 5.2.7, $\beta(A) \le \min(m,n)-1$ for $A \in \mathcal{F}_1(m,n)$, and $\beta(A) \le \min(m,n)$ for $A \in \mathcal{G}_2(m,n)$; in both cases A has neither zero rows nor zero columns. The same is true for the matrix B. Applying assertion (i) of Theorem 5.2.3, we see that in cases (a), (b), and (c)

$$\beta(AB) \le \max\{\min(m,n)+\min(n,p)-n-1,\, 0\}.$$

Hence

$$\beta(AB) \le \min(m,n,p)-1.$$

If m, n, and p satisfy any of the relations (5.2.69), then $\min(m,n,p) \le \min(m,p)$; that is, $\beta(AB) \le \min(m,p)-1$, and therefore $AB \in \mathcal{F}_1(m,p)$. For $A \in \mathcal{G}_2(m,n)$ and $B \in \mathcal{G}_2(n,p)$ we have

$$\beta(AB) \le \max\{\min(m,n)+\min(n,p)-n,\, 0\}.$$

Hence

$$\beta(AB) \le \min(m,p);$$

that is, $AB \in \mathcal{G}_2(m,p)$.

In a similar way, using Theorem 5.2.7, we can derive other corollaries concerning the inclusions of AB into one of the mentioned classes.

THEOREM 5.2.10. *Let* $A_j \in \mathcal{G}_2(n_{j-1}, n_j)$, $1 \le j \le k$, *let* h *matrices among* $A_1, \dots, A_k$, $0 \le h \le k$, *be totally indecomposable, and let* $\Pi_k = A_1 \dots A_k$. *Then*

$$\beta(\Pi_k) \le \max\left\{\frac{1}{2}\left(n_0 + n_k - \sum_{j=1}^{k} |n_j - n_{j-1}|\right) - h,\ 0\right\}. \tag{5.2.70}$$

PROOF. From the inclusion $A_j \in \mathcal{G}_2(n_{j-1}, n_j)$ it follows that A_j has neither zero rows nor zero columns, and therefore the same is true for $\Pi_j = A_1 \cdots A_j$. Using assertion (i) of Theorem 5.2.3, we obtain

$$\beta(\Pi_j) \le \max\{\beta(\Pi_{j-1} + \beta(A_j) - n_{j-1},\ 0\}. \tag{5.2.71}$$

Using this relation, let us show that

$$\beta(\Pi_k) \le \max\left\{\sum_{j=1}^{k} \beta(A_j) - \sum_{j=1}^{k-1} n_j,\ 0\right\}. \tag{5.2.72}$$

We use induction on k. For $k = 2$, from (5.2.71) we obtain

$$\beta(\Pi_2) \le \max\{\beta(A_1) + \beta(A_2) - n_1, 0\}.$$

From the induction assumption and (5.2.71) it follows that

$$\beta(\Pi_k) \le \max\left\{\max\left\{\sum_{j=1}^{k-1} \beta(A_j) - \sum_{j=1}^{k-2} n_j,\ 0\right\} + \beta(A_k) - n_{k-1},\ 0\right\}. \tag{5.2.73}$$

Since $\beta(A_k) \le n_{k-1}$, from the relation

$$\max\left\{\sum_{j=1}^{k-1} \beta(A_j) - \sum_{j=1}^{k-2} n_j, 0\right\} = 0$$

it follows that

$$\max\left\{\max\left\{\sum_{j=1}^{k-1} \beta(A_j) - \sum_{j=1}^{k-2} n_j,\ 0\right\} + \beta(A_k) - n_{k-1},\ 0\right\} = 0;$$

that is, $\Pi_k > 0$. If $\Pi_k \not> 0$, then (5.2.73) implies (5.2.72).

The bounds (5.2.64) and (5.2.65) yield

$$\sum_{j=1}^{k} \beta(A_j) \le \sum_{j=1}^{k} \min\{n_{j-1}, n_j\} - h. \tag{5.2.74}$$

We observe that

$$\min\{n_{j-1}, n_j\} = \frac{1}{2}(n_{j-1} + n_j) - \frac{1}{2}|n_j - n_{j-1}|. \tag{5.2.75}$$

Using (5.2.74), (5.2.75), and (5.2.72), we obtain (5.2.70). □

COROLLARY 5.2.11. *If* $A_j \in \mathcal{F}_1(n_{j-1}, n_j)$, $1 \le j \le k$, *then* $\Pi_k > 0$ *for*

$$k \ge \frac{1}{2}\left(n_0 + n_k - \sum_{j=1}^{k} |n_j - n_{j-1}|\right) - 1.$$

Indeed, it is easily seen that, as follows from (5.2.70), in order for $\Pi_k > 0$ to be true it is sufficient that

$$\frac{1}{2}\left(n_0 + n_k - \sum_{j=1}^{k} |n_j - n_{j-1}|\right) - k \leq 1,$$

because $\beta(A) \geq 2$ for each nonempty matrix A.

The least integer $k \geq 0$ such that $A_1 \cdots A_k > 0$, where $A_i \in \mathcal{F}_1(n,n)$, $1 \leq i \leq k$, is called the exponent of the class $\mathcal{F}_1(n,n)$ of totally indecomposable matrices, and is denoted by $\gamma(\mathcal{F}_1)$.

COROLLARY 5.2.12. *The exponent of the class of totally indecomposable matrices of order n satisfies the inequality*

$$\gamma(\mathcal{F}_1) \leq n - 1. \tag{5.2.76}$$

Indeed, applying Corollary 5.2.11 with $n_1 = \cdots = n_k = n$, we obtain that $A_1 \cdots A_k > 0$ for $k \geq n-1$, which implies (5.2.76).

Problems.

1. Prove, starting from the definitions, that an $m \times n$ matrix $A \geq 0$, $m \leq n$, is weakly totally indecomposable and is not semi-totally indecomposable if and only if there exist permutation matrices Π and Π_1 of orders m and n, respectively, such that $\Pi A \Pi_1 = \begin{pmatrix} A_{11} & \mathbf{0} \end{pmatrix}$, where A_{11} is a totally indecomposable matrix of order n and $\mathbf{0}$ is the $m \times (n-m)$ zero matrix.
2. Prove that an $m \times n$ matrix $A \geq 0$, $m \leq n$, is weakly totally indecomposable if there exists a submatrix of A of order m which is totally indecomposable.
3. Prove that the matrix

$$A = \begin{pmatrix} 1 & 1 & 0 & 0 \\ 0 & 1 & 1 & 0 \\ 0 & 0 & 1 & 1 \end{pmatrix}$$

is semi-totally indecomposable and none of its submatrices of order 3 is totally indecomposable.

5.3. Rectangular nonnegative chainable matrices

5.3.1. Chainable matrices. A matrix $A = (a_{ij}) \geq 0$, $i = 1, \ldots, m$, $j = 1, \ldots, n$, is called chainable if

1. it has no zero rows or columns, and
2. for any pair of its positive elements a_{rt} and a_{pq} there exists a chain whose ends are these elements; that is, there exists a sequence of positive elements $a_{i_1 j_1} a_{i_2 j_2} \ldots a_{i_s j_s}$ such that
   (a) $i_1 = r$ and $j_1 = t$;
   (b) $i_s = p$ and $j_s = q$;
   (c) $i_k = i_{k+1}$ or $j_k = j_{k+1}$, $k = 1, \ldots, s-1$.

A pair of neighbouring elements of the chain is called a link.

The second condition means that for a pair of distinct positive elements a_{rt} and a_{pq} of A, one can pass from one to the other, moving from one positive element to another along the links of the chain in the same way as a rook moves on a chessboard.

For $m = n$, the definition coincides with the definition of a chainable matrix of order n given in Section 5.1.2. For example, the matrices

$$\begin{pmatrix} 1 & 1 & 1 & 0 \\ 0 & 0 & 1 & 1 \\ 1 & 0 & 1 & 1 \end{pmatrix}, \qquad \begin{pmatrix} 1 & 0 & 0 & 0 \\ 0 & 1 & 1 & 1 \\ 0 & 0 & 1 & 0 \end{pmatrix}$$

are, respectively, chainable and nonchainable.

Let $C(m, n)$ be the class of $m \times n$ chainable $(0, 1)$-matrices. Consider properties of the matrices in $C(m, n)$.

PROPERTY 5.3.1. *Let an $m \times n$ matrix A have neither zero rows nor zero columns. The matrix A belongs to the class $C(m, n)$ if and only if there exist no permutation matrices Π and Π_1 of orders m and n, respectively, such that*

$$\Pi A \Pi_1 = \begin{pmatrix} A_1 & \mathbf{0} \\ \mathbf{0} & A_2 \end{pmatrix}. \tag{5.3.1}$$

If $A \in C(m, n)$, then $\Pi A \Pi_1 \in C(m, n)$, because permutation of rows and columns in A preserves the existence of chains with given ends, although the chains themselves are altered. The matrix on the right-hand side of (5.3.1) is not chainable. Therefore, there are no permutation matrices Π and Π_1 which satisfy (5.3.1).

Now let $A \notin C(m, n)$. Then there exist positive elements a_{rs} and a_{pq} which are not the ends of any chain. Consider the set

$$K = \{a_{ij} \colon a_{ij} = 1, \text{ and there exists a chain from } a_{rs} \text{ to } a_{ij}\}.$$

If $a_{lr} = 1$ and $a_{lr} \notin K$, then $a_{lk} \notin K$, $k \in N_n$, and $a_{tr} \notin K$, $t \in N_m$. This means that if the elements of K are located in the rows with indices in P and columns with indices in Q, then the positive elements which do not belong to K are located in the rows with indices in $N_m \setminus P$ and the columns with indices in $N_n \setminus Q$. Therefore, there exist Π and Π_1 such that $\Pi A \Pi_1$ is of the form (5.3.1).

With a $(0, 1)$-matrix $A = (a_{ij})$ of size $m \times n$, we associate the bipartite graph $G(M, N, \tau)$, $M = N_m$, $N = N_n$, such that an edge $(i, j) = (j, i) \in \tau$ if and only if $a_{ij} = 1$, $i \in N_m$, $j \in N_n$ A sequence of edges of the form

$$(i_1, i_2), (i_2, i_3), \ldots, (i_{k-2}, i_{k-1}), (i_{k-1}, i_k), \tag{5.3.2}$$

where $(i_j, i_{j+1}) \in \tau$, $j = 1, \ldots, k-1$, is called a path in $G(M, N, \tau)$. A graph $G(M, N, \tau)$ is called connected if any two vertices i_1, i_k of the graph are joined by a path of the form (5.3.2).

COROLLARY 5.3.1. *A bipartite graph $G(M, N, \tau)$ is connected if and only if the corresponding matrix A is chainable.*

PROPERTY 5.3.2. *If an $m \times n$ matrix A has neither zero rows nor zero columns, then there exist permutation matrices Π and Π_1 such that*

$$\Pi A \Pi_1 = \begin{pmatrix} A_1 & \mathbf{0} & \ldots & \mathbf{0} \\ \mathbf{0} & A_2 & \ldots & \mathbf{0} \\ & & \ldots & \\ \mathbf{0} & \mathbf{0} & \ldots & A_s \end{pmatrix} \tag{5.3.3}$$

where $A_1, \ldots, A_s$ are chainable matrices.

If the matrix A is chainable, then Property 5.3.2 is true for $s = 1$. If A is not chainable, then there exist permutation matrices Π and Π_1 such that $\Pi A \Pi_1$ is of the form (5.3.1). We use induction. By the induction assumption, A_1 and A_2 in the matrix (5.3.1) admit expansions of the form (5.3.3) with chainable matrices at the main diagonal. Since A_1 and A_2 have no common rows or columns, there exist permutations of rows and columns which reduce A to the form (5.3.3).

5.3.2. Minimal chainable matrices. Let a $(0,1)$-matrix $A = (a_{ij})$, $i = 1,\dots,m$, $j = 1,\dots,n$, be chainable, and for each $a_{ij} = 1$ let the matrix $A - E_{ij}$ be not chainable. Here E_{ij} stands for the $(0,1)$-matrix whose only 1 is located at the ith row and jth column. Such an A is called a minimal chainable matrix.

It is clear that if A is a chainable matrix, then there exists a minimal chainable matrix A' such that $A' \leq A$.

PROPERTY 5.3.3. *If a $(0,1)$-matrix A of size $m \times n$ is minimal chainable, then*
(1) $A = (1,1,\dots,1)$ *for* $m = 1$;
(2) $A = (1,1,\dots,1)^T$ *for* $n = 1$;
(3) *for* $m > 1$, $n > 1$ *there exist permutation matrices* Π *and* Π_1 *such that*

$$\Pi A \Pi_1 = \begin{pmatrix} A_1 & \mathbf{0} \\ E & A_2 \end{pmatrix}, \tag{5.3.4}$$

where the matrices A_1 and A_2 are minimal chainable and the matrix E has a single one.

Assertions (1) and (2) are obvious. Let us prove (3). If $a_{ij} = 1$, then $A - E_{ij}$ is not chainable, and by Property 5.3.1 there exist permutation matrices Π and Π_1 such that

$$\Pi(A - E_{ij})\Pi_1 = \begin{pmatrix} A_1 & \mathbf{0} \\ \mathbf{0} & A_2 \end{pmatrix},$$

where A_1 and A_2 are chainable. Moreover, the minimal chainability of the matrix A implies the minimal chainability of the matrices A_1 and A_2. Hence

$$\Pi A \Pi_1 = \begin{pmatrix} A_1 & \mathbf{0} \\ \mathbf{0} & A_2 \end{pmatrix} + \Pi E_{ij} \Pi_1 \tag{5.3.5}$$

where A_1 and A_2 are minimal chainable.

In the right-hand side of this equality, the matrix $\Pi E_{ij} \Pi_1$ has a single element one, which cannot be added to elements of A_1 or A_2 because $\Pi A \Pi_1$ is chainable. Hence, permuting rows and columns, we can reduce the matrix to the form (5.3.4).

A closed chain with no links lying on the same line is called a cycle.

PROPERTY 5.3.4. *If a matrix A is minimal chainable, then none of the chains of A is a cycle.*

If $m = 1$ or $n = 1$, the assertion is obvious.

For $m > 1$ and $n > 1$, by Property 5.3.3 there exist permutation matrices Π and Π_1 such that $\Pi A \Pi_1$ is of the form (5.3.4), where A_1 and A_2 are minimal chainable. We use induction on the matrix order. By the induction assumption, the chains whose elements lie in A_1 only or in A_2 only cannot be cycles. Since E has a single positive element, a chain whose elements belong both to A_1 and to A_2 cannot be a cycle as well.

PROPERTY 5.3.5. *If a $(0,1)$-matrix A has at least two ones in each row and in each column, then A has a cycle.*

We take an element $a_{i_1j_1} = 1$. There exists an element $a_{i_1j_2} = 1$, $j_2 \neq j_1$, since the i_1th row has two ones. There exists an element $a_{i_2j_2} = 1$, since the j_2th column has two ones. Continuing, we arrive at $a_{i_1j_1}$, passing through positive elements of A in a rook-like manner; that is, the matrix A has a cycle.

PROPERTY 5.3.6. *If a matrix A of order n is minimal chainable, then $A = (1)$ for $n = 1$, and for $n > 1$ there exist permutation matrices Π and Π_1 such that*

$$\Pi A \Pi_1 = \begin{pmatrix} 0 & E_1 \\ E_2 & A_2 \end{pmatrix}, \tag{5.3.6}$$

where A_2 is a minimal chainable matrix of order $n-1$ and each of E_1, E_2 has a single one.

The assertion regarding the case $n = 1$ is obvious. To prove the second assertion, we use induction on n. For $n = 2$, the matrix A has one zero; that is, it admits the representation (5.3.6). For $n > 2$, the matrix A has no cycles and therefore, by Property 5.3.5, has a row or a column with a single element one. Let this element one, which is unique in the i_1th row, be in the j_1th column. Removing the i_1th row and the j_1th column from the matrix A, we arrive at the matrix $A(i_1, j_1)$. By the induction assumption the i_2th row and the j_2th column of $A(i_1, j_1)$ have a single element one each. Hence the i_1th row and the j_1th column have a single one each; that is, the matrix A reduces to the form (5.3.6) by permutations of rows and columns.

PROPERTY 5.3.7. *Let a $(0,1)$-matrix $A = (a_{ij})$, $i = 1, \dots, m$, $j = 1, \dots, n$, be chainable, and*

$$\Sigma(A) = \sum_{i,j=1}^{n} a_{ij}.$$

Then the matrix A is minimal chainable if and only if

$$\Sigma(A) = m + n - 1. \tag{5.3.7}$$

Let the matrix A be minimal chainable. We prove (5.3.7) by induction on $m + n$. By Property 5.3.3, (5.3.7) is fulfilled if $m = 1$ or $n = 1$. Let $m, n > 1$. There exist permutation matrices Π and Π_1 such that $\Pi A \Pi_1$ is of the form (5.3.4), where A_1 and A_2 are $s \times r$ and $(m-s) \times (n-r)$ matrices respectively. Since A_1 and A_2 are minimal chainable, we see that

$$\Sigma(A_1) = s + r - 1, \qquad \Sigma(A_2) = m + n - s - r - 1.$$

Hence

$$\Sigma(A) = \Sigma(A_1) + \Sigma(A_2) + \Sigma(E) = m + n - 1,$$

since $\Sigma(E) = 1$.

Let a $(0,1)$-matrix be chainable and $\Sigma(A) = m + n - 1$. We consider the connected bipartite graph $G(M, N, \tau)$ with adjacency matrix A. The graph $G(M, N, \tau)$ is called minimal connected if after removing any edge the graph ceases to be connected. It is clear that $G(M, N, \tau)$ is a minimal connected graph if and only if the matrix A is minimal chainable.

Let, say, $m \le n$. To construct a minimal connected graph $G(M, N, \tau)$, each of n vertices of N has to be joined with some vertex of M. The minimum number of edges needed to do this is equal to n. The minimal connected subgraphs thus obtained can be joined by additional edges to form a connected graph. The minimal number of edges needed to do this is m. Thus, a connected bipartite graph with $m+n-1$ edges is minimal. Hence a chainable matrix A such that $\Sigma(A) = m+n-1$ is minimal chainable.

PROPERTY 5.3.8. *A $(0,1)$-matrix $A = (a_{ij})$, $i = 1, \dots, m$, $j = 1, \dots, n$, is minimal chainable if and only if it has no cycles and the matrix $A + E_{ij}$ has a cycle for each (ij) such that $a_{ij} = 0$.*

Let a matrix A be minimal chainable. Then it has no cycles. If $a_{ij} = 0$, then in the ith row and the jth column we find $a_{ij_1} = 1$, $a_{i_1 j} = 1$, and there exists a chain whose ends are a_{ij_1} and $a_{i_1 j}$. The matrix $A + E_{ij}$ has the element one in the position (i, j). Using this element one and the mentioned chain, we form a cycle in the matrix $A + E_{ij}$.

Let a matrix A have no cycles, and let the matrix $A + E_{ij}$ have a cycle for each $a_{ij} = 0$. This cycle corresponds to a chain with ends $a_{ij_1} = 1$ and $a_{i_1 j} = 1$ in the matrix A. Since A has no cycles, for any $a_{rs} = 1$ and $a_{pq} = 1$ either $a_{rq} = 0$ or $a_{ps} = 0$, and there exist chains whose ends are arbitrary positive elements. It is also clear that the matrix A has no zero elements. Therefore, A is chainable. Now let $a_{kl} = 1$, and let the matrix $A - E_{kl}$ be chainable. Then in the kth row and the lth column we find elements $a_{kl_1} = 1$, $a_{k_1 l} = 1$ which are ends of a chain, allowing us, with the use of $a_{kl} = 1$, to form a cycle in the matrix A. In view of this contradiction, the matrix $A - E_{kl}$ cannot be chainable, and the matrix A is minimal chainable.

5.3.3. Transformations. Let a $(0,1)$-matrix $A = (a_{ij})$, $i = 1, \dots, m$, $j = 1, \dots, n$, be chainable, $a_{i_1 j_1} = 0$, and $\hat{A} = A + E_{i_1 j_1} = (\hat{a}_{ij})$, $i = 1, \dots, m$, $j = 1, \dots, n$. Then $\hat{A}$ has a cycle, say, of the form

$$\hat{a}_{i_1 j_1}, \hat{a}_{i_1 j_2}, \hat{a}_{i_2 j_2}, \dots, \hat{a}_{i_{s-1} j_s}, \hat{a}_{i_s j_s} = \hat{a}_{i_1 j_1}.$$

In the matrix A, we replace one of the elements $a_{i_k j_k} = 1$, $2 \le k \le s$, or $a_{i_k j_{k+1}} = 1$, $1 \le k \le s-1$, by zero, and $a_{i_1 j_1} = 0$ by one. The resulting matrix $\bar{A}$ will be called a transformation of the matrix A.

PROPERTY 5.3.9. *If a $(0,1)$-matrix A is chainable, then its transformation $\bar{A}$ is chainable as well.*

Since $a_{i_k j_k} = 1$ and $a_{i_k j_{k+1}} = 1$ are inner elements of a chain, both the i_kth row and the j_kth (j_{k+1}th, respectively) column have positive elements in addition to $a_{i_k j_k}$ ($a_{i_k j_{k+1}}$). Therefore, the matrix $\bar{A}$ has no zero rows or columns. Assume that $\bar{A}$ is not chainable. Then, by Property 5.3.1, there exist permutation matrices Π and Π_1 such that

$$\Pi \bar{A} \Pi_1 = \begin{pmatrix} \bar{A}_1 & \mathbf{0} \\ \mathbf{0} & \bar{A}_2 \end{pmatrix}.$$

Hence

$$\Pi A \Pi_1 = \begin{pmatrix} A_1 & \mathbf{0} \\ E_2 & A_2 \end{pmatrix} \quad \text{or} \quad \Pi A \Pi_1 = \begin{pmatrix} A_1 & E_2 \\ \mathbf{0} & A_2 \end{pmatrix},$$

where A_1 and $\bar{A}_1$ are of the same order and E_2 contains a single positive element. Under the permutation of rows and columns, the element $a_{i_1 j_1}$ falls into either A_1 or A_2 and therefore cannot lie in the cycle containing the element $a_{i_k j_k}$ ($a_{i_k j_{k+1}}$), which under that permutation falls into E_2. This contradiction proves the required result.

PROPERTY 5.3.10. *If A is a minimal chainable matrix, then the transformation $\bar{A}$ is also minimal chainable.*

Since A is minimal chainable, $\Sigma(A) = m + n - 1$. The matrix $\bar{A}$ has the same number of ones as A. Therefore, $\Sigma(\bar{A}) = m + n - 1$, and $\bar{A}$ is a minimal chainable matrix by Property 5.3.7.

PROPERTY 5.3.11. *If A and B are minimal chainable $(0,1)$-matrices of size $m \times n$, then there exists a sequence of transformations which takes A into B.*

We prove this property by induction on the number

$$d(A, B) = \sum_{i=1}^{m} \sum_{j=1}^{n} |a_{ij} - b_{ij}|,$$

where $A = (a_{ij})$, $B = (b_{ij})$, $i = 1, \dots, m$, $j = 1, \dots, n$. If $d(A, B) = 0$, then $A = B$, and the property is true. Assume that the property holds for all A and B such that $d(A, B) < r$, and let $d(A, B) = r$. Then there exist elements a_{ij} and b_{ij} such that $a_{ij} \neq b_{ij}$. Without loss of generality we set $a_{ij} = 0$, $b_{ij} = 1$. By Property 5.3.8, the matrix $\hat{A} = A + E_{ij}$ has a cycle Γ containing a_{ij}. Consider the sequence of elements of B corresponding to the indices of the cycle Γ. Since the matrix B has no cycles, among those indices we find p and q such that $b_{pq} = 0$. Now we perform a transformation of A to $\bar{A}$ by substituting one for $a_{ij} = 0$ and zero for $a_{pq} = 1$. Then $d(\bar{A}, B) < r$, because the elements in the positions (i, j) and (p, q), which were different in A and B, coincide in $\bar{A}$ and B, whereas the other ones remain unchanged. By the induction assumption, there exists a sequence of transformations which takes $\bar{A}$ into B. Hence there exists a sequence of transformations which takes A into B.

PROPERTY 5.3.12. *If A and B are chainable $(0,1)$-matrices and $\Sigma(A) = \Sigma(B)$, then there exists a sequence of transformations which takes A into B.*

Let $A = (a_{ij})$, $i = 1, \dots, m$, $j = 1, \dots, n$. Performing transformations which replace zero elements of the first row and the first column by ones, we arrive at the matrix $\bar{A}$ whose first row and column consist of ones only. Let the matrix $\bar{B}$ transformed from B possess the same property.

We consider the $(0,1)$-matrix $C = (c_{ij})$, $i = 1, \dots, m$, $j = 1, \dots, n$, whose first row and column consist of ones, whereas the remaining elements are zeros. It is clear that C is a minimal chainable matrix and $C \leq \bar{A}$, $C \leq \bar{B}$.

Let $D = (d_{ij})$ be a chainable matrix with $\Sigma(D) = \Sigma(A)$ such that $C \leq D$. If $d_{ij} = 1$ and $d_{ik} = 0$ for $i > 1$, $j > 1$, then D can be transformed into a matrix $\bar{D} = (\bar{d}_{ij})$ such that $C \leq \bar{D}$ and $\bar{d}_{ik} = 1$, $\bar{d}_{ij} = 0$. The same is true for $d_{ij} = 0$ and $d_{ik} = 1$. The matrix $\bar{D}$ thus obtained is chainable, since $C \leq \bar{D}$ and C is chainable. Applying the above transformation to $\bar{A}$ and $\bar{B}$, we move all ones outside the first row and the first column to the extreme left and the extreme top positions, and arrive at the matrix $\tilde{D} = (\tilde{d}_{ij})$, which is such that if $\tilde{d}_{ij} = 1$, then $\tilde{d}_{i-1,j} = 1$ and $\tilde{d}_{i,j-1} = 1$ for all i and j, and if $\tilde{d}_{ij} = 1$ for $i > 1$ and $j > 1$, then $\tilde{d}_{i-1,n} = 1$.

The transformations of $\bar{A}$ and $\bar{B}$ give the same matrix $\tilde{D}$, since $\Sigma(\bar{A}) = \Sigma(\bar{B})$.

5.3.4. Multiplication of matrices.

PROPERTY 5.3.13. *If matrices A and B of order n are chainable, then the matrix AB is also chainable.*

The matrices A and B are chainable and hence have no zero rows or columns. Thus AB also has no zero rows or columns. Assume that AB is not chainable. Then, by Property 5.3.1, there exist permutation matrices Π and Π_1 such that

$$\Pi AB\Pi_1 = \begin{pmatrix} C_1 & \mathbf{0} \\ \mathbf{0} & C_2 \end{pmatrix}, \tag{5.3.8}$$

where C_1 is an $r \times s$ matrix. We introduce

$$\Pi A = \begin{pmatrix} A_1 \\ A_2 \end{pmatrix}, \qquad B\Pi_1 = \begin{pmatrix} B_1 & B_2 \end{pmatrix},$$

where A_1 and B_1 are $r \times n$ and $n \times s$ matrices.

Let Π_2 be a permutation matrix such that

$$\Pi A\Pi_2 = \begin{pmatrix} A_1 \\ A_2 \end{pmatrix} \Pi_2 = \begin{pmatrix} \hat{A}_{11} & \mathbf{0} \\ \hat{A}_{21} & \hat{A}_{22} \end{pmatrix},$$

where $\hat{A}_{11}$ is an $r \times t$ matrix without zero rows or columns. Furthermore,

$$\Pi_2^T B\Pi_1 = \Pi_2^T \begin{pmatrix} B_1 & B_2 \end{pmatrix} = \begin{pmatrix} \hat{B}_{11} & \hat{B}_{12} \\ \hat{B}_{21} & \hat{B}_{22} \end{pmatrix},$$

where $\hat{B}_{11}$ is a $t \times s$ matrix. Thus,

$$\Pi AB\Pi_1 = \begin{pmatrix} C_1 & \mathbf{0} \\ \mathbf{0} & C_2 \end{pmatrix} = \begin{pmatrix} \hat{A}_{11} & \mathbf{0} \\ \hat{A}_{21} & \hat{A}_{22} \end{pmatrix} \begin{pmatrix} \hat{B}_{11} & \hat{B}_{12} \\ \hat{B}_{21} & \hat{B}_{22} \end{pmatrix}.$$

Hence

$$\hat{A}_{11}\hat{B}_{12} = \mathbf{0}, \qquad \hat{A}_{21}\hat{B}_{11} + \hat{A}_{22}\hat{B}_{21} = \mathbf{0}.$$

Since the matrices are nonnegative, from the last equality it follows that

$$\hat{A}_{21}\hat{B}_{11} = \mathbf{0}, \qquad \hat{A}_{22}\hat{B}_{21} = \mathbf{0}.$$

Since $\hat{A}_{11}$ and $\hat{A}_{22}$ have no zero columns, $\hat{B}_{12} = \mathbf{0}$ and $\hat{B}_{21} = \mathbf{0}$. Since B has no zero columns, the equality $\hat{B}_{21} = \mathbf{0}$ implies that $\hat{B}_{11}$ has no zero columns. Hence $\hat{A}_{21} = \mathbf{0}$. Thus, the matrices A and B reduce to the block form and are not chainable, which is impossible.

COROLLARY 5.3.2. *The set $C(n,n)$ of chainable matrices forms a semigroup with respect to the operation of matrix multiplication.*

COROLLARY 5.3.3. *The set $C(n,n)$ of chainable matrices is a two-sided ideal in the semigroup $H(n,n)$ of matrices of order n without zero rows or columns.*

PROOF. We show that if $A \in C(n,n)$ and $B \in H(n,n) \setminus C(n,n)$, then $AB \in C(n,n)$. Let $AB \notin C(n,n)$. Then there exist permutation matrices Π and Π_1 such that (5.3.8) holds. Similarly, the matrix B satisfies the relation

$$\Pi_2 B\Pi_3 = \begin{pmatrix} B_1 & \mathbf{0} \\ \mathbf{0} & B_2 \end{pmatrix},$$

where Π_2 and Π_3 are permutation matrices. We represent A as

$$A = \begin{pmatrix} A_{11} & A_{12} \\ A_{21} & A_{22} \end{pmatrix},$$

where A_{11} is an $s \times r$ matrix. From the equality

$$\begin{pmatrix} A_{11} & A_{12} \\ A_{21} & A_{22} \end{pmatrix} \begin{pmatrix} B_1 & \mathbf{0} \\ \mathbf{0} & B_2 \end{pmatrix} = \begin{pmatrix} C_1 & \mathbf{0} \\ \mathbf{0} & C_2 \end{pmatrix}$$

it follows that

$$A_{12}B_2 = \mathbf{0}, \qquad A_{21}B_2 = \mathbf{0}.$$

Since B_1 and B_2 have no zero rows or columns, we conclude that $A_{12} = \mathbf{0}$ and $A_{21} = \mathbf{0}$, that is, the matrix A is not chainable, which is impossible. Similar reasoning shows that $BA \in C(n, n)$. □

PROPERTY 5.3.14. *If matrices A and B are chainable, then their Kronecker product $A \otimes B$ is chainable.*

If $A = (a_{ij})$, then $A \otimes B = (a_{ij}B)$. Therefore, $A \otimes B$ has no zero rows or columns, as otherwise either A or B would have them. We set $A \otimes B = (c_{ij})$ and assume that $c_{rs} = 1$, $c_{pq} = 1$. If $c_{rs} = a_{x_1y_1}b_{x_2y_2}$, $c_{pq} = a_{w_1z_1}b_{w_2z_2}$, then $a_{x_1y_1} = 1$, $a_{w_1z_1} = 1$ and $b_{x_2y_2} = 1$, $b_{w_2z_2} = 1$. Therefore, there exist chains in the matrices A and B of the form

$$a_{x_1y_1}, a_{i_1j_1}, a_{j_2j_2}, \dots, a_{i_sj_s}, a_{w_1z_1},$$
$$b_{x_2y_2}, b_{k_1l_1}, b_{k_2l_2}, \dots, b_{k_tl_t}, b_{w_2z_2}$$

respectively. With the use of these chains, in the matrix $A \otimes B$ we construct the chain

$$a_{x_1y_1}b_{x_2y_2}, a_{i_1j_1}b_{x_2y_2}, \dots, a_{w_1z_1}b_{x_2y_2}, a_{w_1z_1}b_{k_1l_1}, a_{w_1z_1}b_{k_2l_2}, \dots, a_{w_1z_1}b_{w_2z_2},$$

whose ends are elements c_{rs} and c_{pq} of the matrix $A \otimes B$.

COROLLARY 5.3.4. *If matrices A and B are totally indecomposable, then the matrix $A \otimes B$ is also totally indecomposable.*

PROOF. It suffices to prove that each element of the matrix $A \otimes B = (c_{ij})$ lies on a positive diagonal (see Property 5.1.10). Let $c_{rs} = a_{i_1j_1}b_{k_1l_1} = 1$, and consequently, $a_{i_1j_1} = 1$, $b_{k_1l_1} = 1$, and these elements lie on positive diagonals, say, $a_{i_1j_1}, a_{i_2j_2}, \dots, a_{i_mj_m}$ and $b_{k_1l_1}, b_{k_2l_2}, \dots, b_{k_nl_n}$, respectively. Then it is clear that c_{rs} lies on a positive diagonal of the form

$$\begin{matrix} a_{i_1j_1}b_{k_1l_1}, & a_{i_1j_1}b_{k_2l_2}, & \dots, & a_{i_1j_1}b_{k_nl_n}, \\ a_{i_2j_2}b_{k_1l_1}, & a_{i_2j_2}b_{k_2l_2}, & \dots, & a_{i_2j_2}b_{k_nn}, \\ & & \cdots & \\ a_{i_mj_m}b_{k_1l_1}, & a_{i_mj_m}b_{k_2l_2}, & \dots, & a_{i_mj_m}b_{k_nl_n}, \end{matrix}$$

Thus, the corollary is proved. □

Problems.

1. Indicate which of the following matrices are chainable:

$$\begin{pmatrix} 0 & 1 \\ 1 & 0 \end{pmatrix}, \quad \begin{pmatrix} 1 & 1 \\ 1 & 1 \end{pmatrix}, \quad \begin{pmatrix} 1 & 0 & 1 \\ 0 & 1 & 0 \\ 1 & 0 & 0 \end{pmatrix}, \quad \begin{pmatrix} 1 & 1 & 0 \\ 1 & 0 & 1 \\ 0 & 1 & 1 \end{pmatrix}.$$

2. Let Γ be a directed graph with n vertices and adjacency matrix $A = A(\Gamma)$. Prove that the graph Γ is connected if and only if the matrix $A + \mathbf{I}$ is chainable, where $\mathbf{I}$ is the identity matrix.

5.4. Extension of partial diagonals

5.4.1. Partial diagonals. Recall that a partial diagonal of length t of a nonnegative matrix $A = (a_{ij})$, $i = 1, \dots, m$, $j = 1, \dots, n$, is a sequence of positive elements

$$a_{i_1 j_1}, a_{i_2 j_2}, \dots, a_{i_t j_t}, \tag{5.4.1}$$

where $i_\mu \neq i_\nu$, $j_\mu \neq j_\nu$ for $\mu \neq \nu$ (see Section 1.2.1).

If $m < n$ and $t = m$, then a partial diagonal is called a row diagonal, and if $m = n = t$, then it is called a diagonal of the matrix A (see Section 1.2). A partial diagonal (5.4.1) can be extended to a partial diagonal of length $t+d$, $d > 0$, if there exist positive elements $a_{i_{t+1} j_{t+1}}, \dots, a_{i_{t+d}, j_{t+d}}$ of A such that the sequence

$$a_{i_1 j_1}, a_{i_2 j_2}, \dots, a_{i_t j_t}, a_{i_{t+1} j_{t+1}}, \dots, a_{i_{t+d}, j_{t+d}} \tag{5.4.2}$$

is a partial diagonal of A of length $t + d$.

The following properties of partial diagonals can be deduced from the results of Section 5.2.

PROPERTY 5.4.1. *If a matrix $A \geq 0$ has a partial diagonal of length t, then its term rank $\rho(A)$ satisfies the inequality*

$$\rho(A) \geq t. \tag{5.4.3}$$

PROPERTY 5.4.2. *An $m \times n$ nonnegative matrix A, $m \leq n$, has a partial diagonal of length $t \leq m$ if and only if any covering of the positive elements of A contains at least t lines (rows and columns).*

This assertion follows from the König Theorem (Theorem 1.2.6).

PROPERTY 5.4.3. *An $m \times n$ nonnegative matrix A, $m \leq n$, has a partial diagonal of length $t \leq m$ if and only if A contains no zero $p \times q$ submatrix such that $p + q = m + n - t + 1$, $p, q \geq 1$.*

This assertion is usually called the König–Frobenius Theorem.

PROPERTY 5.4.4. *A totally indecomposable $m \times n$ matrix A, $m \leq n$, has a row diagonal.*

Indeed, a matrix A is totally indecomposable, $2 \leq m \leq n$, if by a permutation of rows and columns it cannot be represented in the form

$$\begin{pmatrix} A_{11} & A_{12} \\ \mathbf{0} & A_{22} \end{pmatrix}, \tag{5.4.4}$$

where A_{11} is of order p, $1 \leq p \leq m-1$. This is equivalent to the fact that A contains no zero $p \times (m-p)$ submatrix, $1 \leq p \leq m-1$. It remains to use Property 5.4.3.

PROPERTY 5.4.5. *For a nonnegative matrix A of order n which has a positive diagonal, there exist permutation matrices Π and Π_1 such that*

$$\Pi A \Pi_1 = \begin{pmatrix} A_1 & \mathbf{0} & \dots & \mathbf{0} \\ * & A_2 & \dots & \mathbf{0} \\ & & \dots & \\ * & * & \dots & A_k \end{pmatrix}, \tag{5.4.5}$$

where $A_1, \dots, A_k$ are totally indecomposable square matrices and each $*$ stands for a nonnegative matrix.

Since A has a diagonal, there exist permutation matrices Π' and Π'_1 such that $\Pi' A \Pi'_1$ has a positive main diagonal. For $\Pi' A \Pi'_1$ there exists a permutation matrix $\tilde{\Pi}$ such that $\tilde{\Pi}\Pi' A \Pi'_1 \tilde{\Pi}^{-1}$ is of the form (5.4.5) with $\Pi = \tilde{\Pi}\Pi'$, $\Pi_1 = \Pi'_1\tilde{\Pi}^{-1}$; here $A_1, \dots, A_k$ are indecomposable square matrices and the positive main diagonal remains unchanged. Since each matrix A_i, $1 \le i \le k$, is indecomposable and has a positive diagonal, it is totally indecomposable (see Property 5.1.2).

PROPERTY 5.4.6. *Each nonzero element of a nonnegative square matrix A with nonzero diagonal belongs to some diagonal if and only if there exist permutation matrices Π and Π_1 such that*

$$\Pi A \Pi_1 = \begin{pmatrix} A_1 & \mathbf{0} & \dots & \mathbf{0} \\ \mathbf{0} & A_2 & \dots & \mathbf{0} \\ & & \dots & \\ \mathbf{0} & \mathbf{0} & \dots & A_k \end{pmatrix}, \tag{5.4.6}$$

where $A_1, \dots, A_k$ are totally indecomposable matrices. In particular, for $k = 1$ the matrix A itself is totally indecomposable.

Let each nonzero element of the matrix A belong to some diagonal. Then A has at least one diagonal and admits the representation (5.4.5). Using the Laplace theorem on permanents, we find from (5.4.5) that

$$\operatorname{per} A = \operatorname{per} A_1 \operatorname{per} A_2 \cdots \operatorname{per} A_k. \tag{5.4.7}$$

This means that the diagonal products containing elements of matrices labelled $*$ provide no positive contribution to $\operatorname{per} A$. Hence all matrices labelled $*$ are zero matrices, and the matrix $\Pi A \Pi_1$ is of the form (5.4.6).

Now let the matrix A admit the representation (5.4.6). In view of (5.4.7) it suffices to show that each positive element of the matrix A_i lies on a diagonal of this matrix, $1 \le i \le k$. The matrix A_i is totally indecomposable; therefore $\operatorname{per} A_i(r, s) > 0$ for any element $a_{rs}^{(i)} > 0$ of it (see Property 5.1.1). This means that $a_{rs}^{(i)}$ lies on some diagonal.

COROLLARY 5.4.1. *Any totally indecomposable matrix of order n has at least two diagonals.*

This follows from the fact that a totally indecomposable matrix of order n has at least two positive elements in each row.

5.4.2. Extension of partial diagonals.

PROPERTY 5.4.7. *For an $m \times n$ totally indecomposable matrix A, $m \leq n$, each partial diagonal of A of length t, $1 \leq t < m$, can be extended to a row diagonal of A if and only if A has no zero $p \times q$ submatrices such that*

$$p + q = n - t + 1, \quad 1 \leq p \leq m - t, \quad 1 \leq q \leq n - t. \tag{5.4.8}$$

We begin by proving sufficiency. Without loss of generality, assume that

$$A = \begin{pmatrix} A_1 & A_2 \\ A_3 & A_4 \end{pmatrix},$$

where A_1 is a matrix of order t with a diagonal. Assume that the $(m - t) \times (n - t)$ matrix A_4 has no row diagonals. Then, by Property 5.4.3, A_4 has a $p \times q$ zero submatrix such that

$$p + q = (m - t) + (n - t) - (m - t) + 1 = n - t + 1,$$

and $1 \leq p \leq m - t$, $1 \leq q \leq n - t$. This contradicts condition (5.4.8). Thus, A_4 has a row diagonal, which, combined with the row diagonal of A_1, forms a row diagonal of A.

Note that in the proof of sufficiency we did not use the fact that A is totally indecomposable.

Now let a matrix A have a $p \times q$ zero submatrix satisfying (5.4.8). Assume that each row diagonal of length t can be extended to a row diagonal of A. Without loss of generality we assume that A is of the form

$$A = \begin{pmatrix} \mathbf{0} & A_2 \\ A_3 & A_4 \end{pmatrix},$$

where the dimension of the $p \times q$ zero submatrix $\mathbf{0}$ satisfies (5.4.8).

Assume that the $(m - p) \times (n - q)$ matrix A_4 has a partial diagonal of length t. It is clear that an extension of this diagonal must contain p additional elements in $n - q - 1$ different last columns; that is, $p \leq n - q - t$, which contradicts (5.4.8).

It remains to prove that A_4 has a partial diagonal of length t. If this were not the case, then there would exist a $k \times l$ zero submatrix such that

$$k + l = (m - p) + (n - q) - t + 1 = m.$$

This contradicts the fact that A is totally indecomposable.

COROLLARY 5.4.2. *For a totally indecomposable matrix A of order n, any partial diagonal of length t, $1 \leq t < n$, can be extended to a diagonal of the A if and only if A contains no $p \times q$ zero submatrix such that $p \geq 1$, $q \geq 1$, and $p + q = n - t + 1$.*

5.4.3. Extension of diagonals and coverings.

PROPERTY 5.4.8. *Let an $m \times n$ matrix A, $m \leq n$, have a partial diagonal of length k and have no diagonals of greater length. Then each diagonal of length t, $1 \leq t \leq k$, can be extended to a diagonal of length k if and only if each covering of A consisting of r rows and s columns, $r \geq 1$, $s \geq 1$, $r + s = k + t - 1$, is associated with an $r \times s$ submatrix which can be covered by less than t lines (rows and columns).*

Let

$$A = \begin{pmatrix} A_1 & A_2 \\ A_3 & A_4 \end{pmatrix},$$

where A_1 is a matrix of order t with a diagonal. Assume that the diagonal of A_1 cannot be extended to a partial diagonal of A of length k. Then the matrix A_4 has no partial diagonals of length $k-t$; that is, A_4 is covered by $k-t-1$ lines. Without loss of generality we assume that this covering consists of the first p rows and the first q columns of A_4 (the cases $p=0$ and $q=0$ are allowed). Thus, the first $t+p$ rows and $t+q$ columns of the matrix A cover the whole matrix A, and $t+p \geq 1$, $t+q \geq 1$, $(t+p)+(t+q) = k-t-1+2t = k+t-1$. But it is clear that the main $(t+p) \times (t+q)$ submatrix of A cannot be covered by less than t lines.

Now we assume that each partial diagonal of length t can be extended to a diagonal of length k and there exists a covering consisting, say, of the first r rows and the first s columns, $r \geq 1$, $s \geq 1$, $r+s = k+t-1$, such that the main $r \times s$ submatrix A' cannot be covered by less than t lines. Then A' has a partial diagonal Δ of length t. If

$$A = \begin{pmatrix} A_1 & A_2 \\ A_3 & A_4 \end{pmatrix},$$

where A_1 is a matrix of order t, then the matrix

$$A_4 = \begin{pmatrix} X & Y \\ Z & \mathbf{0} \end{pmatrix}$$

can be covered by $r-t$ rows and $s-t$ columns, that is, by $r-t+s-t = k-t-1$ lines. Thus, A_4 has no partial diagonal of length $k-t$, and hence, Δ cannot be extended to a partial diagonal of length k, which is impossible.

5.4.4. The uniqueness of extension for partial diagonals. We are interested in conditions under which any partial diagonal of length t of an $m \times n$ $(0,1)$-matrix A can be uniquely extended to a row diagonal of A. If $m < n$, then the uniqueness does not occur in nontrivial cases.

PROPERTY 5.4.9. *Let $m < n$, and let an $m \times n$ nonnegative matrix A admit the representation $A = \begin{pmatrix} A_1 & A_2 \end{pmatrix}$, where A_1 has a row diagonal and $A_2 \neq \mathbf{0}$. Then for the matrix A the uniqueness of extension of partial diagonals does not take place.*

Without loss of generality we assume that A_1 has the row diagonal

$$\Delta = (a_{11}, a_{22}, \ldots, a_{mm}),$$

that is, $a_{ii} > 0$, $1 \leq i \leq m$. Let $1 \leq i \leq m$, $j > m$, and $a_{ij} > 0$. Then the partial diagonal $(a_{11}, \ldots, a_{i-1,i-1}, a_{i+1,i+1}, \ldots, a_{t+1,t+1})$ can be extended both to the diagonal Δ and to the diagonal

$$\Delta' = (a_{11}, a_{i-1,i-1}, a_{ij}, a_{i+1,i+1}, \ldots, a_{mm}).$$

Property 5.4.9 shows that the question on uniqueness of extension of partial diagonals is of interest for square matrices only.

We begin with the case $t = 1$.

PROPERTY 5.4.10. *For a totally indecomposable $(0,1)$-matrix A of order $n \geq 2$, any partial diagonal of length one can be uniquely extended to a diagonal of A if and only if an appropriate permutation of rows and columns reduces A to*

$$A = \mathbf{I}_n + C_n, \tag{5.4.9}$$

where $\mathbf{I}_n$ and C_n are permutation matrices of order n corresponding to the identity permutation and the unicyclic permutation $C = (1\,2\,\ldots\,n)$.

The matrix $A = \mathbf{I}_n + C_n$ is totally indecomposable, and $\operatorname{per}(\mathbf{I}_n + C_n) = 2$. Each row and column has two elements 1, and each of these elements lies on a unique diagonal.

Now let a matrix A be totally indecomposable and let every one of A lie on a unique diagonal. Each row contains at least two ones; hence there exist at least two diagonals. Assume that one of them is the main diagonal and the other corresponds to the permutation matrix Π associated with a permutation s. Assume that s is not unicyclic and contains a cycle, say, $(1, 2, \ldots, l)$. Then, for example, a_{nn}, in addition to the main diagonal, lies on the diagonal $a_{12}a_{23}, \ldots, a_{l-1,l}, a_{l1}, a_{l+1,l+1}, \ldots, a_{nn}$, which is impossible. Therefore, s is a unicyclic permutation. It is clear that after applying the same appropriate permutation to rows and columns of A the permutation matrix Π is transformed to the permutation matrix corresponding to the unicyclic permutation C, while the main diagonal remains unchanged. Finally, no element of A except those lying on the two mentioned diagonals can be equal to one. Assume that there is an element $a_{ij} = 1$ that does not lie on these two diagonals. Without loss of generality, assume that $i < j$. Then a_{ij} lies on the diagonal

$$(a_{12}, a_{23}, \ldots, a_{i-1,i}, a_{ij}, a_{j,j+1}, \ldots, a_{n-1,n}, a_{n1}, a_{i+1,i+1}, \ldots, a_{j-1,j-1}).$$

At the same time, a_{12} lies on the same diagonal, which contradicts the uniqueness of diagonal extension.

We formulate one more criterion for the uniqueness of extension of a diagonal of length one.

PROPERTY 5.4.11. *Any partial diagonal of length one of a* $(0, 1)$*-matrix* $A = (a_{ij})$, $i, j = 1, \ldots, n$, $n \geq 2$, *can be uniquely extended to a diagonal of* A *if and only if*

(1) $\displaystyle\sum_{i=1}^{n} a_{ij} = \sum_{j=1}^{n} a_{ij} = 2,$

(2) $\operatorname{per} A = 2.$

In view of Lemmas 1.3.21 and 1.3.16, condition (1) implies that A is permutationally equivalent to the direct sum of matrices of the form $\mathbf{I}_{n_i} + C_{n_i}$, $i = 1, \ldots, q$. The second condition shows that $q = 1$, and in the matrix $\mathbf{I}_n + C_n$ any diagonal of length one admits a unique extension to a diagonal of the matrix.

Let A satisfy the uniqueness condition for the extension of a diagonal of length one. If in some row one of the elements equal to one lies on some diagonal, then the other one should lie on another diagonal. Therefore by permutations of rows and columns the matrix A reduces to $\mathbf{I}_n + C_n + A'$. As we have seen in the proof of Property 5.4.10, A' must be a zero matrix. Then both the conditions hold.

Note that in Property 5.4.11 A is not required to be decomposable.

Now we turn to the question on the possibility of extending any partial diagonal of length two to a unique diagonal of a matrix.

PROPERTY 5.4.12. *In a totally indecomposable* $(0, 1)$*-matrix* A *of order* $n > 3$, *any partial diagonal of length* 2 *can be extended to at least two different diagonals of* A.

The matrix $A = (a_{ij})$ is totally indecomposable. Therefore, each element 1 lies on a diagonal of A. Without loss of generality we set $a_{11} = 1$ and represent A as

$$A = \begin{pmatrix} 1 & X \\ Y & A_1 \end{pmatrix} = (a_{ij}), \qquad i, j = 1, \ldots, n,$$

where A_1 is of order $n-1$. We choose a partial diagonal of two elements, one of which is $a_{11} = 1$ and the other element $a_{ij} = 1$ belongs to the matrix A_1. For an extension of this partial diagonal to be unique, it is necessary that the element $a_{ij} = 1$ lie on only one diagonal of A. Then A_1 either is totally indecomposable and possesses the property of uniqueness of extension of partial diagonals of length one, or is a direct sum of totally indecomposable matrices possessing the same property. The direct sum cannot contain terms of order 1, because otherwise A would contain a $1 \times (n-2)$ zero submatrix, and hence would does not satisfy the uniqueness condition for the extension of a diagonal of length 2.

Thus, each term of the direct sum is a square matrix of the form determined by Property 5.4.10. Each row and each column of A contains at least three ones, as otherwise it would again contain a $1 \times (n-2)$ zero submatrix.

Therefore

$$X = (1, \ldots, 1) = Y^T.$$

Now it is obvious that for the matrix $A = (a_{ij})$ the equality

$$\begin{pmatrix} a_{11} & a_{12} \\ a_{21} & a_{22} \end{pmatrix} = \begin{pmatrix} 1 & 1 \\ 1 & 1 \end{pmatrix}$$

is true. Hence any pair of elements one on the main diagonal outside of this submatrix can be extended in at least two ways.

PROPERTY 5.4.13. *For a totally indecomposable* $(0,1)$*-matrix* A *of order* n*, any partial diagonal of length* t*,* $2 \leq t < n-1$*, can be extended to at least two different diagonals of* A.

In the case where $t = 2$ and hence $n > 3$, Property 5.4.13 follows from Property 5.4.12. We use induction on n. Let $t > 2$. Assume that the assertion is true for all $n-1 > t$, and prove it for $n > t+1$. In a matrix A of order n, for $n > t+1$ let each partial diagonal of length t be uniquely extended to a diagonal of A. The matrix A is totally indecomposable; therefore it has a positive diagonal. Without loss of generality we assume that it is the main diagonal, and represent A in the form

$$A = \begin{pmatrix} 1 & X \\ Y & A_1 \end{pmatrix},$$

where the $(n-1) \times (n-1)$ matrix A_1 has the property that each partial diagonal of length $t-1$ can be uniquely extended to a diagonal of A_1 and each one of A_1 lies on a diagonal. This one can be extended to a partial diagonal of length $t-1$ with the use of $t-2 < n-3$ elements on the main diagonal of A_1. Since $n-1 > t$, this partial diagonal of length $t-1$ admits an extension to a diagonal of A_1. Thus, A_1 is a direct sum of totally indecomposable matrices. If a term of this sum is of dimensions $s \times s$, then $s > t$. If $s \leq t$, then there exists an $(n-1-s) \times s$ zero submatrix of A such that $(n-1-s)+s = n-1 > n-t$ and $n-1-s \geq n-1-t > 0$, which contradicts the condition that a partial diagonal of length t can be extended (see (5.4.8)). On the other hand, the inequality $s > t$ is impossible because each partial diagonal of length $t-1$ should be uniquely extendible to a diagonal of A_1, which contradicts the induction assumption.

Note that the condition $n > t+1$ cannot be weakened. Set $n = t+1$ and choose a matrix whose elements are all equal to 1. It is clear that each diagonal of length t is uniquely extendible to a diagonal of this matrix, but no other totally indecomposable matrix possesses this property (see Property 5.4.7).

PROPERTY 5.4.14. *Let a* $(0,1)$*-matrix* A *of order* n *have a positive diagonal, for a fixed* t, $1 \leq t < n$, *let each one of* A *lie on a partial diagonal of length* t, *and let any partial diagonal of length* t *have a unique extension to a diagonal of* A. *Then the following cases are possible.*

(1) *If* $n = t + 1$, *then* A *is a direct sum of totally indecomposable matrices with all elements equal to one.*
(2) *If* $n > t + 1$, *then two cases are possible:*
  (a) *for* $t = 1$, *either* $A = \mathbf{I}_n + C_n$ *or* $A = \mathbf{I}_n$,
  (b) *for* $t \geq 2$, $A = \mathbf{I}_n$.

The matrices mentioned above have the property of unique extendibility of the partial diagonals of length t. Show that the above list of matrices is exhaustive.

Each element 1 lies on a diagonal of length t and hence can be extended to a diagonal of A. Therefore, by permutations of rows and columns the matrix A reduces to a direct sum of s totally indecomposable matrices. For $n = t + 1$ a partial diagonal of length $t = n - 1$ consists of diagonals of $s - 1$ components and a partial diagonal of length $r - 1$ of a component of order r. From the uniqueness of extension of the partial diagonal of length $r - 1$ it follows that the component of order r consists of 1's only. Taking a partial diagonal of any of s components, we see that each of these components consists of ones only.

Now let $n > t+1$ and $A \neq \mathbf{I}_n$. Then A has a totally indecomposable component A_1 of order $r \geq 2$. If $n > t + 1$ and $A = \mathbf{I}_n$, then $r = 1$. It is clear that A_1 has at least two diagonals. We represent A in the form

$$A = \begin{pmatrix} A_1 & \mathbf{0} \\ \mathbf{0} & A_2 \end{pmatrix},$$

where $\mathbf{0}$ is a zero matrix and A_2 is of order $n - r$. Each partial diagonal of length $t - (n - r) \geq 1$ should be extended in A_1 to a diagonal of A_1 in a unique way. In view of the condition $r > t - (n - r) + 1$, for $t - (n - r) \geq 2$ such an extension is impossible by Property 5.4.13. For $t - (n - r) = 1$, by Property 5.4.10 $A_1 = \mathbf{I}_r + C_r$. If $t = 1$, then $r = n$, and A has a single component of the form $\mathbf{I}_n + C_n$.

For $t \geq 2$, consider a partial diagonal of A consisting of $t - 2$ elements of the matrix A_2. The condition $t - (n - r) = 1$ of unique extendibility of partial diagonals implies that A_2 is of order $t - 1$. We extend a partial diagonal of length $t - 2$ to a diagonal of length t by adding two elements a_{11} and a_{23} in A_1. Since $A_1 = \mathbf{I}_{n-t+1} + C_{n-t+1}$, the resulting partial diagonal of length t cannot be extended to a diagonal of A. Therefore, for $t \geq 2$ the only possible case is $A = \mathbf{I}_n$.

Now let any element of a matrix $A \geq 0$ of order $n \geq 2$ have 2^{d-1} distinct extensions to a diagonal of A. Such an extension will be called d-variant. For $d = 1$ the extension is unique. We give a characterization of a class of $(0,1)$-matrices of order n such that each nonzero element has a d-variant extension.

PROPERTY 5.4.15. *Let a* $(0,1)$*-matrix* $A = (a_{ij})$, $i, j = 1, \ldots, n$, *satisfy the condition*

$$\sum_{j=1}^{n} a_{ij} = \sum_{i=1}^{n} a_{ij} = 2.$$

Then each element of A has a d-variant extension if and only if there exist permutation matrices Π and Π' of order n such that

$$\Pi A \Pi' = (\mathbf{I}_{n_1} + C_{n_1}) \oplus (\mathbf{I}_{n_2} + C_{n_2}) \oplus \cdots \oplus (\mathbf{I}_{n_d} + C_{n_d}), \tag{5.4.10}$$

that is, $\Pi A \Pi'$ is a direct sum of matrices of the form

$$\mathbf{I}_{n_i} + C_{n_i}, \quad n_i > 1, \quad 1 \leq i \leq d.$$

By Theorem 1.3.23 there exist two permutation matrices Π and Π' such that $\Pi A \Pi'$ is a direct sum of the form (5.4.10) with r summands.

Note that if a matrix A is totally indecomposable, then $d = 1$ and the hypotheses of Property 5.4.11 are satisfied; that is, in A any element equal to one is uniquely extendible to a diagonal of A.

Problem.

1. Let s and s' be permutations of order n, and let Π and Π' be the corresponding permutation matrices. Let $l_1, \dots, l_k$ be the cycle lengths of the permutation $s^{-1}s'$. Prove that the formula

$$\operatorname{per}(\alpha\Pi + \beta\Pi') = \prod_{i=1}^{k} (\alpha^{l_i} + \beta^{l_i})$$

is true for any real α and β.

5.5. Prime Boolean matrices

5.5.1. Properties of prime Boolean matrices. The set of all $(0,1)$-matrices of order n with Boolean operations of addition and multiplication is called the Boolean and denoted by $\mathcal{B}_n$ (see Section 1.1.12). Let $\mathcal{P}_n$ stand for the set of all permutation matrices of order n.

A matrix $A \in \mathcal{B}_n$ is called prime (more precisely, algebraically prime) if $A \notin \mathcal{P}_n$ and the equality $A = BC$ for $B, C \in \mathcal{B}_n$ implies that either $B \in \mathcal{P}_n$ or $C \in \mathcal{P}_n$.

If there exists at least one representation $A = BC$, where $B \notin \mathcal{P}_n$ and $C \notin \mathcal{P}_n$, then the matrix $A \in \mathcal{B}_n$ is called factorable. Thus, a matrix $A \in \mathcal{B}_n$ is prime if and only if it is not factorable.

By the definition, the set $\mathcal{B}_1$ does not contain prime matrices. It is easy to see that $\mathcal{B}_2$ has no prime matrices as well. For $n = 3$, the matrix

$$\begin{pmatrix} 0 & 1 & 1 \\ 1 & 0 & 0 \\ 1 & 1 & 0 \end{pmatrix}$$

is prime.

For $(0,1)$-matrices $A = (a_{ij})$, $B = (b_{ij})$, $i = 1, \dots, m$, $j = 1, \dots, n$, we say that B is included in A, and write $B \subseteq A$ or $A \supseteq B$, if $a_{ij} \geq b_{ij}$ for all $1 \leq i \leq m$, $1 \leq j \leq n$ (that is, if $A \geq B$). If $m = 1$ or $n = 1$, then we obtain the definition of inclusion for rows and columns respectively. The concept of inclusion for rows and columns of matrices allows us to formulate a necessary condition for primality of a matrix.

PROPERTY 5.5.1. *If $A \in \mathcal{B}_n$ is a prime matrix, then none of its columns (rows) includes another column (row).*

Let $A_1, \ldots, A_n$ be the columns of A; that is, $A = \begin{pmatrix} A_1 & \ldots & A_n \end{pmatrix}$. Without loss of generality we assume that $A_1 \supseteq A_2$. Then $A_1 \boxplus A_2 = A_1$ (the symbol $\boxplus$ stands for the Boolean addition) and therefore $A = AC$, where

$$C = \begin{pmatrix} 1 & 0 & & & 0 \\ 1 & 1 & & & \\ & & 1 & & \\ & & & \ddots & \\ 0 & & & & 1 \end{pmatrix};$$

that is, the matrix A is factorable.

The next property follows immediately from Property 5.5.1.

PROPERTY 5.5.2. *If $A \in \mathcal{B}_n$ is a prime matrix, then it has no zero rows (columns) and rows (columns) consisting of ones only.*

We recall that matrices $A, A' \in \mathcal{B}_n$ are said to be permutationally equivalent if there exist permutation matrices $\Pi, \Pi' \in \mathcal{P}_n$ such that $A' = \Pi A \Pi'$.

PROPERTY 5.5.3. *A matrix $A \in \mathcal{B}_n$ is prime if and only if any matrix A' permutationally equivalent to A is prime.*

Let $A = \Pi A' \Pi'$ be a prime matrix, but $A' = B'C'$, where $B', C' \notin \mathcal{P}_n$. Then $A = \Pi B'C'\Pi'$, where $\Pi B', C'\Pi' \notin \mathcal{P}_n$, which contradicts the primality of A.

In what follows, the concept of r-indecomposability of a matrix is essential. A matrix $A \in \mathcal{B}_n$ is called r-indecomposable, $0 \leq r < n$, if it does not contain an $s \times t$ zero submatrix, $s, t \geq 1$, $s + t = n - r + 1$.

From the definition it follows that a matrix $A \in \mathcal{B}_n$ is 0-indecomposable if and only if its permanent differs from zero, that is, there exists a matrix $\Pi \in \mathcal{P}_n$ such that $\Pi \subseteq A$. 1-indecomposability coincides with total indecomposability. In this case, if $A = (a_{ij})$, $i, j = 1, \ldots, n$, and $A(i,j)$ is the result of removing the ith row and the jth column from the matrix A, then $\operatorname{per} A(i,j) > 0$, $1 \leq i, j \leq n$. The following characteristic property holds for r-indecomposable matrices.

PROPERTY 5.5.4. *A matrix $A \in \mathcal{B}_n$ is r-indecomposable if and only if each row (column) of A has at least $r+1$ ones; the Boolean sum of each pair of rows (columns) of A has at least* $\min(r+2, n)$ *ones; and, in general, the Boolean sum of each k rows (columns) of A has at least* $\min(r+k, n)$ *ones, $1 \leq k \leq n$.*

If $k < n - r$ and the Boolean sum of k arbitrarily chosen columns has less than $k + r$ ones, then there is an $s \times t$ zero submatrix, $s = n - k - r + 1$, $t = k$, and $s + t = n - r + 1$; that is, the matrix A is not r-indecomposable. On the other hand, if for any k columns their Boolean sum has at least $k + r$ ones, then the $s \times t$ zero submatrix on these k rows satisfies the condition $s + t \leq n - r$; that is, the matrix A is r-indecomposable.

From Property 5.5.4 we immediately obtain the following useful assertion.

PROPERTY 5.5.5. *A matrix $A \in \mathcal{B}_n$ is totally indecomposable if and only if the Boolean sum of any k rows (columns) of A has at least* $\min(k+1, n)$ *ones, $1 \leq k \leq n$. In particular, any row (column) contains at least two ones.*

The concept of r-indecomposability will be used below in order to formulate sufficient conditions for primality of a matrix.

Let $\mathbf{J}$ be an $s \times t$ matrix whose elements are all equal to one. Let $A \in \mathcal{B}_n$, and let $\mathbf{J}$ be a submatrix of A, $1 \leq s, t \leq n$. We say that $\mathbf{J}$ is a proper submatrix if $s, t > 1$. A proper submatrix $\mathbf{J}$ for $s = 2$, $t = 2$ is denoted by $\mathbf{J}_{22}$.

Let $A \in \mathcal{B}_n$, $\Pi \in \mathcal{P}_n$, and $\Pi \subseteq A$. We say that the matrix Π is isolated in A, if none of the $\mathbf{J}_{22}$-submatrices of A contain two ones belonging to Π.

LEMMA 5.5.1. *Let a matrix $A \in \mathcal{B}_n$ include an identity matrix $\mathbf{I}$. If the matrix $\mathbf{I}$ is isolated in A and $A = BC$, where $B, C \in \mathcal{B}_n$, then there exists a matrix $\Pi \in \mathcal{P}_n$ such that $\mathbf{I} \subseteq B\Pi$ and $\mathbf{I} \subseteq \Pi^T C$, where Π^T is the transposed matrix Π.*

PROOF. Note that if $B_{\cdot k}$ and $C_{k\cdot}$ denote the kth column and the kth row of the matrices B and C, respectively, then

$$A = \sum_{k=1}^{n} B_{\cdot k} C_{k\cdot}.$$

It is obvious that $B_{\cdot k} C_{k\cdot} \subseteq A$, $1 \leq k \leq n$. Since $\mathbf{I} \subseteq A$, each one of $\mathbf{I}$ belongs to $B_{\cdot k} C_{k\cdot}$ for some k. The matrix $\mathbf{I}$ is isolated in A; therefore $B_{\cdot k} C_{k\cdot}$ has a unique one shared with $\mathbf{I}$.

The multiplication of the matrix A from the left by a permutation matrix transposes rows, and the multiplication from the right transposes columns. We choose a permutation matrix Π so that the kth one of $\mathbf{I}$ belongs to the matrix $(B\Pi)_{\cdot k}(\Pi^T C)_{k\cdot}$, $1 \leq k \leq n$. Hence $\mathbf{I} \subseteq B\Pi$ and $\mathbf{I} \subseteq \Pi^T C$. □

Let $A \in \mathcal{B}_n$ and $\Pi \subseteq A$, where $\Pi \in \mathcal{P}_n$. We say that a submatrix $\mathbf{J}$ of A is tangent to Π if $\mathbf{J}$ contains a one of Π.

THEOREM 5.5.2. *Let $A \in \mathcal{B}_n$ be an r-indecomposable matrix, $r \geq 1$, and let a permutation matrix $\Pi \subseteq A$ be isolated in A. If the number of submatrices $\mathbf{J}_{22}$ of A which are tangent to Π is strictly less than r, then the matrix A is prime.*

PROOF. The hypotheses of the theorem remain valid for permutationally equivalent matrices. Therefore, without loss of generality, we may assume that $\Pi = \mathbf{I}$, where $\mathbf{I}$ is the identity matrix. The ones of $\mathbf{I}$ will be diagonal ones. There exist columns of A which do not contain two nondiagonal ones belonging to some submatrix $\mathbf{J}_{22}$ which is tangent to $\mathbf{I}$. If this were not the case, at least $n > r$ submatrices $\mathbf{J}_{22}$ would be tangent to $\mathbf{I}$, which contradicts the hypotheses of the theorem.

Let $A_{\cdot j}$ be some column of the matrix $A = (a_{ij})$, $i, j = 1, \dots, n$, which does not contain two nondiagonal ones belonging to a matrix $\mathbf{J}_{22}$ which is tangent to $\mathbf{I}$. We introduce

$$W_j = \{k \colon a_{kj} = 1,\ k \neq j,\ 1 \leq k \leq n\}.$$

If $|A_{\cdot j}|$ is the number of ones in the column $A_{\cdot j}$, then $|A_{\cdot j}| \geq r + 1$, due to the r-indecomposability of A. Hence $|W_j| \geq r$. For a given column $A_{\cdot j}$, consider the submatrix W of A whose rows and columns have indices in the set W_j. The submatrix W includes the identity submatrix of order $|W_j|$. Since the permutation $\mathbf{I}$ is isolated and all elements of $A_{\cdot j}$ belonging to the set W_j are equal to one, all coordinates of $A_{j\cdot}$ in this set are equal to zero. Thus each column of A whose index is in W_j has at least r ones in positions with indices in $\{1, \dots, n\} \setminus (W_j \cup \{j\})$. The permutation $\mathbf{I}$ is isolated; therefore each row whose index is in W_j has at least r zeros in the positions with indices in $\{1, \dots, n\} \setminus (W_j \cup \{j\})$. Thus, each row with index in W_j has at most $n - (|W_j| - 1) - r$ ones. From r-indecomposability of A it follows that $n - |W_j| - r + 1 \geq r + 1$. Therefore, $n \geq |W_j| + 2r \geq 3r$.

Now let us prove that the matrix A is prime. Assume that $A = BC$, where $B, C \in \mathcal{B}_n$, and show that either B or C is a permutation matrix. The case where both B and C are permutation matrices is impossible, because A is indecomposable. Since the matrix $\mathbf{I}$ is isolated, by Lemma 5.5.1 we can assume that $\mathbf{I} \subseteq B$, $\mathbf{I} \subseteq C$.

Let

$$K = \{k\colon |B_{\cdot k}| \leq 1,\ 1 \leq k \leq n\}.$$

Since $\mathbf{I} \subseteq B$, we see that $|B_{\cdot k}| = 1$ for all $k \in K$. If $|K| = n$, then $B = \mathbf{I}$ is a permutation matrix, and the theorem is proved. Assume that $|K| < n$. Let $K \cup \bar{K} = \{1, \dots, n\}$ and $K \cap \bar{K} = \emptyset$. Assume that $r \leq |K| < n$ and therefore $\bar{K} \neq \emptyset$. Since $\mathbf{I}$ is isolated, each matrix $B_{\cdot k}C_{k\cdot}$ is tangent to $\mathbf{I}$ at a unique element 1. If $k \in \bar{K}$, then $|B_{\cdot k}| \geq 2$, so that $B_{\cdot k}C_{k\cdot}$ contains a $|B_{\cdot k}| \times |C_{k\cdot}|$ submatrix $\mathbf{J}$ that is tangent to $\mathbf{I}$ at the kth one only. Thus, for $k \in \bar{K}$ each one in $C_{k\cdot}$ differing from $c_{kk} = 1$ provides a way for a new matrix $\mathbf{J}_{22}$ to emerge which is tangent to $\mathbf{I}$ only at the kth one. By the hypothesis, less than r submatrices $\mathbf{J}_{22}$ of A are tangent to $\mathbf{I}$. Therefore, the $|\bar{K}| \times n$ submatrix of C with rows from $\bar{K}$ must have less than r nondiagonal ones and therefore less than $|\bar{K}|+r$ elements 1 altogether. This means that this $|\bar{K}| \times n$ submatrix has at most $|\bar{K}|+r-1$ nonzero columns, and therefore, this submatrix has at least $n-|\bar{K}|-r+1 = |K|-r+1 > 0$ zero columns. Multiplying the rows of the matrix B with indices in $\bar{K}$ by C, we obtain a zero submatrix in A of size at least $|\bar{K}| \times (|K|-r+1)$. Since $|\bar{K}|+|K|-r+1 = n-r+1$, this contradicts r-indecomposability of A. Therefore, the case $r \leq |K| < n$ is impossible. Thus, if B is not a permutation matrix and, hence, $|K| < n$, then the number of columns containing a single element 1 is less than r. Similarly, if C is not a permutation matrix, then the number of rows containing a single elements 1 is less than r.

If $|K| < r$, then C should be a permutation matrix. Assume the contrary. Then there are more than $n-r$ rows of C each containing a single one, and at most $|K| \leq r-1$ of them have indices in K. Hence more than $n-r-(r-1) = n-2r+1$ rows with indices in $\bar{K}$ have more than a single one, so that each of them has a nondiagonal one. These ones generate more than $n-2r+1$ submatrices $\mathbf{J}_{22}$ of A which are tangent to $\mathbf{I}$. Since we have proved that $n \geq 3r$, at least $n-2r+1 \geq r+1$ submatrices $\mathbf{J}_{22}$ are tangent to $\mathbf{I}$. This contradicts the hypothesis of the theorem. □

As corollaries to Theorem 5.5.2, we give more obvious sufficient conditions for primality of a matrix.

PROPERTY 5.5.6. *If the matrix A' obtained from a totally indecomposable matrix $A \in \mathcal{B}_n$ by replacing all submatrices $\mathbf{J}_{22}$ by zero submatrices satisfies the condition* per $A' > 0$, *then the matrix A is prime.*

Indeed, A is an r-indecomposable matrix with $r = 1$ including a permutation matrix which is tangent to $0 = r-1 < r$ submatrices $\mathbf{J}_{22}$. By Theorem 5.5.2, A is prime.

PROPERTY 5.5.7. *If a matrix $A \in \mathcal{B}_n$ is r-indecomposable for some $r \geq 1$ and has less than r submatrices $\mathbf{J}_{22}$, then A is prime.*

In order to prove Property 5.5.7, it suffices to demonstrate that A contains an isolated permutation matrix Π and use Theorem 5.5.2.

We start with the case $n = 3$. For $r = 2$, we see that $A = \mathbf{J}_{33}$, and therefore, we have to analyze only the case $r = 1$. In this case, each row and each column contains at least two ones, and the matrix is equivalent to the above example of a prime matrix for $n = 3$. The matrix A is totally indecomposable, and there exists a permutation submatrix Π which is isolated because A contains no submatrices $\mathbf{J}_{22}$. Thus, for $n = 3$ the property is true.

We use induction on n. Let $A \in \mathcal{B}_n$, $n \geq 4$, and assume that the matrices of order $n' < n$ satisfying the hypotheses of the property contain an isolated permutation matrix. We prove that A contains an isolated permutation matrix as well. In view of the r-decomposability there exists a permutation matrix $\Pi \subset A$. If A contains no submatrices $\mathbf{J}_{22}$, then Π is isolated, and the property is proved. Assume that A has $r - 1 > 0$ submatrices $\mathbf{J}_{22}$; that is, $r \geq 2$. The matrix A in each column has at least $r + 1$ ones; therefore, the total number of ones in A is at least $n(r + 1)$. All ones of A cannot simultaneously belong to the submatrices $\mathbf{J}_{22}$, because A would have at least $\lceil n(r+1)/4 \rceil \geq r$ submatrices $\mathbf{J}_{22}$, in contradiction with the hypothesis. Therefore, there exists an element a_{ij} of A which is equal to one but does not belong to any of the submatrices $\mathbf{J}_{22}$. Furthermore, A possesses a one in some row or column overlapping some submatrix $\mathbf{J}_{22}$, since otherwise A would then be a direct sum of the submatrix containing all ones in the submatrices $\mathbf{J}_{22}$ and the submatrix containing the ones lying not in the submatrices $\mathbf{J}_{22}$, in contradiction with the total indecomposability of A. Choosing such an a_{ij}, we delete the ith row and the jth column of A. The resulting matrix A' of order $n - 1$ is $(r - 1)$-indecomposable and has less than $r - 1$ submatrices $\mathbf{J}_{22}$. By the induction assumption, A' contains a permutation matrix Π' isolated in A'. Consider the permutation matrix Π derived from Π' by adding the element $a_{ij} = 1$. None of the submatrices $\mathbf{J}_{22}$ contains two ones of the matrix Π, since such a submatrix would be contained in A', which contradicts the fact that Π' is isolated in A'. On the other hand, the element $a_{ij} = 1$ is tangent to none of the submatrices $\mathbf{J}_{22}$ of A. Thus, Π is isolated in A, and the property is proved.

From Property 5.5.6 we immediately obtain the following property.

PROPERTY 5.5.8. *If a matrix $A \in \mathcal{B}_n$ is totally indecomposable and has no submatrices $\mathbf{J}_{22}$, then A is prime.*

PROPERTY 5.5.9. *Let a matrix $A \in \mathcal{B}_n$ be of the form*

$$A = \begin{pmatrix} \mathbf{I} & & & & & A_d \\ A_1 & \mathbf{I} & & & & 0 \\ & A_2 & \mathbf{I} & & & \\ & & A_3 & \mathbf{I} & & \\ & & & \ddots & \ddots & \\ 0 & & & & A_{d-1} & \mathbf{I} \end{pmatrix},$$

where $d \geq 3$ and $\mathbf{I}$ are identity matrices, generally speaking, of different dimensions. If A is totally indecomposable, then it is prime.

The property follows from Theorem 5.5.2 with $r = 1$ and $\Pi = \mathbf{I}$.

LEMMA 5.5.3. *Let a matrix $A \in \mathcal{B}_n$ be totally indecomposable and factorable; that is, let there exist nonpermutation matrices $B, C \in \mathcal{B}_n$ such that $A = BC$. Then there exists k, $1 \leq k \leq n$, such that the number of ones in the kth column of B is at least 2 and the number of ones in the kth row of C is at least 2; that is, $|B_{\cdot k}| \geq 2$, $|C_{k \cdot}| \geq 2$.*

PROOF. The conditions $|B_{\cdot k}| \le 1$ or $|C_{k\cdot}| \le 1$ mean, respectively, that all ones of the matrix $B_{\cdot k}C_{k\cdot}$ are placed in the same row or in the same column. This means that all ones of the matrix

$$BC = \sum_{k=1}^{n} B_{\cdot k}C_{k\cdot}$$

are covered by n lines and the covering contains those rows for which $|B_{\cdot k}| \le 1$ and those columns for which $|C_{k\cdot}| \le 1$. If both conditions hold, then the covering includes either the row or the column satisfying the conditions. Since the matrix A is totally indecomposable, in order to cover it by n lines, we can use either only rows or only columns. The matrix A has no zero rows or columns; therefore B has no zero rows and C no zero columns. Thus, if to cover A we use the rows, that is, $|B_{\cdot k}| \le 1$, $1 \le k \le n$, then B is a permutation matrix, and if to cover A we use the columns, then C is a permutation matrix. This contradicts the hypothesis of the lemma. The lemma is proved. □

THEOREM 5.5.4. *Let a matrix $A \in \mathcal{B}_n$ be totally indecomposable, and let each row and each column of A have exactly k ones. If the number of ones in each proper submatrix* **J** *of A is less than k, then A is prime.*

PROOF. We assume that the hypotheses of the theorem hold but A is not prime. Then there exist matrices $B, C \notin \mathcal{P}_n$ such that $A = BC$. If $|A|$ is the number of ones of A, then

$$nk = |A| = \left|\sum_{j=1}^{n} B_{\cdot j}C_{j\cdot}\right| \le \sum_{j=1}^{n} |B_{\cdot j}C_{j\cdot}|.$$

Each product $B_{\cdot j}C_{j\cdot}$, $1 \le j \le n$, induces a submatrix **J** contained in A. From Lemma 5.5.3 it follows that at least one such proper submatrix exists for some j', $1 \le j' \le n$. By the hypothesis, $|B_{\cdot j'}C_{j'\cdot}| < k$, because $|B_{\cdot j'}C_{j'\cdot}|$ does not exceed the number of ones of the corresponding submatrix **J**. Since $B_{\cdot j'}C_{j'\cdot}$ is a submatrix of A, by the hypothesis $|B_{\cdot j}C_{j\cdot}| \le k$ for all other $1 \le j \le k$. Thus,

$$nk \le \sum_{j=1}^{n} |B_{\cdot j}C_{j\cdot}| < nk.$$

This contradiction proves the theorem. □

As a corollary to Theorem 5.5.4, we formulate a property of incidence matrices of (v, k, l)-configurations (in other words, symmetric block designs, see Chapter 1). Assume that $v > k > l > 0$, and consider the incidence matrix $A \in \mathcal{B}_n$ of (v, k, l)-configurations. Each row and each column of this matrix contains exactly k ones, while each pair of different rows and each pair of different columns has exactly l coinciding elements equal to one.

PROPERTY 5.5.10. *If a matrix $A \in \mathcal{B}_n$ is the incidence matrix of a (v, k, l)-configuration and $k > l^2$, then A is prime.*

If a matrix is partially decomposable (that is, not totally indecomposable), then it can be represented in the form

$$A = \begin{pmatrix} A_1 & A_2 \\ \mathbf{0} & A_3 \end{pmatrix},$$

where the matrices A_1, A_2, A_3, $\mathbf{0}$ are of dimensions $s \times s$, $s \times t$, $t \times t$, $t \times s$, respectively, $s, t > 1$, $s + t = n$. If A is the incidence matrix of a (v, k, l)-configuration and $\sigma(A_i)$ is the sum of elements of the matrix A_i, then, summing over the first s columns and the last t rows, we find that $\sigma(A_1) + \sigma(A_3) = nk$. On the other hand, $\sigma(A) = \sigma(A_1) + \sigma(A_2) + \sigma(A_3) = nk$. Hence $\sigma(A_2) = 0$, and therefore the matrix A_2 is zero. Thus any two rows taken, respectively, from the first s and the last t rows have no common elements equal to one. This contradicts the condition $l > 0$. Thus, A is totally indecomposable. Each proper submatrix $\mathbf{J}$ of A is at most $l \times l$ and hence contains at most l^2 ones. Since $k > l^2$, the matrix A is prime by Theorem 5.5.4.

In conclusion we note that if ${}^sA$ is the support of a nonnegative matrix A, then the equality $A = BC$ implies that ${}^sA = {}^sB^sC$. Hence if ${}^sA$ is prime, then the matrix A is prime as well.

5.5.2. Multiplicative rank and scope of a Boolean matrix. The multiplicative rank of an $m \times n$ Boolean matrix A is the least k such that A admits a representation in the form of a product of $m \times k$ and $k \times n$ Boolean matrices.

THEOREM 5.5.5. *Let a matrix $A \in \mathcal{B}_n$ be totally indecomposable. The matrix A is prime if and only if each matrix $A' \in \mathcal{B}_n$ obtained by deleting some ones of any proper submatrix $\mathbf{J}$ of A is of multiplicative rank n (the ones are removed completely, or partially, or not at all).*

PROOF. We begin by proving the sufficiency. Let A' be of multiplicative rank n. Assume that A is not prime. Then

$$A = BC = \sum_{j=1}^{n} B_{\cdot j} C_{j \cdot},$$

and there is an i, $1 \leq i \leq n$, such that by Lemma 5.5.3 there exists a proper submatrix $\mathbf{J}$ corresponding to $B_{\cdot i} C_{i \cdot}$. The matrix

$$A' = \sum_{j \neq i} B_{\cdot j} C_{j \cdot}$$

is the result of removing the ones of the matrix $B_{\cdot i} C_{i \cdot}$ from A. On the other hand, A' is a sum of products of $n-1$ columns of B and $n-1$ rows of C; therefore, its multiplicative rank does not exceed $n-1$, which is impossible.

Now let A be a prime matrix, and let A' be the result of removing from A the ones of a proper submatrix $\mathbf{J}$ placed at rows and columns determined by Boolean vectors X and Y, so that $XY \in \mathcal{B}_n$ is the matrix $\mathbf{J}$ and $|X| \geq 2$, $|Y| \geq 2$. We show that A' has multiplicative rank n. Assume that there exist an $n \times (n-1)$ matrix B' and an $(n-1) \times n$ matrix C' such that $A' = B'C'$. Then A can be represented in the form

$$A = \begin{pmatrix} B' & X \end{pmatrix} \begin{pmatrix} C' \\ Y \end{pmatrix},$$

where none of the multipliers is a permutation matrix. This contradicts the hypothesis that A is a prime matrix. $\square$

PROPERTY 5.5.11. *If a matrix $A \in \mathcal{B}_n$ includes an isolated permutation matrix, then A has multiplicative rank n.*

If there exists an isolated permutation matrix included in A, then without loss of generality we assume that it is the identity matrix $\mathbf{I}$. If $A = BC$, $B, C \in \mathcal{B}_n$, then by Lemma 5.5.1 there exists a permutation matrix Π such that $\mathbf{I} \subseteq B\Pi$, $\mathbf{I} \subseteq \Pi^T C$. This means that the multiplicative rank of A is equal to n.

PROPERTY 5.5.12. *Let a matrix $A \in \mathcal{B}_n$ be totally indecomposable and have exactly k ones in each row and each column. Furthermore, let the number of ones in each proper submatrix $\mathbf{J}$ of A not exceed k. Then A is prime if and only if any matrix $A' \in \mathcal{B}_n$ obtained by deleting all ones of a proper submatrix $\mathbf{J}$ of A with exactly k ones has multiplicative rank n.*

The necessity of the conditions of Property 5.5.12 follows from Theorem 5.5.5. Assume that the conditions of the property hold but the matrix is not prime; that is, $A = BC$, $B, C \in \mathcal{B}_n$. Then

$$nk = |A| \leq \sum_{j=1}^{n} |B_{\cdot j} C_{j \cdot}| \leq nk,$$

since the number of ones in each submatrix $\mathbf{J}$ associated with $B_{\cdot j} C_{j \cdot}$ does not exceed k. From these inequalities it follows that $|B_{\cdot j} C_{j \cdot}| = k$ for all $1 \leq j \leq n$. Therefore, any pair $B_{\cdot i} C_{i \cdot}$ and $B_{\cdot j} C_{j \cdot}$, $i \neq j$, contains no common ones. Now consider the proper submatrix $\mathbf{J}$ associated with $B_{\cdot i} C_{i \cdot}$. It does exist, due to Lemma 5.5.3. Let A' be obtained from A by removing all ones of $B_{\cdot i} C_{i \cdot}$. Then the multiplicative rank of A' does not exceed $n - 1$. This contradiction completes the proof.

For a matrix $A \in \mathcal{B}_n$, the scope $s(A)$ of the matrix A is the set of n-columns consisting of zero n-column and all sums of columns of A. It is clear that $s(A)$ is the least subset of the set of all Boolean n-columns which includes the columns of A and is closed with respect to the Boolean addition.

PROPERTY 5.5.13. *If $A, B \in \mathcal{B}_n$, then $s(A) \subseteq s(B)$ if and only if there exists a matrix $C = (c_{kj}) \in \mathcal{B}_n$ such that $A = BC$.*

If

$$A = BC = \sum_{k=1}^{n} B_{\cdot k} C_{k \cdot},$$

then

$$\sum_{k=1}^{n} \sum_{j=1}^{n} c_{kj} B_{\cdot k} = \sum_{k=1}^{n} A_{\cdot k};$$

that is, $s(A) \subseteq s(B)$. In turn, if $s(A) \subseteq s(B)$, then the above equality for the sums of columns of A and B holds with coefficients c_{kj} which determine C.

Let $2^{[n]}$ denote the set of all n-columns.

PROPERTY 5.5.14. *For a matrix $A \in \mathcal{B}_n$, the equality $s(A) = 2^{[n]}$ holds if and only if A is a permutation matrix.*

If A is a permutation matrix, then it is clear that $s(A) = 2^{[n]}$. If a matrix A satisfies the relation $s(A) = 2^{[n]}$, then the set $s(A)$ must include all columns containing a single one. Therefore, A is a permutation matrix.

We consider the family of sets

$$S_n = \{s(A) \colon A \in \mathcal{B}_n\}.$$

On this family we define the natural partial order with respect to inclusion. We say that an element of S_n is maximal if it does not coincide with the set $2^{[n]}$ of all n-columns and it is not a proper subset of any element of S_n, excluding $2^{[n]}$.

A matrix $A \in \mathcal{B}_n$ is called elementary if it is the direct sum of the matrix

$$\begin{pmatrix} 1 & 0 \\ 1 & 1 \end{pmatrix}$$

and the identity matrix of order $n-2$.

PROPERTY 5.5.15. *For $A \in \mathcal{B}_n$, the scope of the matrix A is maximal in S_n if and only if A is either prime or permutationally equivalent to the elementary matrix.*

Assume that A is a prime matrix and

$$s(A) \subseteq s(B) \neq 2^{[n]}, \qquad B \in \mathcal{B}_n.$$

By Property 5.5.14, $A = BC$, where $C \in \mathcal{P}_n$. Since $B = AC^{-1}$, we see that $s(B) \subseteq s(A)$, and hence $s(A) = s(B) \neq 2^{[n]}$. Thus, $s(A)$ is maximal in S_n.

The elementary matrix has the maximal scope in S_n. To see this, it suffices to show that none of its columns belongs to the scope of another column. But this immediately follows from the structure of the elementary matrix.

Now let the scope of a matrix A be maximal in S_n. In particular, this means that A cannot have a zero column or a pair of identical columns. Assume that some column of A includes another column. Without loss of generality we assume that $A_{\cdot 1} \supseteq A_{\cdot 2}$. Consider the matrix A' obtained from A by replacing the first column by $A_{\cdot 1} \setminus A_{\cdot 2}$. It is clear that such a replacement increases the scope of the matrix; that is, $s(A) \subseteq s(A')$, To prove this, we note that $A_{\cdot 1}$ lies outside the scope of the other columns of A, and therefore, so does $A'_{\cdot 1}$. It is clear that the columns $A'_{\cdot 1}, A_{\cdot 2}, \ldots, A_{\cdot n}$ generate $s(A)$, while $s(A)$ does not contain, for example, the column $A'_{\cdot 1}$ which lies in $s(A')$. Since $s(A)$ is maximal, we conclude that $s(A') = 2^{[n]}$; that is, A' is a permutation matrix. Therefore, A is an elementary matrix.

Now let $A_{\cdot i} \not\subseteq A_{\cdot j}$ for all $i \neq j$. Assume that $A = BC$, $B, C \in \mathcal{B}_n$. We prove that either B or C is a permutation matrix. If B is not a permutation matrix, then $s(A) \subseteq s(B) \neq 2^{[n]}$. From the hypothesis that $s(A)$ is maximal it follows that $s(A) = s(B)$; hence there exists a matrix $C' \in \mathcal{B}_n$ such that $B = AC'$. Therefore, $A = AC''$, where $C'' = CC'$. Since A has no columns including each other, any column of C'' contains a single one and the columns of C'' are pairwise different. Therefore, C'', and hence C, are permutation matrices. Since $s(A) \neq 2^{[n]}$, A is not a permutation matrix. Therefore, A is a prime matrix.

PROPERTY 5.5.16. *Let $A \in \mathcal{B}_n$, and let A not be a permutation matrix. If none of the rows (columns) includes another row (column) of A, then there exist matrices $B, C \in \mathcal{B}_n$ such that $A = BC$, where B (C) is prime.*

Let the hypothesis hold for the columns of A. Take a matrix $B \in \mathcal{B}_n$ such that the scope $s(B)$ is maximal in S_n and $s(A) \subseteq s(B)$. Then there exists a matrix $C \in \mathcal{B}_n$ such that $A = BC$, where by virtue of Property 5.5.15 the matrix B is either prime or elementary. Since none of the columns of A includes another column, the matrix B possesses this property as well, so that B is a prime matrix. The transpose of a prime matrix is prime; therefore, if the condition on rows holds for A, then the property can be proved by passing to the matrix A^T.

Problems.

1. Prove that the matrix

$$\begin{pmatrix} 1 & 1 & 0 & 0 \\ 0 & 1 & 1 & 0 \\ 0 & 0 & 1 & 1 \\ 1 & 0 & 0 & 1 \end{pmatrix}$$

is prime.

2. Prove that a $(0,1)$-matrix A of order 3 is prime if and only if there exist permutation matrices Π and Π' such that

$$\Pi A \Pi' = \begin{pmatrix} 0 & 1 & 1 \\ 1 & 0 & 0 \\ 1 & 1 & 0 \end{pmatrix}.$$

3. Let a $(0,1)$-matrix A of order n satisfy the matrix equation

$$ACA^T = D,$$

where A^T is the transposed matrix A, C and D are diagonal matrices of order n with real-valued elements, and D is nonsingular. Prove that A is a permutation matrix.

5.6. Prime nonnegative matrices

5.6.1. Factorability of a matrix. Let $\mathcal{R}_n$ be the semigroup of nonnegative matrices of order n. In this section we denote the support of a matrix $A \in \mathcal{R}_n$ by a. The columns of A and a are denoted by $A_1, \dots, A_n$ and $a_1, \dots, a_n$.

A matrix A is called a monomial if the support of A is a permutation matrix. A matrix $A \in \mathcal{R}_n$ is invertible if $A^{-1} \in \mathcal{R}_n$.

PROPERTY 5.6.1. *A matrix $A \in \mathcal{R}_n$ is invertible if and only if it is a monomial.*

It is obvious that if A is a monomial, then it is invertible, and the inverse matrix is a monomial as well. Now let a matrix $A \in \mathcal{R}_n$ be invertible. Then neither A nor A^{-1} has zero rows or columns. Let $A = (a_{ij})$, $A^{-1} = (\bar{a}_{ij})$, $i, j = 1, \dots, n$. It is clear that

$$\sum_{k=1}^{n} a_{ik}\bar{a}_{kj} = \delta_{ij},$$

where δ_{ij} is the Kronecker delta. In the ith row of A, we choose a nonzero element $a_{ik} > 0$. Then $\bar{a}_{ki} > 0$ and $\bar{a}_{kj} = 0$, $j \neq i$; that is, in the kth row all elements but $\bar{a}_{kj}$ are zero. Setting $i = 1, \dots, n$, we see that A^{-1} is a monomial. Hence A is a monomial.

A matrix $A \in \mathcal{R}_n$ is called prime (more precisely, algebraically prime) if it is not a monomial and the equality $A = BC$ for some $B, C \in \mathcal{R}_n$ implies that either B or C is a monomial. A matrix A which is neither prime nor monomial is called factorable.

PROPERTY 5.6.2. *If for a matrix $A \in \mathcal{R}_n$ there exist columns a_i and a_k, $i \neq k$, of the support a such that $a_i \supseteq a_k$, then A is factorable.*

By permuting columns of A the assertion can be reduced to the case where $a_1 \supseteq a_2$. Then there exists $\varepsilon > 0$ such that $B_1 = A_1 - A_2\varepsilon \geq 0$ and $b_1 = a_1$. Set $B_i = A_i$, $i = 2, \dots, n$, and consider the matrix $B = \begin{pmatrix} B_1 & B_2 & \dots & B_n \end{pmatrix} \in \mathcal{R}_n$. Let us demonstrate that B is not a monomial.

We consider two cases, where $B_2 = 0$ and $B_2 \neq 0$. If $B_2 = 0$, then B is not a monomial. If $B_2 \neq 0$, then there is an r, $1 \leq r \leq n$, such that the rth coordinate B_{r2} of the column B_r is positive. Since $a_1 \supseteq a_2 = b_2$, the coordinate $A_{r1} > 0$, and therefore, $B_{r1} > 0$. Since $B_{r1} > 0$ and $B_{r2} > 0$, B is not a monomial.

Consider the matrix

$$C = \begin{pmatrix} 1 & 0 \\ \varepsilon & 1 \end{pmatrix} \oplus \mathbf{I}_{n-2},$$

where $\mathbf{I}_{n-2}$ is the identity matrix of order $n-2$ and $\oplus$ stands for the direct sum. The matrix C is not a monomial. Nevertheless, it is easy to see that $A = BC$; that is, the matrix A is factorable.

PROPERTY 5.6.3. *If a matrix A is prime, then its support a has zeros and ones in each row and in each column.*

Since the zero column is included in any other column and any column is included in the column whose elements are all ones, the support of a prime matrix, by Property 5.6.2, cannot contain zero columns or columns consisting of ones only. The same is true for rows.

Now we formulate a sufficient condition for primality of a matrix $A \in \mathcal{R}_n$.

THEOREM 5.6.1. *Let $A \in \mathcal{R}_n$ be totally indecomposable, and let the condition on the columns of the support*

$$a_i^T a_k \leq 1$$

hold for all $i \neq k$, $1 \leq i, k \leq n$. Then the matrix A is prime.

PROOF. From the total indecomposability it follows that A is not a monomial. Assume that $A = BC$, where $B, C \in \mathcal{R}_n$. Let J denote the set of indices of those columns of B which have at most one positive element. Then for any $j \in N_n \setminus J$, $N_n = \{1, \ldots, n\}$, there exists at most one $i \in N_n$ such that the element C_{ji} of the matrix C is positive. Assume that $C_{ji} > 0$, $C_{jk} > 0$, $i \neq k$, $j \in N_n \setminus J$. Then the ith column of A is

$$A_i = \sum_{l=1}^{n} B_l C_{li} \geq B_j C_{ji}.$$

Hence $a_i \supseteq b_j$. Similarly, $a_k \supseteq b_j$. Since b_j has at least two positive elements, it follows that $a_i^T a_k \geq 2$, which is impossible.

We set $|J| = q$ and demonstrate that the inequalities $0 < |J| < n$ are impossible. Let

$$I = \{i \colon C_{ji} = 0 \text{ for all } j \in N_n \setminus J\}, \qquad |I| = r.$$

We see that

$$A_i = \sum_{j \in J} B_j C_{ji}$$

for each $i \in I$. Thus,

$$\begin{aligned} \sum_{i \in I} A_i &= \sum_{i \in I} \left(\sum_{j \in J} B_j C_{ji} + \sum_{j \in N_n \setminus J} B_j C_{ji} \right) \\ &= \sum_{j \in J} B_j \sum_{i \in I} C_{ji} + \sum_{j \in N_n \setminus J} B_j \sum_{i \in I} C_{ji}. \end{aligned}$$

Since

$$\sum_{i \in I} C_{ji} = 0$$

for all $j \in N_n \setminus J$, we obtain

$$\sum_{i \in I} A_i = \sum_{j \in J} B_j \sum_{i \in I} C_{ji}.$$

The vector on the left-hand side of this equality has at least $r+1$ positive elements due to the total indecomposability of A, whereas the vector on the right-hand side has at most q positive elements. Hence $r < q$.

We set $J' = N_n \setminus J$, $I' = N_n \setminus I$. Then I' is the set of those i for which there exists $j \in J'$ such that $C_{ji} > 0$. Since $|I'| = n - r > n - q = |J'|$, there exists at least one $j \in J'$ such that $C_{ji} > 0$, $C_{jk} > 0$, $i \neq k$. We have seen that this is impossible. Thus, the assumption $0 < |J| < n$ is wrong.

If $|J| = n$, then every column of the matrix B has at most one positive element. Since A is totally indecomposable, neither A nor B has zero rows. Therefore, B is a monomial. If $|J| = 0$, then, by the above, every row of C has at most one positive element. Since C has no zero columns, C is a monomial. Thus, the matrix A is prime. □

5.6.2. A criterion for primality. We give a necessary and sufficient condition for a matrix $A \in \mathcal{R}_n$ to be prime (see Property 5.6.6 below).

PROPERTY 5.6.4. *If $A = (a_{ij}) \in \mathcal{R}_n$ is prime, then there exist r, $1 \leq r \leq n$, and a totally indecomposable prime matrix $M \in \mathcal{R}_r$ such that*

$$\Pi A \Pi' = M \oplus D,$$

where Π, Π' are permutation matrices of order n, and $D \in \mathcal{R}_{n-r}$ is a nonsingular diagonal matrix.

We prove this property by induction on n. For $n = 1$, this property is trivial since there are no prime matrices. Let the property be true in $\mathcal{R}_{n-1}$. We take a prime matrix $A \in \mathcal{R}_n$. If A is totally indecomposable, then the property holds with $r = n$, and the matrix D is absent. If A is not totally indecomposable, then there exist permutation matrices Π and Π' such that

$$\Pi A \Pi' = \begin{pmatrix} A_1 & A_2 \\ \mathbf{0} & A_3 \end{pmatrix},$$

where A_1 is of order s, $0 < s < n$, and $\mathbf{0}$ is a zero matrix.

Show that A_2 is also a zero matrix. If $A_2 \neq \mathbf{0}$, then there exists $A_{ij} > 0$, $1 \leq i \leq s$, $s+1 \leq j \leq n$. Then A_1 is not a monomial, since otherwise for the columns of the support the inclusion $a_j \supseteq a_k$, $1 \leq k \leq s$, would hold, which, by Property 5.6.2, would contradict the primality of A. If $\mathbf{I}_s$ is the identity matrix of order s, then

$$\Pi A \Pi' = \begin{pmatrix} \mathbf{I}_s & A_2 \\ \mathbf{0} & A_3 \end{pmatrix} \begin{pmatrix} A_1 & \mathbf{0} \\ \mathbf{0} & \mathbf{I}_{n-s} \end{pmatrix},$$

where Π, Π' are permutation matrices and the matrices in the right-hand side are not monomials because A_1 is not a monomial and $A_2 > 0$. This equality contradicts the primality of A. Hence $A_2 = \mathbf{0}$ and

$$\Pi A \Pi' = \begin{pmatrix} A_1 & \mathbf{0} \\ \mathbf{0} & A_3 \end{pmatrix}.$$

If either A_1 or A_3 is factorable, then $\Pi A \Pi'$ is factorable as well, since

$$\begin{pmatrix} A_1'A_2' & \mathbf{0} \\ \mathbf{0} & A_3 \end{pmatrix} = \begin{pmatrix} A_1' & \mathbf{0} \\ \mathbf{0} & A_3 \end{pmatrix} \begin{pmatrix} A_2' & \mathbf{0} \\ \mathbf{0} & \mathbf{I}_{n-s} \end{pmatrix}.$$

If neither A_1 nor A_3 is factorable, then the equality

$$\Pi A \Pi' = \begin{pmatrix} A_1 & \mathbf{0} \\ \mathbf{0} & \mathbf{I}_{n-s} \end{pmatrix} \begin{pmatrix} \mathbf{I}_s & \mathbf{0} \\ \mathbf{0} & A_3 \end{pmatrix}$$

in view of the primality of A allows for two cases:

1. A_1 is prime and A_3 is a monomial;
2. A_1 is a monomial and A_3 is prime.

In the first case, by the induction assumption, the rows and columns of A_1 can be permuted so that the matrix $M \oplus D_1$ is obtained, where M is totally indecomposable in $\mathcal{R}_r$, $1 \leq r \leq s$, and $D_1 \in \mathcal{R}_{s-r}$ is a nonsingular diagonal matrix. By a permutation of rows and columns of A_3, it also reduces to a nonsingular diagonal matrix $D_2 \in \mathcal{R}_{n-s}$. Thus,

$$\Pi A \Pi' = M \oplus D$$

with appropriate permutation matrices Π and Π', where $D = D_1 \oplus D_2 \in \mathcal{R}_{n-s}$ is a nonsingular diagonal matrix. The second case is treated similarly.

PROPERTY 5.6.5. *If a matrix $A_1 \in \mathcal{R}_r$ is prime and $A_2 \in \mathcal{R}_{n-r}$ is a monomial, then the matrix $A = A_1 \oplus A_2 \in \mathcal{R}_n$ is prime.*

Assume that $A = BC$, where $B, C \in \mathcal{R}_n$. We represent B and C in the form

$$B = \begin{pmatrix} B_1 \\ B_2 \end{pmatrix}, \qquad C = \begin{pmatrix} C_1 & C_2 \end{pmatrix},$$

where B_1 is an $r \times n$ matrix and C_1 is an $n \times r$ matrix. Let Π be a permutation matrix such that

$$B\Pi = \begin{pmatrix} B_{11} & B_{12} \\ B_{21} & B_{22} \end{pmatrix}, \qquad \Pi^{-1}C = \begin{pmatrix} C_{11} & C_{12} \\ C_{21} & C_{22} \end{pmatrix},$$

where B_{11} is an $r \times s$ matrix, C_{11} is an $s \times r$ matrix, $B_{12} = \mathbf{0}$, and B_{11} has no zero columns. We obtain

$$\begin{pmatrix} A_1 & \mathbf{0} \\ \mathbf{0} & A_2 \end{pmatrix} = BC = \begin{pmatrix} B_{11}C_{11} & B_{11}C_{12} \\ B_{21}C_{11} + B_{22}C_{21} & B_{21}C_{12} + B_{22}C_{22} \end{pmatrix}.$$

Hence $B_{11}C_{12} = \mathbf{0}$. Since B_{11} has no zero columns, we conclude that $C_{12} = \mathbf{0}$. Thus,

$$A_1 = B_{11}C_{11}, \qquad A_2 = B_{22}C_{22},$$

and $s < n$, because A has no zero columns. Therefore, $0 < s < n$.

Now we prove that $r = s$. Assume that $s < r$, and introduce the matrices

$$B_{11}' = \begin{pmatrix} B_{11} & \mathbf{0} \end{pmatrix}, \qquad C_{11}' = \begin{pmatrix} C_{11} \\ \mathbf{0} \end{pmatrix},$$

where $B_{11}', C_{11}' \in \mathcal{R}_r$. For these matrices, the relation $A_1 = B_{11}'C_{11}'$ holds, which contradicts the primality of A_1. Similarly, if $s > r$, then $n - r < n - s$, and we introduce the matrices

$$B_{22}' = \begin{pmatrix} B_{22} & \mathbf{0} \end{pmatrix}, \qquad C_{22}' = \begin{pmatrix} C_{22} \\ \mathbf{0} \end{pmatrix},$$

where $B'_{22}, C'_{22} \in \mathcal{R}_{n-r}$. Then $A_2 = B'_{22}C'_{22}$, which contradicts the fact that A_2 is a monomial. From the equalities

$$A = BC = \begin{pmatrix} B_{11}C_{11} & \mathbf{0} \\ B_{21}C_{11} + B_{22}C_{21} & B_{22}C_{22} \end{pmatrix} = \begin{pmatrix} A_1 & \mathbf{0} \\ \mathbf{0} & A_2 \end{pmatrix}$$

it follows that $B_{21}C_{11} = \mathbf{0}$, $B_{22}C_{21} = \mathbf{0}$. Since $A_1 = B_{11}C_{11}$ and A_1 is prime, the matrix C_{11} is either prime or monomial; hence it has no zero rows. Thus, from $B_{21}C_{11} = \mathbf{0}$ it follows that $B_{21} = \mathbf{0}$. Since A_2 is a monomial, B_{22} is a monomial as well, and the relation $B_{22}C_{21} = \mathbf{0}$ implies that $C_{21} = \mathbf{0}$. Hence

$$B\Pi = B_{11} \oplus B_{22}, \qquad \Pi^{-1}C = C_{11} \oplus C_{22}.$$

Since A_2 and B_{22} are monomials, the equality $A_2 = B_{22}C_{22}$ implies that C_{22} is a monomial as well. From the primality of A_1 and the equality $A_1 = B_{11}C_{11}$ it follows that either B_{11} or C_{11} is a monomial; hence either $B\Pi$ or $\Pi^{-1}C$ is a monomial. Therefore, either B or C is a monomial.

PROPERTY 5.6.6. *A matrix $A \in \mathcal{R}_n$ is prime if and only if for some r, $1 \leq r \leq n$, there exists a totally indecomposable prime matrix $M \in \mathcal{R}_r$ such that*

$$\Pi A \Pi' = M \oplus D,$$

where Π, Π' are permutation matrices of order n and D is a diagonal nonsingular matrix of order $n - r$.

Necessity follows from Property 5.6.4, and sufficiency from Property 5.6.5.

We say that a matrix A is perfectly decomposable if there exist permutation matrices Π and Π' such that

$$\Pi A \Pi' = A_1 \oplus \cdots \oplus A_s, \qquad s \geq 1,$$

where the matrices $A_1, \ldots, A_s$ are totally indecomposable. In particular, a totally indecomposable matrix is perfectly decomposable (with $s = 1$).

PROPERTY 5.6.7. *Any prime matrix $A \in \mathcal{R}_n$ is perfectly decomposable.*

Property 5.6.7 follows from Property 5.6.6.

PROPERTY 5.6.8. *There are no prime matrices in $\mathcal{R}_2$.*

To prove this, we use Property 5.6.2. There are 16 supports of matrices of $\mathcal{R}_2$; all of them, except two monomials, possess the inclusion property for rows and columns and hence are factorable in $\mathcal{R}_2$.

CHAPTER 6

Sequences of Nonnegative Matrices

Introduction

The main objective of this chapter is to analyze the behaviour of powers of nonnegative matrices as the exponent grows to infinity. This problem is of considerable importance in various fields of mathematics and its applications. In particular, for sequences of stochastic matrices this problem is related to the asymptotic behaviour of Markov chains with a finite number of states. It is also of interest in the analysis of ergodic properties of finite probabilistic automata.

In Section 6.1, we consider combinatorial properties of the directed graphs corresponding to nonnegative matrices. We analyze the structure of components and strong components of such graphs, and divide the vertices of a digraph into two kinds, essential and inessential ones; on this base, we give the canonical form of a nonnegative matrix. The reduction to this form is performed by means of permutation of rows and columns according to a certain permutation matrix Π and its transpose Π^T. To conclude the section, we give a series of number-theoretical lemmas to be used in the subsequent sections.

In Section 6.2, we deal with indecomposable periodic matrices. For such matrices of period d, we demonstrate that the vertices of the corresponding directed graph are partitioned into d subclasses, referred to as cyclic. Such a partition allows us to give the canonical form of an indecomposable periodic matrix of period d. An indecomposable aperiodic matrix is primitive. In line with this, the digraph of a primitive matrix contains a family of elementary circuits whose lengths are relatively prime. Such a structure of the digraph allows us to obtain bounds for the exponent of a matrix.

In the first two sections of Chapter 6, we mainly use the papers [**53, 54, 82**].

In Section 6.3, the basic combinatorial properties of nonnegative matrices are applied to analysis of the so-called tournament matrices, which are adjacency matrices for tournaments. A tournament is a directed graph where two distinct vertices are necessarily joined with exactly one arc. A tournament is irreducible if and only if it is strongly connected. For $n \geq 4$ a tournament matrix is primitive if and only if the tournament is irreducible. For $n \geq 5$, the exponent γ of a tournament matrix lies within the bounds $3 \leq \gamma \leq n+2$. A tournament is said to be transitive if its vertices can be renumbered so that the upper triangle submatrix of the tournament matrix consists of ones only and the other elements are zeros. Any tournament matrix can be reduced to the canonical form, where block submatrices are placed on the main diagonal, ones are above them, and zeros below. Each block submatrix corresponds to either an irreducible or a transitive subtournament, and neighbouring submatrices cannot simultaneously correspond to transitive subtournaments. The canonical form of a tournament matrix allows us to estimate the convergence index and period in terms of the corresponding characteristics of the block matrices. In Section 6.3, results of [**79**] are used.

In Section 6.4, the semigroup approach is applied to the analysis of powers of nonnegative matrices. We use the concept of associated operator studied in [**59, 82**]. For a matrix A and any subset of vertices L, the associated operator $F_A(L)$ determines the set of vertices of the digraph $\Gamma(A)$ into each of which an arc, emanating from at least one vertex of the set L, enters. To the product of matrices there correspond sequential actions of the operators associated with the multipliers. Thus, the associated operator for the matrix A^h is $F_A^h(L)$. In the case of a primitive matrix, for any $L \subset N_n = \{1, \dots, N\}$ and some $h > 0$ the equality $F_A^h(L) = N_n$ holds. This fact allows us to obtain a series of bounds for the exponent of a primitive matrix depending on characteristics of the matrix. With the use of the associated operator, we study another characteristic, called the index of a nonnegative matrix. The index of a matrix A is equal to the least k for which the matrix A^k has the greatest number of positive elements. Similarly we define the indices of the rows of A. If A is primitive, then, of course, the index of A coincides with the exponent. The associated operator F_A is nonsingular if and only if $\operatorname{per} A > 0$. For a nonsingular operator F_A, we give estimates of indices of the rows of A.

In Section 6.5, we consider sequences of powers of $(0,1)$-matrices of order n. We study the convergence index and the period of a $(0,1)$-matrix A. We introduce the concept of q-universality of a strong component of the digraph $\delta(A)$. On the basis of this concept, we formulate conditions under which the limit of the sequence $\{A^k,\ k = 1, 2, \dots\}$ of powers of the matrix A exists.

In the general case, the limit of this sequence is an idempotent matrix A^* such that $\mathbf{0} \le A^* \le \mathbf{J}$, where $\mathbf{0}$ is the zero matrix and $\mathbf{J}$ is the matrix whose all elements are ones.

If the sequence $\{A^k,\ k = 1, 2, \dots\}$ is not convergent, then it oscillates with some period p equal to the least common multiple of the periods of the submatrices corresponding to the strong components that are not q-universal. With the use of these results, for simple homogeneous Markov chains properties of powers of the transition matrices are expressed in terms of properties of sequences of powers of their supports and hence in terms of properties of the corresponding directed graphs. This section is based on [**85**].

In Section 6.6, we study products of nonnegative matrices. We introduce the concept of r-primitivity of a sequence of nonnegative matrices. For a sequence of matrices $\{A_k,\ k = 1, 2, \dots\}$ of order n satisfying the condition of r-primitivity, we consider the product

$$T_{pk} = (t_{ij}(p,k)) = A_{p+1}A_{p+2}\cdots A_{p+k}, \qquad i, j = 1, \dots, n,$$

for any $p \ge 0$ and $k = 1, 2, \dots$. We prove the theorem on weak ergodicity, which asserts that the rows of the matrix T_{pk} are proportional in limit as $k \to \infty$. Further, we formulate the concept of strong r-primitivity of a sequence of nonnegative matrices. For a strong r-primitive sequence of matrices, we prove a theorem on strong ergodicity. This theorem establishes existence of the limits

$$\lim_{k \to \infty} \frac{t_{ij}(p,k)}{\sum_{q=1}^{n} t_{iq}(p,q)} = v_j, \qquad i, j = 1, \dots, n,$$

where $v^T = (v_1, \ldots, v_n)$, $v_1 + \cdots + v_n = 1$, and v^T is the left eigenvector of the matrix A corresponding to the maximal eigenvalue λ.

The results on ergodicity of sequences of nonnegative matrices are used to study the weak and strong ergodicity of Markov chains with a finite number of states. In this section we use results of the book [**98**].

6.1. Directed graphs of nonnegative matrices

6.1.1. Strong components of a directed graph. We consider a nonnegative matrix $A = (a_{ij})$, $i, j = 1, \ldots, n$. On the set $N_n = \{1, \ldots, n\}$ we define a multivalued mapping $\Gamma\colon N_n \to N_n$ such that

$$\Gamma(i) = \{j\colon a_{ij} > 0\}, \qquad i \in N_n.$$

This mapping defines a directed graph whose vertices are labelled with elements of the set N_n. Vertices i and j are joined by an arc (i, j) if $\Gamma(i) = j$, that is, if $a_{ij} > 0$. We denote this digraph by $\Gamma(A)$. Note that this digraph coincides with the digraph whose adjacency matrix is the support of the matrix A.

If vertices i and j in A are joined by a chain $i, i_1, \ldots, i_{l-1}, j$ of length l, we say that j is attainable from i in l steps, and write $i \to j$. If $i = j$, then the chain is called a circuit passing through the vertex i. A circuit without coinciding vertices is called elementary. Two circuits are not distinguished if they differ from one another only by a cyclic shift of vertices.

A digraph $\Gamma(A)$ is called strongly connected if $i \to j$ for any i and j. It is clear that a matrix A is indecomposable or irreducible if and only if $\Gamma(A)$ is strongly connected.

On the set N_n, we define the equivalence relation $\sim$. We say that $i \sim j$, $i, j \in N$, if either $i = j$ or $i \to j$, $j \to i$. In the latter case, we write $i \leftrightarrow j$.

To the equivalence relation there corresponds a partition of the set of vertices into equivalence classes $V_1, \ldots, V_m$ such that

$$N_n = V_1 \cup \cdots \cup V_m, \quad V_\mu \cap V_\nu = \emptyset, \quad \mu \neq \nu.$$

The restriction of the mapping Γ to an equivalence class V_k determines a strong component of the digraph $\Gamma(A)$, $1 \leq k \leq m$. Thus, if $\Gamma_1, \ldots, \Gamma_m$ are strong components of $\Gamma(A)$, then

$$\Gamma_k(i) = \Gamma(i) \cap V_k, \quad i \in V_k, \quad 1 \leq k \leq m.$$

The strong components of $\Gamma(A)$ are maximal strongly connected subdigraphs of $\Gamma(A)$. If $\Gamma(A)$ is strongly connected, then it has a single strong component.

A strong component is said to be transient if it consists of a single vertex and contains no loops. Otherwise the component is called nontransient. Each nontransient component contains at least one circuit, allowed to be a loop.

For any nonnegative matrix A there exists a permutation matrix Π such that A reduces to the normal form

$$\Pi A \Pi^T = \begin{pmatrix} A_{11} & & & 0 \\ A_{21} & A_{22} & & \\ A_{m1} & A_{m2} & \ldots & A_{mm} \end{pmatrix}. \tag{6.1.1}$$

The digraph $\Gamma(\Pi A \Pi^T)$ differs from the digraph $\Gamma(A)$ in the following: the vertices of $\Gamma(\Pi A \Pi^T)$ are enumerated so that the matrices $A_{11}, \ldots, A_{mm}$ correspond to the strong components $\Gamma_1, \ldots, \Gamma_m$.

We consider the digraph $\Gamma(A+A^T)$ and assume that $\tilde{V}_1, \dots, \tilde{V}_{\tilde{m}}$ are the equivalence classes determiined by this digraph. We define the components of the digraph $\Gamma(A)$ as the subdigraphs $\tilde{\Gamma}_1, \dots, \tilde{\Gamma}_{\tilde{m}}$ determined by the conditions

$$\tilde{\Gamma}_k(i) = \Gamma(i) \cap \tilde{V}_k, \quad i \in \tilde{V}_k, \quad 1 \le k \le \tilde{m}.$$

It is clear that the components of $\Gamma(A)$ are unions of several strong components.

If Γ_μ, Γ_ν are strong components of $\Gamma(A)$ and there exists an arc (i,j) such that $i \in V_\mu$, $j \in V_\nu$, $\mu \ne \nu$, then we say that Γ_μ is joined to Γ_ν. In any component of $\Gamma(A)$ containing more than one strong component, any strong component is joined to at least one other strong component. No strong components can be joined if they belong to different components.

For two strong components Γ_μ and Γ_ν belonging to the same component of $\Gamma(A)$, we say that the vertices of Γ_μ precede those of Γ_ν, and write $\Gamma_\mu < \Gamma_\nu$, if for any $i \in V_\mu$ there exists a chain joining i to any vertex $j \in V_\nu$. In particular, in view of strong connectedness, $\Gamma_\mu < \Gamma_\nu$.

Let $\delta_{\mu\nu}$ denote the maximum number of arcs in a chain joining two different strong components Γ_μ and Γ_ν such that $\Gamma_\mu < \Gamma_\nu$. We define the diameter of a component $\tilde{\Gamma}_k$ as

$$\delta_k = \max_{\mu,\nu}\{\delta_{\mu\nu} : \Gamma_\mu < \Gamma_\nu;\ V_\mu, V_\nu \subset \tilde{V}_k\}.$$

Let $\theta_k(i,j)$ be the minimum of lengths of all chains joining the vertices i and j, where $i,j \in V_k$. Then the diameter of a strong component Γ_k is defined as

$$d_k = \max_{i \ne j,\ i,j \in V_k} \theta_k(i,j), \qquad 1 \le k \le m.$$

It follows from the strong connectedness of Γ_k that

$$d_k \le |V_k| - 1.$$

Let a strong component Γ_k contain only elementary circuits of lengths $l_1^{(k)}, \dots, l_t^{(k)}$ which satisfy the condition $0 < l_1^{(k)} < \dots < l_t^{(k)}$. Let $P_k = \gcd(l_1^{(k)}, \dots, l_t^{(k)})$ be the greatest common divisor of these lengths. For a transient strong component, we set $P_k = 1$.

We define the circuit diameter D_k of a strong component Γ_k, $1 \le k \le m$, by the equality

$$D_k = \max_{i,j \in V_k} l_k(i,j),$$

where $l_k(i,j)$ is the minimum length of the chains joining the vertex i with the vertex j and passing through at least one vertex of each elementary circuit of length $l_1^{(k)}, \dots, l_t^{(k)}$.

The next lemma gives a bound for the circuit diameter expressed in terms of the diameter of a strong component.

LEMMA 6.1.1. *For the circuit diameter D_k of a strong component Γ_k, we have the bounds*

$$d_k \le D_k \le d_k + \sum_{\mu=1}^{t-1}(n_k - l_\mu^{(k)}), \qquad 1 \le k \le m, \tag{6.1.2}$$

where $n_k = |V_k|$ and d_k is the diameter of Γ_k.

PROOF. From the definition of d_k it follows that there exist vertices $i, j \in V_k$ such that d_k coincides with the minimum of the lengths of the chains joining i and j. The length of the corresponding chain passing through the vertices of elementary circuits is no less than d_k. Hence $d_k \leq D_k$.

Let us construct a chain from a vertex i to a vertex j of a strong component Γ_k which passes through at least one vertex of each of the elementary circuits of lengths $l_1^{(k)}, \dots, l_t^{(k)}$. From the vertex i a chain of length not exceeding $n_k - l_\mu^{(k)}$ goes out and reaches some vertex of an elementary circuit of length $l_\mu^{(k)}$, $1 \leq \mu \leq t$. The minimum length of a chain from a vertex of an elementary circuit of length $l_\mu^{(k)}$ to a vertex of another elementary circuit of length $l_\nu^{(k)}$, $\mu \neq \nu$, does not exceed $n_k - l_\nu^{(k)}$. Thus, the minimum length of a chain from i to j does not exceed

$$d_k + \sum_{\nu=1,\ \nu\neq\mu}^{t} (n_k - l_\nu^{(k)}) \leq d_k + \sum_{\mu=1}^{t-1} (n_k - l_\mu^{(k)}). \tag{6.1.3}$$

Since i and j are chosen arbitrarily, the lemma is proved. □

COROLLARY 6.1.2. *If all elementary circuits of Γ_k are of length P_k, then*

$$D_k = d_k;$$

that is, the circuit diameter coincides with the diameter of Γ_k, $1 \leq k \leq m$.

It is easy to see that any vertex of Γ_k lies on at least one elementary circuit. Therefore, a chain joining two vertices of Γ_k necessarily passes through at least one vertex belonging to an elementary circuit of length P_k. Hence the coincidence of the diameter of Γ_k with the circuit diameter follows.

6.1.2. Classification of vertices of a directed graph. Let $\Gamma(A)$ be the digraph of a nonnegative matrix $A = (a_{ij})$, and let $A^t = (a_{ij}^{(t)})$, $i, j = 1, \dots, n$. If for two vertices i and j of $\Gamma(A)$ the relation $i \to j$ holds, then there exists t such that $a_{ij}^{(t)} > 0$. In this case we say that i transits to j. If $i \leftrightarrow j$, then we say that i and j communicate, or they are communicating vertices.

A vertex i of the digraph $\Gamma(A)$ is called inessential if there exists j such that $i \to j$ but $j \not\to i$. A vertex i which transits to no other vertex is also considered inessential. In the latter case the ith row of the matrix A is zero.

A vertex i of a digraph is called essential if there exists at least one vertex j such that $i \to j$, and, moreover, for any j such that $i \to j$ the relation $j \to i$ holds; that is, $i \leftrightarrow j$. The set of essential vertices of $\Gamma(A)$ is thus partitioned into essential classes. Vertices i and j belong to the same essential class if and only if $i \leftrightarrow j$. Furthermore, all inessential vertices of $\Gamma(A)$ (if they exist) which communicate with some vertex are partitioned into inessential classes. A class includes all inessential vertices which communicate with one another. Furthermore, each inessential vertex which transits to no other vertex forms an inessential class.

As an example, we consider the matrix

$$A_c = \left(\begin{array}{c|ccccccc} & 1 & 2 & 3 & 4 & 5 & 6 & 7 \\ \hline 1 & 1 & 1 & 0 & 0 & 0 & 0 & 0 \\ 2 & 1 & 1 & 0 & 0 & 0 & 0 & 0 \\ 3 & 0 & 1 & 1 & 1 & 0 & 0 & 0 \\ 4 & 0 & 0 & 1 & 1 & 0 & 0 & 0 \\ 5 & 0 & 0 & 0 & 0 & 1 & 0 & 0 \\ 6 & 0 & 0 & 0 & 1 & 0 & 1 & 0 \\ 7 & 0 & 0 & 0 & 0 & 0 & 1 & 1 \end{array}\right) \tag{6.1.4}$$

In the corresponding digraph $\Gamma(A_c)$, the classes of essential vertices are $\{1,2\}$, $\{5\}$. The inessential classes are $\{3,4\}$, $\{6\}$, $\{7\}$.

LEMMA 6.1.3. *If a matrix $A \geq 0$ contains no zero rows, then it is impossible to transit from an essential vertex of A to an inessential vertex.*

PROOF. Let i be an essential vertex and $i \longrightarrow j$. Since the matrix A has no zero rows, there exists k such that $j \longrightarrow k$. It is clear that $i \longrightarrow k$. The vertex i is essential; therefore $k \longrightarrow i$. In view of transitivity, $k \longrightarrow j$. Since k is arbitrary, j is essential. □

A property defined for all vertices of $\Gamma(A)$ is called a class property if the fact that a certain vertex of a class possesses it implies that all vertices of the class possess it. The negation of a class property is a class property.

COROLLARY 6.1.4. *The property of a vertex of a digraph $\Gamma(A)$ to be essential or inessential is a class property.*

Note that the assertion converse to Lemma 6.1.3 is not true. If $i \longrightarrow j$ and j is an essential vertex, then i is allowed to be inessential if $j \not\twoheadrightarrow i$. Thus, if a suitable chain exists, then it is possible to transit from a class of inessential vertices to a class of essential vertices, but the transition from an essential vertex to an inessential vertex is impossible.

6.1.3. The canonical form of a matrix. We change the vertex labels in the digraph $\Gamma(A)$ so that first come, arranged in an arbitrary order, the vertices of essential classes with consecutive numeration inside a class, followed by the vertices of inessential classes, numbered also consecutively within a class. Moreover, an inessential class precedes another inessential class if its vertices do not transit to vertices of succeeding inessential classes. As a result the digraph $\Gamma(A)$ becomes the digraph $\Gamma(\Pi A \Pi^T)$, where Π is a permutation matrix and

$$\Pi A \Pi^T = \left(\begin{array}{cccc|c} A_1 & & & 0 & \\ & A_2 & & & \\ & & \ddots & & \mathbf{0} \\ 0 & & & A_k & \\ \hline & & B & & C \end{array}\right) \tag{6.1.5}$$

Each of the diagonal matrices $A_1, \ldots, A_k$ corresponds to one of k classes of essential vertices; the matrix C corresponds to inessential classes. To the left and to the right of the matrices $A_1, \ldots, A_k$, zero elements are placed. The matrix B has nonzero elements if some inessential vertices transit to vertices in essential classes. The matrix C is also of block structure:

$$C = \begin{pmatrix} C_1 & & & 0 \\ & C_2 & & \\ & & \ddots & \\ D & & & C_r \end{pmatrix}, \tag{6.1.6}$$

where the diagonal matrices $C_1, \ldots, C_r$ correspond to the classes of inessential vertices, and D has nonzero elements if some vertices of inessential classes transit to vertices of other inessential classes.

In view of the block structure of the matrices, their powers are, respectively, of the form

$$\Pi A^\nu \Pi^T = \left(\begin{array}{cccc|c} A_1^\nu & & & 0 & \\ & A_2^\nu & & & \\ & & \ddots & & \mathbf{0} \\ 0 & & & A_k^\nu & \\ \hline & & B^\nu & & C^\nu \end{array}\right),$$

$$C^\nu = \begin{pmatrix} C_1^\nu & & & 0 \\ & C_2^\nu & & \\ & & \ddots & \\ D^\nu & & & C_r^\nu \end{pmatrix}.$$

In the case where all vertices of the digraph $\Gamma(A)$ are essential, the subdigraphs $\Gamma(A_1), \ldots, \Gamma(A_k)$ are strong components of $\Gamma(A)$. For $k = 1$, the digraph $\Gamma(A)$ is strongly connected and the matrix A is indecomposable.

LEMMA 6.1.5. *If a matrix $A \geq 0$ of order n has no zero rows, then the digraph $\Gamma(A)$ contains at least one class of essential vertices.*

PROOF. Assume that all vertices are inessential. Then from the hypothesis that all rows of A are nonzero it follows that for any vertex i, $1 \leq i \leq n$, there exists at least one j such that $i \to j$, $j \not\to i$.

We take a chain of the form $i_1 \to i_2 \to \cdots \to i_n \to i_{n+1}$ (starting from some vertex i_1) such that $i_{k+1} \not\to i_k$, and therefore $i_{k+1} \not\to i_1, \ldots, i_{k-1}$. The sequence $i_1, \ldots, i_{n+1}$ has n distinct elements, and hence has repetitions. This contradicts the possibility of constructing a chain where no vertex transits to a preceding vertex. Therefore, there exist μ and ν, $\mu \neq \nu$, such that $i_\mu = i_\nu$. Therefore, the subchain $i_\mu \to i_{\mu+1} \to \cdots \to i_{\nu-1} \to i_\nu \to i_\mu$ belongs to a class of essential vertices. □

For a matrix $A = (a_{ij})$, $i, j = 1, \ldots, n$, let a vertex i of the digraph $\Gamma(A)$ satisfy the condition $i \to i$. The period $d(i)$ of the vertex i is the greatest common divisor of those k which satisfy the relation $a_{ii}^{(k)} > 0$, $1 \leq i \leq n$, where $A^k = (a_{ij}^{(k)})$, $i, j = 1, \ldots, n$.

It is clear that the period of a vertex i is equal to the greatest common divisor of the lengths of the circuits passing through i. In particular, if $a_{ii} > 0$, then $d(i) = 1$.

LEMMA 6.1.6. *If for a pair of vertices i and j of the digraph $\Gamma(A)$ of a matrix $A = (a_{ij})$, $i, j = 1, \ldots, n$, we have $i \leftrightarrow j$, then $d(i) = d(j)$.*

PROOF. From the hypotheses of the lemma it follows that there exist t_1 and t_2 such that $a_{ij}^{(t_1)} > 0$, $a_{ji}^{(t_2)} > 0$, where $A^t = (a_{ij}^{(t)})$, $i, j = 1, \dots, n$, $t = 1, 2, \dots$ For any positive integer t such that $a_{jj}^{(t)} > 0$, we see that

$$a_{ii}^{(t_1+t_2+t)} \geq a_{ij}^{(t_1)} a_{jj}^{(t)} a_{ji}^{(t_2)} > 0.$$

Since the hypothesis $a_{jj}^{(t)} > 0$ implies that $a_{jj}^{(2t)} > 0$, argueing as above we find that

$$a_{ii}^{(t_1+t_2+2t)} > 0.$$

Thus, the period $d(i)$ divides the numbers $t_1 + t_2 + t$ and $t_1 + t_2 + 2t$, and therefore, their difference t. Thus, $d(i)$ divides any t such that $a_{jj}^{(t)} > 0$ and, consequently, divides $d(j)$. Hence $d(i) \leq d(j)$. Similarly one proves that $d(j) \leq d(i)$. □

Lemma 6.1.6 implies that the following object is well defined. The period d of a class of communicating vertices of the digraph $\Gamma(A)$ is the period of an arbitrary vertex of this class. If $d = 1$, then the class is called aperiodic or acyclic. In particular, if a class contains a vertex i such that $d(i) = 1$, then the class is aperiodic, and the vertex i itself is called aperiodic.

6.1.4. Number-theoretical lemmas. In what follows, in order to express the lengths of chains joining vertices of a strong component of a digraph in terms of the lengths of elementary circuits of this component, we need a series of lemmas from number theory.

LEMMA 6.1.7. *A set X of integers that contains at least one nonzero element and is closed with respect to addition and subtraction contains the smallest positive element x such that all elements of X are divisible by this element.*

Indeed, it is easy to see that if $y \in X$, $y \neq 0$, then $y - y = 0 \in X$, and therefore $0 - y = -y \in X$. Consequently, X contains at least one positive element. Let x be the smallest positive element in X. Then $nx \in X$ and $-nx = 0 - nx \in X$, $n = 1, 2, \dots$. We take an arbitrary element $z \in X$ and assume that $z = qx + r$, $0 \leq r < x$. Then $r = z - qx \in X$, and therefore $r = 0$ because the inequality $r > 0$ contradicts the minimality of x.

LEMMA 6.1.8. *Let $d = \gcd(a_1, \dots, a_k)$ be the greatest common divisor of integers $a_1, \dots, a_k$. Then there exist integers $b_1, \dots, b_k$ such that*

$$d = a_1 b_1 + \cdots + a_k b_k.$$

PROOF. For given $a_1, \dots, a_k$ and varying integers $x_1, \dots, x_k$, consider the set X consisting of $a_1 x_1 + \cdots + a_k x_k$. This set is closed with respect to the operations of addition and subtraction, and by Lemma 6.1.7 it consists of multiples of some minimal element $\nu = a_1 b_1 + \cdots + a_k b_k$. Since d divides $a_1, \dots, a_k$, we see that d divides ν as well; that is, $0 < d \leq \nu$. Since d is the greatest common divisor of $a_1, \dots, a_k$, we conclude that $\nu = d = a_1 b_1 + \cdots + a_k b_k$. □

The greatest common divisor d of an infinite numerical sequence $a_1, a_2 \dots$ is defined as the limit

$$d = \lim_{k \to \infty} d_k,$$

where $d_k = \gcd(a_1, \dots, a_k)$. This limit exists because the sequence $d_1, d_2, \dots$ does not increase and is bounded from below. The limit is attained at some finite k.

LEMMA 6.1.9. *An infinite set $Y = \{a_1, a_2, \dots\}$ of positive integers closed with respect to addition contains all but a finite number of the positive integers that are divisible by the greatest common divisor d of the elements of this set.*

PROOF. By dividing all elements of Y by d, we reduce the proof to the case where $d = 1$. For a sufficiently large k, $\gcd(a_1, \dots, a_k) = 1$. Therefore, by Lemma 6.1.8, there exist integers $b_1, \dots, b_k$ such that $a_1 b_1 + \dots + a_k b_k = 1$. If S_1 and S_2 are the absolute values of the sum of positive and negative terms of the above sum, respectively, then $S_1 - S_2 = 1$. We choose q such that $q > S_2(S_2 - 1)$ and represent $q = aS_2 + b$, $0 \le b \le S_2$, where a, b are integers and $a \ge S_2 - 1$. We see that $q = aS_2 + b(S_1 - S_2)$, and therefore, $q = (a - b)S_2 + bS_1$. Hence $q \in Y$. Thus, any sufficiently large number belongs to the set Y. □

COROLLARY 6.1.10. *If $a_1, \dots, a_k$ are positive integers and $\gcd(a_1, \dots, a_k) = 1$, then any sufficiently large positive q can be represented as*

$$q = a_1 b_1 + \dots + a_k b_k,$$

where $b_1, \dots, b_k$ are nonnegative integers.

For positive integers $a_1 < \dots < a_k$ such that $\gcd(a_1, \dots, a_k) = 1$, the Frobenius function $g(a_1, \dots, a_k)$ is defined as the greatest positive integer which cannot be represented as a linear combination of $a_1, \dots, a_k$ with nonnegative integer coefficients $x_1, \dots, x_k$.

We give without proof the following assertion.

LEMMA 6.1.11. *The inequalities*

$$a_1[(a_k - 2)/(a_k - a_1)] \le g(a_1, \dots, a_k) + 1 \le (a_1 - 1)(a_k - 1)$$

are true, where we assume that $g(a_1, \dots, a_k) = -1$ for $a_1 = 1$.

Problems.

1. Reduce the matrix in (6.1.4) to the canonical form (6.1.5).
2. For the matrix defined by (6.1.4), find the matrices $A_1, \dots, A_k, B, C$ and $C_1, \dots, C_r, D$ in the representations (6.1.5) and (6.1.6).
3. For the matrix in (6.1.4), find the periods of the classes of communicating vertices.

6.2. Irreducible and primitive matrices

6.2.1. Irreducible periodic matrices. We recall that a nonnegative matrix $A = (a_{ij})$, $i, j = 1, \dots, n$, is called indecomposable or irreducible if for powers $A^t = (a_{ij}^{(t)})$ the following property holds: for any i and j there exists a positive integer $t = t(i, j)$ such that $a_{ij}^{(t)} > 0$. This means that in the digraph $\Gamma(A)$ any pair of vertices i, j satisfies the condition $i \leftrightarrow j$. An indecomposable matrix A is called periodic (or cyclic) if the period of any vertex of the corresponding digraph $\Gamma(A)$ is equal to $d > 1$.

If $d = 1$, then the matrix is called aperiodic (or acyclic). These notions are well defined, because for an indecomposable matrix A all vertices of $\Gamma(A)$ belong to the same essential class.

LEMMA 6.2.1. *Let $A = (a_{ij})$ be an indecomposable periodic matrix of period d, and let $A^t = (a_{ij}^{(t)})$, $i, j = 1, \dots, n$. Then for each $i = 1, \dots, n$ there exists an integer $t_0 = t_0(i)$ such that $a_{ii}^{(kd)} > 0$ for all $k \ge t_0$.*

PROOF. We consider the set of positive integers $Y = \{kd\}$, $k = 1, 2, \ldots$, such that $a_{ii}^{(kd)} > 0$. This set is nonempty and closed with respect to addition, because the conditions $a_{ii}^{(kd)} > 0$ and $a_{ii}^{(sd)} > 0$ imply that

$$a_{ii}^{((k+s)d)} \geq a_{ii}^{(kd)} a_{ii}^{(sd)} > 0.$$

By Lemma 6.1.9, any sufficiently large number that is divisible by d belongs to the set Y. The lemma is proved. □

THEOREM 6.2.2. *Let $A = (a_{ij})$ be an indecomposable matrix of period d, and let $A^t = (a_{ij}^{(t)})$, $i, j = 1, \ldots, n$. If i is a fixed vertex of the digraph $\Gamma(A)$, $1 \leq i \leq n$, then for any j, $1 \leq j \leq n$, there exists a unique nonnegative integer r_j, $0 \leq r_j < d$, such that*

(1) *the condition $a_{ij}^{(s)} > 0$ implies that $s \equiv r_j \pmod d$, and*
(2) *there exists a positive $t(j)$ such that $a_{ij}^{(kd+r_j)} > 0$ for all $k \geq t(j)$.*

PROOF. For $d = 1$, the theorem is obvious.

Let $d \geq 2$. From indecomposability of A it follows that there exist m, m', $m \neq m'$, and p such that

$$a_{ij}^{(m)} > 0, \quad a_{ij}^{(m')} > 0, \quad a_{ji}^{(p)} > 0.$$

Hence

$$a_{ii}^{(m+p)} \geq a_{ij}^{(m)} a_{ji}^{(p)} > 0, \qquad a_{ii}^{(m'+p)} \geq a_{ij}^{(m')} a_{ji}^{(p)} > 0.$$

Therefore, d divides $m + p$ and $m' + p$, hence also $m - m'$; that is, $m - m' \equiv 0 \pmod d$. Therefore, there exists r_j, $0 \leq r_j < d$, such that $m \equiv m' \equiv r_j \pmod d$. The first part of the theorem is thus proved.

Since $i \to j$, from the first part of the theorem it follows that there exists a positive integer m satisfying the condition $a_{ij}^{(md+r_j)} > 0$. Let t_0 be the number in Lemma 6.2.1 satisfying the condition $a_{ii}^{(kd)} > 0$ for all $k \geq 0$. We set $t(j) = m + t_0$. Then $s = k - m \geq t_0$ and $kd + r_j = md + sd + r_j$ for all $k \geq t(j)$. Thus, for all $k \geq t(j)$ we obtain

$$a_{ij}(kd + r_j) \geq a_{ii}^{(sd)} a_{ij}^{(md+r_j)} > 0.$$

The theorem is proved. □

Let $A = (a_{ij})$, $i, j = 1, \ldots, n$, be an indecomposable matrix of period d. For a fixed vertex i of the strongly connected digraph $\Gamma(A)$, we consider the set of vertices which correspond, according to Theorem 6.2.2, to the same smallest nonnegative residue r modulo d. This set is called the subclass of the unique essential class of vertices of $\Gamma(A)$, and is denoted by C_r, $0 \leq r < d$. It is clear that the subclasses $C_0, C_1, \ldots, C_{d-1}$ form a partition of the set of vertices of the digraph $\Gamma(A)$:

$$C_0 \cup C_1 \cup \ldots C_{d-1} = \{1, 2, \ldots, n\}, \quad C_j \cap C_k = \emptyset, \quad j \neq k, \quad 0 \leq j < k \leq d-1.$$

Let us show that the partitioning into subclasses does not depend on the choice of the vertex i.

LEMMA 6.2.3. *In Theorem 6.2.2, the choice of an arbitrary fixed vertex instead of i results in a cyclic permutation of the indices of the subclasses.*

PROOF. Let ν be an arbitrary fixed vertex, $\nu \neq i$. Let r_ν be the smallest nonnegative residue such that $a_{i\nu}^{(kd+r_\nu)} > 0$ for all $k \geq t(\nu)$. Similarly, let r'_j be the smallest nonnegative residue such that $a_{\nu j}^{(md+r'_j)} > 0$ for all $m \geq t(j)$. Therefore,

$$a_{ij}(md + r_nu + kd + r'_j) \geq a_{i\nu}^{(kd+r_\nu)} a_{\nu j}^{(md+r'_j)} > 0$$

for sufficiently large k and m.

If r_j is the least nonnegative residue under the classification of the vertex j with respect to the fixed vertex i in Theorem 6.2.2, then

$$md + r_\nu + kd + r'_j \equiv r_j \pmod{d}.$$

Hence

$$r_\nu + r'_j \equiv r_j \pmod{d}, \qquad j = 0, 1, \dots, d-1.$$

Thus, the residues $r_0, r_1, \dots, r_{d-1}$ under the classification of the vertices with respect to the fixed vertex i are obtained from the residues $r'_0, r'_1, \dots, r'_{d-1}$ under the classification with respect to the fixed vertex ν by adding r_ν. This means that in passing from the fixed vertex i to the fixed vertex ν we renumber the subclasses $C_0, C_1, \dots, C_{d-1}$ according to the permutation s^{-r_ν}, where s is the cycle $(0\,1\,\dots\,(d-1))$. □

6.2.2. The canonical form of a periodic matrix. For a class of essential vertices of $\Gamma(A)$, we define the classes C_r for all nonnegative integers r by setting $C_r = C_{r_j}$, where $r \equiv r_j \pmod{d}$ and r_j is defined under classification with respect to a fixed vertex i. From indecomposability of the matrix $A = (a_{ij})$ it follows that for any power $m > 0$ there exists j such that $a_{ij}^{(m)} > 0$, where $A^m = (a_{ij}^{(m)})$, $i, j = 1, \dots, n$. If this were not the case, the matrices $A^m, A^{m+1}, \dots$ would have zero rows, which is impossible. By Theorem 6.2.2, for $a_{ij}^{(m)} > 0$ the relation $m \equiv r_j \pmod{d}$ holds, and therefore, $j \in C_{r_j}$. Furthermore, there exists k such that $a_{ik}^{(m+1)} > 0$. Since $m + 1 \equiv r_j + 1 \pmod{d}$, we conclude that $k \in C_{r_j+1}$.

Thus, taking into consideration consecutive powers of the matrix A, we determine cyclic subclasses which contain the indices of columns of positive elements of the ith row. None of the d cyclic subclasses is empty.

Considering the ith rows of consecutive powers of the matrix A, we see that there occurs a sequential transition from one cyclic subclass to another. Classifying the vertices $i = 1, \dots, n$, we find that an indecomposable matrix A of period d has the canonical form A_c.

We renumber rows and columns of A so that the indices of vertices of $\Gamma(A)$ belonging to subclasses $C_0, \dots, C_{d-1}$, $d \geq 2$, go in consecutive order. Then

$$(6.2.1) \qquad A_c = \Pi A \Pi^T = \begin{pmatrix} \mathbf{0} & Q_{01} & \mathbf{0} & \dots & \mathbf{0} \\ \mathbf{0} & \mathbf{0} & Q_{12} & \dots & \mathbf{0} \\ & & & \dots & \\ \mathbf{0} & \mathbf{0} & \mathbf{0} & \ddots & Q_{d-2,d-1} \\ Q_{d-1,0} & \mathbf{0} & \mathbf{0} & \dots & \mathbf{0} \end{pmatrix},$$

where Π is a permutation matrix and the submatrix $Q_{i,i+1}$ is located at the intersection of those rows and columns that are elements of the classes C_i and C_{i+1}, $0 \leq i < d-1$, respectively. Matrix (6.2.1) is the canonical form to which a periodic matrix A reduces.

For example, the indecomposable matrix A which has three cyclic subclasses $C_{01} = \{1, 2\}$, $C_{12} = \{3, 4, 5\}$, $C_{20} = \{6, 7\}$, has the canonical form

$$A_c = \left(\begin{array}{c|cc|ccc|cc} & 1 & 2 & 3 & 4 & 5 & 6 & 7 \\ \hline 1 & 0 & 0 & 1 & 0 & 1 & 0 & 0 \\ 2 & 0 & 0 & 1 & 1 & 0 & 0 & 0 \\ \hline 3 & 0 & 0 & 0 & 0 & 0 & 1 & 0 \\ 4 & 0 & 0 & 0 & 0 & 0 & 1 & 1 \\ 5 & 0 & 0 & 0 & 0 & 0 & 0 & 1 \\ \hline 6 & 1 & 1 & 0 & 0 & 0 & 0 & 0 \\ 7 & 1 & 1 & 0 & 0 & 0 & 0 & 0 \end{array}\right).$$

6.2.3. Primitive matrices. A nonnegative matrix A is called primitive if there exists a positive integer γ such that $A^\gamma > 0$. The minimum of these γ is called the exponent of the matrix A.

It is clear that in the digraph $\Gamma(A)$ of a primitive matrix whose exponent is equal to γ for any vertices i and j there exists a chain of length γ joining i with j. Moreover, there are infinitely many chains of lengths $\gamma, \gamma + 1, \dots$ joining an arbitrary vertex i with an arbitrary vertex j.

THEOREM 6.2.4. *A nonnegative matrix A is primitive if and only if it is indecomposable and aperiodic.*

PROOF. Let a matrix $A = (a_{ij})$, $i, j = 1, \dots, n$, be indecomposable and aperiodic, that is, be of period $d = 1$. Therefore, there is a unique subgraph C_0 coinciding with the whole set of vertices of the digraph $\Gamma(A)$. From Theorem 6.2.2 it follows that there exists $N(i,j) > 0$ such that $a_{ij}^{(k)} > 0$ for all $k \geq N(i,j)$. We set $N_0 = \max_{i,j} N(i,j)$. Then $a_{ij}^{(k)} > 0$ for any i and j, provided that $k \geq N_0$. Hence $A^k > 0$; that is, A is primitive.

Let a matrix A be primitive. Then it is indecomposable, because the digraph $\Gamma(A)$ is strongly connected. In addition, $a_{ij}^{(k)} > 0$ and $a_{ii}^{(k+1)} > 0$ for any vertex i and sufficiently large k; that is, for i there are two circuits whose lengths k and $k+1$ are relatively prime. Hence $d = 1$ and the matrix A is aperiodic. □

Note that if a matrix A is primitive and $A^\gamma > 0$, then $A^{\gamma+p} > 0$ for all $p = 0, 1, 2, \dots$. Therefore, any power of a primitive matrix is primitive as well.

LEMMA 6.2.5. *Let $\{C_1, \dots, C_t\}$ be the set of all elementary circuits of the digraph $\Gamma(A)$ of a primitive matrix A, and let $l_1, \dots, l_t$ be the corresponding circuit lengths. Then the length l of any circuit C of $\Gamma(A)$ is of the form*

$$l = x_1 l_1 + \cdots + x_t l_t,$$

where $x_1, \dots, x_t$ are nonnegative integers.

PROOF. We set $C_1^* = C$ and define the transition from the circuit C_i^* to the circuit C_{i+1}^*, $i = 1, \dots, t$, as follows. If C_i^* includes the elementary circuit C_i, we eliminate it, except the last vertex, and obtain the circuit C_{i+1}^*. Performing this procedure for $i = 1, \dots, t$, we obtain a single vertex, which is the origin of the circuit C of zero length. If x_i is the number of occurrences of the elementary circuit C_i in C_i^*, then the total length of C takes the value given above. □

THEOREM 6.2.6. *If $\Gamma(A)$ is the digraph corresponding to a primitive matrix A, then it contains a set of elementary circuits whose lengths are relatively prime.*

PROOF. If γ is the exponent of a primitive matrix A, then in the digraph $\Gamma(A)$ there are circuits whose lengths take values $l_n = \gamma + n$, $n = 0, 1, \ldots$. If $l_1, \ldots, l_t$ are the lengths of all elementary circuits of $\Gamma(A)$, then, by Lemma 6.2.5, there exist $x_1^{(n)}, \ldots, x_t^{(n)}$ such that

$$l_n = x_1^{(n)} l_1 + \cdots + x_t^{(n)} l_t = \gamma + n, \qquad n = 0, 1, \ldots.$$

Assume that $\gcd(l_1, \ldots, l_t) = d$. Then the numbers $\gamma, \gamma+1, \gamma+2$ are divisible by d, which is possible only for $d = 1$. □

Let S be a subset of vertices of the digraph $\Gamma(A)$. We say that a chain in $\Gamma(A)$ is tangent to S if there exists a vertex which belong both to S and to the chain.

LEMMA 6.2.7. *Let S be a subset of vertices of the digraph $\Gamma(A)$ corresponding to a primitive matrix A of order n, and let i be an arbitrarily chosen vertex in $\Gamma(A)$. Then there exists a chain of length $l \leq n - k$ leaving the vertex i which is tangent to the set S, where k is the number of elements of S.*

PROOF. If $i \in S$, then the lemma is obvious. Let $i \notin S$. By the strong connectivity of $\Gamma(A)$, there is at least one chain emanating from i which is tangent to S. Let $i_1, \ldots, i_l$ be such a chain of the least length. This chain has no repeating vertices, and i_l is the only vertex such that $i_l \in S$. Hence $l \leq n - k$. □

THEOREM 6.2.8. *Let $\{C_1, \ldots, C_k\}$ be the set of all elementary circuits of the digraph $\Gamma(A)$ of a primitive matrix A. Then for any vertices i and j there exists a chain from i to j which is tangent to each of the circuits $C_1, \ldots, C_k$, and the length L of this chain satisfies the inequality*

$$L \leq n(k+1) - 1 - \sum_{i=1}^{k} l_i.$$

PROOF. The number of distinct vertices which form each elementary circuit is equal to the length of the circuit. Therefore, in view of Lemma 6.2.7, a chain of length at most $n - l_1$, which is tangent to the circuit C_1, leaves an arbitrary vertex i. One more chain of length at most $n - l_2$ that is tangent to C_2 goes out of the terminal vertex of this chain. Continuing these arguments, we find a chain of length at most $n - l_k$ which is tangent to C_k and continues the chain constructed before. Finally, there is a path from a vertex of C_k to a vertex j whose length is at most $n - 1$. The chain thus constructed joins an arbitrary vertex i with an arbitrary vertex j, and its length L satisfies the inequality

$$L \leq (n - l_1) + (n - l_2) + \cdots + (n - l_k) + (n - 1).$$

The theorem is thus proved. □

The following assertion is known as the Schur Lemma.

LEMMA 6.2.9. *If $l_1, \ldots, l_k$ are relatively prime positive integers, $l_1 \leq \cdots \leq l_k$, then the equation*

$$l_1 x_1 + \cdots + l_k x_k = l$$

has a solution in nonnegative integers for any $l \geq (l_1 - 1)(l_k - 1)$. The bound in the last inequality is attained for $k = 2$.

This assertion follows immediately from Lemma 6.1.11.

6.2.4. The exponent of a primitive matrix. We estimate the exponent of a primitive matrix (see also

THEOREM 6.2.10. *The exponent* γ *of a primitive matrix* A *of order* n *satisfies*

$$\gamma \leq n^2 - 2n + 2.$$

PROOF. In view of Theorem 6.2.6, the digraph $\Gamma(A)$ contains the set of elementary circuits, whose lengths are relatively prime. Eliminating, if necessary, several circuits from this set, we arrive at the set $\{C_1, \dots, C_k\}$ of elementary circuits whose relatively prime lengths $\{l_1, \dots, l_k\}$ form a minimal set (that is, eliminating any of them leaves the rest not relatively prime). We assume that $l_1 \leq \cdots \leq l_k$. By Theorem 6.2.8, for any i and j there exists a chain from i to j of length

$$L \leq n(k+1) - 1 - \sum_{i=1}^{k} l_i$$

which is tangent to each of the circuits $C_1, \dots, C_k$. If we adjoin $C_1, \dots, C_k$ to the tangential points taking the circuit C_i exactly x_i times, $1 \leq i \leq k$, we increase the length of the chain up to $L + l$, where $l = l_1 x_1 + \cdots + l_k x_k$. By Lemma 6.2.9, this equation has a solution for $l = (l_1 - 1)(l_k - 1)$. Therefore, for a chain of minimum length γ joining i with j we have the bound

$$\gamma \leq L + l = L + \sum_{i=1}^{k} l_i x_i \leq n(k+1) - 1 - \sum_{i=1}^{k} l_i + (l_1 - 1)(l_k - 1).$$

For $k = 2$,

$$\gamma \leq 3n - 2l_1 - 2l_2 + l_1 l_2 = 3n + (l_1 - 2)(l_k - 2) - 4.$$

Since $l_1 \leq n - 1$ and $l_2 \leq n$, we arrive at the bound $\gamma \leq n^2 - 2n + 2$. For $k = 2$ the theorem is thus proved.

For $k \geq 3$, the minimality of the set of relatively prime numbers $l_1, \dots, l_k$ yields

$$\begin{aligned} \sum_{i=1}^{k} l_k &\geq l_1 + (l_1 + 2) + (l_1 + 4) + \cdots + (l_1 + 2(k-2)) + l_k \\ &= (k-1)l_1 + (k-1)(k-2) + l_k. \end{aligned}$$

Hence we obtain

$$\gamma \leq n(k+1) + (l_1 - 2)(l_k - k) - 2k - (k-1)(k-2).$$

Since

$$l_1 \leq n - k + 2, \qquad l_k \leq n,$$

we have

$$\gamma \leq n^2 - (k-1)n + (k-2).$$

For $n \geq 1$, we conclude that

$$n^2 - (k-1)n + k - 2 < n^2 - 2n + 2,$$

which completes the proof. □

The bound for the exponent given in the theorem is attainable. For example, in the case $n = 3$ for the primitive matrix B we obtain

$$B = \begin{pmatrix} 0 & 1 & 1 \\ 1 & 0 & 0 \\ 1 & 1 & 0 \end{pmatrix}, \quad B^4 = \begin{pmatrix} 1 & 1 & 2 \\ 1 & 0 & 1 \\ 1 & 1 & 1 \end{pmatrix}, \quad B^5 = \begin{pmatrix} 2 & 1 & 2 \\ 1 & 1 & 1 \\ 1 & 1 & 2 \end{pmatrix}.$$

Let us give one more bound for the exponent of a primitive matrix A.

THEOREM 6.2.11. *Let l be the least length of the circuits in the digraph $\Gamma(A)$ of a primitive matrix A of order n. Then for the exponent γ of A we have*

$$\gamma \leq (n-2)l + n.$$

PROOF. We consider the shortest circuit $\nu_1, \nu_2, \ldots, \nu_l, \nu_1$ in $\Gamma(A)$. Choose arbitrary vertices i and j of $\Gamma(A)$, and estimate the length of a chain joining these vertices. Due to the primitivity of A, there exists a chain of length at most $n-l$ joining the vertex i with some vertex ν_k, $1 \leq k \leq l$. Let this chain be $i, i_1, \ldots, i_{n-l-1}, \nu_k$. It is obvious that

$$a_{i,i_1} > 0, a_{i_1 i_2} > 0, \ldots, a_{i_{n-l-1}\nu_k} > 0.$$

Since the matrix A^l is primitive, in the digraph $\Gamma(A^l)$ there exists a chain $\nu_k, j_1, j_2, \ldots, j_{n-2}, j$ of length at most $n-1$ which goes from ν_k to j, and therefore

$$a^{(l)}_{\nu_k j_1} > 0, a^{(l)}_{j_1 j_2} > 0, \ldots, a^{(l)}_{j_{n-2} j} > 0.$$

From the above chains of inequalities we obtain $a^{(n-l)}_{i\nu_k} > 0$ and $a^{(l(n-1))}_{\nu_k j} > 0$. Hence

$$a^{(n-l+l(n-1))}_{ij} \geq a^{(n-l)}_{i\nu_k} a^{(l(n-1))}_{\nu_k j} > 0.$$

Since i and j are arbitrary, we have $A^{n-l+l(n-1)} > 0$. Therefore, the bound for the exponent given in the theorem is true. □

COROLLARY 6.2.12. *If in a primitive matrix $A = (a_{ij})$, $i, j = 1, \ldots, n$, there exists i, $1 \leq i \leq n$, such that $a_{ii} > 0$, then for the exponent of this matrix we have*

$$\gamma \leq 2(n-1).$$

This corollary is just the preceding theorem with $l = 1$.

Note that Theorem 6.2.10 follows from Theorem 6.2.11, due to the obvious inequality $l \leq n-1$.

6.3. Tournament matrices

6.3.1. Primitive tournaments, the exponent. A directed graph T_n with vertices $v_1, \ldots, v_n$ is called a tournament if each pair of distinct vertices v_i and v_j is joined by either the arc (v_i, v_j) or the arc (v_j, v_i), and there exist no arcs of the form (v_i, v_i), $1 \leq i \leq n$. If a contest of n players takes place, where each player plays matches against each other only once and draws are not allowed, then, representing the players as vertices of the graph and passing an arc from the winner to the loser, we arrive at a directed graph which is a tournament. If M_n is the adjacency matrix of a tournament T_n and M_n^T is its transpose, then

$$M_n + M_n^T = \mathbf{J}_{nn} - \mathbf{I}_n.$$

A tournament T_n with vertex set V_n is said to be reducible if there exists a partition of V_n into two subsets $V_n^{(1)}$ and $V_n^{(2)}$ such that T_n contains all arcs (v, v') where $v \in V_n^{(1)}$ and $v' \in V_n^{(2)}$. Otherwise, the tournament is said to be irreducible.

It is not difficult to see that a tournament is irreducible if and only if it is strongly connected.

THEOREM 6.3.1. *Let l, n be integers, $3 \leq l \leq n$. Then at least one elementary circuit of length l passes through each vertex of an irreducible tournament T_n.*

PROOF. Let V denote the set of vertices of the tournament T_n under consideration, let E be the set of arcs, and let $v \in V$. We consider the partition

$$V = U \cup \{v\} \cup W,$$

where

$$U = \{x \in V\colon (v, x) \in E\}, \qquad W = \{x \in V\colon (x, v) \in E\}.$$

Since T_n is irreducible, there exist vertices $u \in U$, $w \in W$ such that $(u, w) \in E$. Therefore, there is an elementary circuit (v, u, w, v) of length three in T_n. By induction, assume that there is an elementary circuit

$$C = (v = v_0, v_1, \dots, v_{l-1}, v_l = v)$$

of length $l < n$ in T_n. Let us show that T_n contains an elementary circuit of length $l+1$. It is clear that such a circuit exists if for some i, $1 \leq i \leq l$, there exists a vertex v' such that $(v_{i-1}, v'), (v', v_i) \in E$. Assume that such a vertex does not exist. Then, starting from the vertex v_1, we easily show that $(v', u) \in E$ for $u \in U \backslash C$ and $v' \in C$. In the same way, taking successively $v_{l-1}, v_{l-1}, \dots, v_1$, we see that $(w, v') \in E$ for $w \in W \setminus C$ and $v' \in C$. Since $l < n$, one of the sets $U_1 = U \setminus C$ or $W_1 = W \setminus C$ is nonempty. Let, for example, W_1 be nonempty. Then, by the irreducibility of T_n, in E there exists an arc (u, w) from a vertex $u \in U$ to some vertex $w \in W_1$. As we have seen, $u \notin C$. Let s denote an integer such that $v_s, v_{s+1}, \dots, v_{l-1} \in W$ and $v_{s-1} \in U$. In view of the above, $(v_{s-1}, u), (w, v_{s+1}) \in E$, and we obtain an elementary circuit

$$C' = (v = v_0, v_1, \dots, v_{s-1}, u, w, v_{s+1}, \dots, v_l = v)$$

of length $l + 1$. Similarly we obtain a circuit of length $l + 1$ passing through v in the case where the set U_1 is nonempty. The theorem is proved. □

We say that a tournament T_n is primitive if the corresponding matrix M_n is primitive. The exponent of a tournament is the exponent of the matrix M_n.

The diameter of an irreducible tournament is the least positive integer d such that for any ordered pair of vertices v_i and v_j in T_n there exists a chain from v_i to v_j of length at most d. A criterion of primitivity of a tournament and a bound for its exponent are given in the following theorem.

THEOREM 6.3.2. *For $n \geq 4$, a tournament T_n is primitive if and only if it is irreducible. For $n \geq 5$, the exponent γ and the diameter d of a tournament satisfy the inequalities*

$$d \leq \gamma \leq d + 3.$$

PROOF. The inequality $\gamma \geq d$ is obvious, since a pair of vertices v_i, v_j exists in T_n such that the minimum length of a chain joining them is equal to d. To prove the inequality $\gamma \leq d + 3$, first we observe that for $n = 3$ no tournament is primitive. In essence, there exists a single irreducible tournament T_3 which contains a circuit of length three, say, of arcs (v_1, v_2), (v_2, v_3), and (v_3, v_1). It is clear that T_3 is not primitive, because a path of length three from v_1 to v_3 does not exist if the condition $l \equiv 2 \pmod 3$ does not hold. There are four types of an irreducible tournament T_4, and it is not difficult to check that all of them are primitive. We prove that, for $n \geq 5$, for any ordered pair of vertices v_i and v_j of an irreducible tournament there exists a chain from v_i to v_j of length $d+3$. Since $n \neq 1$, in view of the strong connectedness of T_n there exists at least one chain $P(v_i, v_j)$ going from v_i to v_j. Let $P'(v_i, v_j)$ be the shortest such chain, and let its length be equal to l. Since $0 \leq l \leq d \leq n-1$, we obtain $3 \leq d-l+3 \leq n+2$. If $3 \leq d-l+3 \leq n$, then by Theorem 6.3.1 there exists a simple circuit $P_1(v_i, v_i)$ of length $d-l+3$ passing through the vertex v_i. Then the chain

$$P(v_i, v_j) = P_1(v_i, v_i) \cup P'(v_i, v_j)$$

is of the required length $d+3$.

Assume that $d-l+3 = n+\nu$, $\nu = 1, 2$. Then $3 \leq n+\nu-3 \leq n-1$, because $n \geq 5$. By Theorem 6.3.1, there exist simple circuits $P_2(v_i, v_i)$ and $P_3(v_i, v_i)$ of lengths three and $n+\nu-3$ respectively. Then the chain

$$P(v_i, v_j) = P_2(v_i, v_i) \cup P_3(v_i, v_i) \cup P'(v_i, v_j)$$

is of length $3+(n+\nu-3)+l = d+3$. The proof of the theorem is completed. □

COROLLARY 6.3.3. *If $n \geq 5$ and T_n is an irreducible tournament of exponent γ, then*

$$3 \leq \gamma \leq n+2.$$

Indeed, the diameter d of an irreducible tournament T_n satisfies the obvious inequalities $3 \leq d \leq n-1$. Hence we obtain the required bound for the exponent.

THEOREM 6.3.4. *If $n \geq 6$ and $3 \leq \gamma \leq n+2$, then there exists an irreducible tournament T_n whose exponent is equal to γ.*

PROOF. We take positive integers n and k such that $3 \leq k \leq n-1$, $n \geq 6$. Consider a tournament T_n with vertices $v_1, \dots, v_n$ which has the arcs (v_1, v_n), $(v_n, v_{n-1}), \dots, (v_3, v_2), (v_2, v_1)$ forming a complete circuit, and the arcs (v_j, v_i), $k \leq i < j \leq n$. In the remaining arcs, the vertices are indexed in increasing order.

This tournament contains a simple complete circuit; therefore it is irreducible and, consequently, primitive. The longest of the shortest paths between the vertices in T_n is $P(v_n, v_1)$ of length k; therefore, for $k \geq 3$ the diameter d of the tournament is equal to k. In the tournament T_n, the chain $P(v_k, v_1)$ cannot be of length k or $k+1$; therefore the exponent cannot take the value k or $k+1$. Similarly, the chain $P(v_{k+1}, v_1)$ cannot be of length $k+2$, and therefore, the exponent cannot take the value $k+2$. By Theorem 6.3.2, $k \leq \gamma \leq k+3$; therefore, $\gamma = k+3$. From the above it follows that by choosing an appropriate k for $n \geq 6$ we can obtain a tournament T_n whose exponent γ satisfies the condition $6 \leq \gamma \leq n+2$.

If $k = 2, 3$, then we construct another tournament T'_n differing from T_n only in that the arc joining v_{k+1} and v_n is directed to v_n. Similar argumens show that $\gamma = k+2$. To complete the proof of the theorem, it remains to show that there exists a tournament T_n for $n \geq 6$ whose exponent is equal to three. For $n = 6$, consider the tournament whose matrix is

$$M_6 = \begin{pmatrix} 0 & 1 & 0 & 0 & 0 & 1 \\ 0 & 0 & 1 & 0 & 1 & 0 \\ 1 & 0 & 0 & 1 & 0 & 0 \\ 1 & 1 & 0 & 0 & 1 & 0 \\ 1 & 0 & 1 & 0 & 0 & 1 \\ 0 & 1 & 1 & 1 & 0 & 0 \end{pmatrix}.$$

By direct multiplications we obtain that $\gamma = 3$. If $n > 6$, then we construct the tournament T_n which contains an arc (v_i, v_j) if and only if $0 < j - i \leq [n/2]$, where the subtraction is modulo n or $n+1$ depending on whether n is odd or even. The exponent of this tournament is equal to three. □

6.3.2. Index, period, and convergence index. A tournament T_n is called transitive if its vertices $v_1, \dots, v_n$ can be renumbered so that an arc (v_i, v_j) belongs to the tournament if and only if $1 \leq i < j \leq n$. Each tournament T_n possesses a canonical expansion into subtournaments $T^{(1)}, \dots, T^{(l)}$ such that

(a) there is an arc from each vertex of $T^{(i)}$ to each vertex of $T^{(j)}$ for $1 \leq i < j \leq l$,
(b) each subtournament $T^{(i)}$, $1 \leq i \leq l$, is either irreducible or transitive, and
(c) consecutive subtournaments $T^{(i)}$ and $T^{(i+1)}$ cannot be simultaneously transitive.

If T_n is irreducible or transitive, then $l = 1$.

The index of a nonnegative matrix A is defined as the least k for which the matrix A^k has the greatest number of positive elements. Let $k = \operatorname{ind} M$ and $k_i = \operatorname{ind} M_i$, $1 \leq i \leq l$, denote the indices of the matrix M of the tournament T_n and the matrices $M_1, \dots, M_l$ of the subtournaments $T^{(1)}, \dots, T^{(l)}$ in the canonical expansion.

THEOREM 6.3.5. *The indices of the matrix of the tournament itself and the matrices of subtournaments of the canonical expansion satisfy the inequality*

$$k \leq \max\{k_1, \dots, k_l\}.$$

PROOF. Without loss of generality we represent the matrix of a tournament T_n as

$$M = \begin{pmatrix} M_1 & & & 1 \\ & M_2 & & \\ & & \ddots & \\ 0 & & & M_l \end{pmatrix},$$

where the elements above and below the diagonal of blocks are equal to 1 and 0 respectively. It is obvious that the Boolean power of this matrix (when the operations on elements are over the Boolean algebra B_2; see Section 1.1) is of the form

$$M^t = \begin{pmatrix} M_1^t & & & 1 \\ & M_2^t & & \\ & & \ddots & \\ 0 & & & M_l^t \end{pmatrix}, \qquad t = 1, 2, \dots. \tag{6.3.1}$$

Let

$$J = \{j \colon T^{(j)} \text{ is transitive}\},$$

and let $\omega(A)$ be the number of nonzero elements of a matrix A. Then

$$\omega(M^t) = \sum_{j\in J} \omega(M_j^t) + \sum_{i\notin J} \omega(M_i^t) + C,$$

where C is the number of ones above the diagonal of the blocks.

If $m = \max\{k_1, \dots, k_l\}$, then $\omega(M_i^t) \leq \omega(M_i^m)$ for all $i \notin J$ if $t \geq 1$, since each subtournament $T^{(i)}$, $i \notin J$, is either a primitive tournament or a circuit of length three. Furthermore, for $j \in J$ we see that $\omega(M_j^{t+1}) \leq \omega(M_j^t)$ as $t \geq 1$. Therefore, $\omega(M_j^t) \leq \omega(M_j^m)$ for all $j \in J$ as $t \geq m$. Thus, $\omega(M^t) \leq \omega(M^m)$ for $t \geq m$. This means that the index of the matrix A does not exceed $\max\{k_1, \dots, k_l\}$. The theorem is thus proved. □

The results below concern the convergence index and the period of the matrix M of a tournament T_n. We recall that the convergence index $t = t(M)$ and the period $p = p(M)$ of the matrix M are defined as the least positive integers such that $M^{t+p} = M^t$. If M is primitive, then $t = \gamma$ and $p = 1$, where γ is the exponent of M. For M corresponding to a transitive tournament, $t = n$ and $p = 1$. For an irreducible tournament T_3, we have $t = 1$ and $p = 3$.

We consider the canonical expansion of a tournament T_n. Let the components of this expansion be $T^{(1)}, \dots, T^{(l)}$, and, as above, let J be the set of indices of the transitive components.

We denote by α the number of vertices in the maximal transitive component $T^{(j)}$, and by β the greatest index of the nontransitive subtournaments $T^{(i)}$. We set α and β equal to zero when they are not defined.

THEOREM 6.3.6. *If M is the matrix of a tournament T_n, then for the convergence index of M the bound*

$$t(M) = \max(\alpha, \beta) \leq n + 2 \tag{6.3.2}$$

holds, and the period is given by the formula

(6.3.3)

$$p(M) = \begin{cases} 3 & \text{if there is an irreducible subtournament } T^{(i)} \text{ with 3 vertices;} \\ 1 & \text{otherwise.} \end{cases}$$

PROOF. From (6.3.1) it follows that

$$t(M) = \max\{t(M_1), \dots, t(M_l)\}.$$

Furthermore,

$$\begin{aligned} \alpha &= \max\{t(M_j) : j \in J\}, \\ \beta &= \max\{k_i : i \notin J\} = \max\{t(M_i) : i \notin J\}. \end{aligned}$$

Hence the equality in (6.3.2) follows. It is obvious that the greatest value of $t(M)$ is attained in the case where there is a unique component in the expansion of T_n. In this case, the matrix M is either irreducible or transitive. In the former case, $t(M) = \gamma(M) \leq n + 2$ by Corollary 6.3.3, and in the latter case, $t(M) = n$. Hence the inequality in (6.3.2) follows.

Finally, note that for the transitive components $T^{(j)}$ and the irreducible components $T^{(i)}$ with at least four vertices the relation $p(M_j) = p(M_i) = 1$ is true. The period of an irreducible component with three vertices is equal to three. Hence formula (6.3.3) follows. □

6.4. Associated operator

6.4.1. Properties of the associated operator. Let $A = (a_{ij})$ be a nonnegative indecomposable matrix of order n, let $L \subseteq N_n$, and let $h \geq 0$ be an integer. We introduce the mapping $F_A^h\colon 2^{N_n} \to 2^{N_n}$ such that

$$F_A^h(L) = \bigcup_{i \in L} \{j\colon a_{ij}^{(h)} > 0\} \tag{6.4.1}$$

for $L \neq \emptyset$, where $A^h = (a_{ij}^{(h)})$, $i, j = 1, \ldots, n$, $h = 1, 2, \ldots$. We also set

$$F_A^h(\emptyset) = \emptyset, \quad F_A^0(L) = L, \quad F_A^1(L) = F_A(L), \tag{6.4.2}$$

$$F_A^h(i) = F_A^h(\{i\}). \tag{6.4.3}$$

The mapping F_A^h will be called the associated operator corresponding to the matrix A^h.

We consider the properties of the associated operator.

PROPERTY 6.4.1. *Let $\Gamma(A)$ be the digraph corresponding to a matrix $A \geq 0$. Then $F_A(L)$ is the set of vertices where the arcs emanating from the vertices of the set L end.*

Indeed, if $\tilde{A} = (\tilde{a}_{ij})$, $i, j = 1, \ldots, n$, is the support of a matrix A, then $\tilde{A}$ is the adjacency matrix of $\Gamma(A)$. It is clear that

$$F_A(L) = \bigcup_{i \in L} \{j\colon a_{ij} > 0\} = \bigcup_{i \in L} \{j\colon \tilde{a}_{ij} = 1\}, \tag{6.4.4}$$

which is the required result.

PROPERTY 6.4.2. *If $L \subseteq M \subseteq N_n$, then*

$$F_A(L) \subseteq F_A(M). \tag{6.4.5}$$

PROPERTY 6.4.3. *If $L, M \subseteq N_n$, then*

$$F_A(L \cup M) = F_A(L) \cup F_A(M). \tag{6.4.6}$$

Both properties follow immediately from (6.4.4).

PROPERTY 6.4.4. *If the product of matrices AB is defined and $L \subseteq N_n$, then*

$$F_{AB}(L) = F_B(F_A(L)). \tag{6.4.7}$$

Indeed, it is easy to see that

$$\begin{aligned} F_{AB}(L) &= \bigcup_{i \in L} \left\{ j\colon \sum_{k=0}^{n} a_{ik} b_{kj} > 0 \right\} \\ &= \bigcup_{k \in \bigcup_{i \in L} \{l\colon a_{il} > 0\}} \{j\colon b_{kj} > 0\} = \bigcup_{k \in F_A(L)} F_B(k), \end{aligned}$$

which proves (6.4.7)

PROPERTY 6.4.5. *For any integer $h \geq 0$ and any $L \subseteq N_n$,*

$$F_A^{h+1}(L) = F_A(F_A^h(L)) = F_A^h(F_A(L)). \tag{6.4.8}$$

From (6.4.1) it follows that

$$F_A^h(F_A(L)) = \bigcup_{i\in\bigcup_{k\in L}\{l:\ a_{kl}>0\}} \{j\colon a_{ij}^{(h)} > 0\}. \tag{6.4.9}$$

From (6.4.6) and (6.4.9) we obtain

$$F_A^h(F_A(L)) = \bigcup_{k\in L}\ \bigcup_{i\in\{l:\ a_{kl}>0\}} \{j\colon a_{ij}^{(h)} > 0\}.$$

Hence

$$F_A^h(F_A(L)) = \bigcup_{k\in L}\left\{j\colon \sum_{i,l:\ i=l} a_{kl}a_{ij}^{(h)} > 0\right\} = \bigcup_{k\in L}\{j\colon a_{kj}^{(h+1)} > 0\}.$$

Thus,

$$F_A^h(F_A(L)) = F_A^{h+1}(L).$$

The equality

$$F_A(F_A^h(L)) = F_A^{h+1}(L)$$

is proved similarly.

The property below immediately follows from (6.4.7) and (6.4.8).

PROPERTY 6.4.6. *For* $L \subseteq N_n$

$$F_A^h(L) = F_{A^h}(L). \tag{6.4.10}$$

PROPERTY 6.4.7. *For* $L \subseteq N_n$, *the set* $F_A^h(L)$ *is the set of vertices of the digraph* $\Gamma(A)$ *where the directed paths of length* h *emanating from the vertices of the set* L *end.*

In view of the equality

$$a_{ij}^{(h)} = \sum_{1\le i_1,i_2,\dots,i_{h-1}\le n} a_{ii_1}a_{i_1i_2}\cdots a_{i_{h-1}j}, \tag{6.4.11}$$

the inequality $a_{ij}^{(h)} > 0$ holds if and only if $\Gamma(A)$ contains a directed path of length h from the vertex i to the vertex j. Now the required result follows from (6.4.1).

PROPERTY 6.4.8. *If* $F_A^h(i) = N_n$, *then* $a_{i1}^{(h)}\cdots a_{in}^{(h)} > 0$; *that is, the ith row of the matrix* $A^h = (a_{ij}^{(h)})$, $i, j = 1,\dots,n$, *is positive.*

Indeed, in this case (6.4.1) implies that

$$F_A^h(i) = \{j\colon a_{ij}^{(h)} > 0\} = N_n, \tag{6.4.12}$$

which means that $a_{i1}^{(h)}\dots a_{in}^{(h)} > 0$.

PROPERTY 6.4.9. *A matrix* A *of order* n *is primitive if and only if there exists* $h > 0$ *such that*

$$F_A^h(L) = N_n \tag{6.4.13}$$

for all $L \subseteq N_n$.

In particular, (6.4.13) means that

$$F_A^h(1) = F_A^h(2) = \cdots = F_A^h(n) = N_n, \tag{6.4.14}$$

and by Property 6.4.8 all rows of A^h are positive, and therefore $A^h > 0$. Conversely, from the condition $A^h > 0$ it follows that (6.4.14) is true, which implies (6.4.13).

6.4.2. The exponent of a matrix. In what follows we consider the operator associated with the same nonnegative matrix A. Therefore we write $F_A = F$. In addition, in view of (6.4.10), the associated operator F^h corresponds to the matrix A^h.

LEMMA 6.4.1. *If a matrix $A = (a_{ij})$ of order $n \geq 2$ is indecomposable, then $F(i) \neq \emptyset$ for any $i \in N_n$.*

Indeed, it is easy to see that $F(i) = \{j\colon a_{ij} > 0\} \neq \emptyset$, $1 \leq i \leq n$, since an indecomposable matrix contains no zero rows.

LEMMA 6.4.2. *If A is an indecomposable matrix of order $n \geq 2$, $L \neq \emptyset$, and $L \neq N_n$, then $F(L)$ contains at least one element of N_n not belonging to L, and $F(N_n) = N_n$.*

PROOF. Suppose that $F(L) \subseteq L$. Then from Property 6.4.4 it follows that $F^h(L) \subseteq L$ for any positive integer h. Let $i \in L$, $j \notin L$. Then

$$j \notin F^h(L) = \bigcup_{i \in L} \{l\colon a_{il}^{(h)} > 0\}$$

for all $h = 1, 2, \ldots$. This means that the graph $\Gamma(A)$ is not strongly connected, which contradicts the indecomposability of A. If we assume that there exists an element $j \in N_n$ such that

$$j \notin F(N_n) = \bigcup_{i \in N_n} \{l\colon a_{il} > 0\},$$

then A has a zero column, which also contradicts its indecomposability. □

LEMMA 6.4.3. *If a matrix A of order n is indecomposable, then for all h, $0 \leq h \leq n-1$, the set $\{i\} \cup F(i) \cup \cdots \cup F^h(i)$ contains at least $h+1$ elements of N_n.*

PROOF. We prove the lemma by induction on h. For $h = 0$ the lemma reduces to the obvious inequality $|\{i\}| \geq 1$. Assume that the lemma is true for some h, $0 \leq h < n-1$. Then $L = \{i\} \cup F(i) \cup \cdots \cup F^h(i)$ contains at least $h+1$ elements of N_n. Two cases are possible.

We begin with the case $L = N_n$. In this case, $F(N_n) = N_n$ by Lemma 6.4.2, and therefore

$$F(L) = F(N_n) = N_n = F(i) \cup \cdots \cup F^{h+1}(i).$$

Hence

$$n = |\{i\} \cup F(i) \cup \cdots \cup F^{h+1}(i)| > h+1.$$

In the second case, where $L \neq N_n$, the set

$$F(L) = F(i) \cup \cdots \cup F^{h+1}(i)$$

contains, by Lemma 6.4.2, at least one element of N_n which does not belong to L. Since $i \in L$, the set

$$\{i\} \cup F(L) = \{i\} \cup F(i) \cup \cdots \cup F^{h+1}(i)$$

contains all elements of L and at least one more element of N_n which does not belong to L; that is, it contains at least $h+2$ elements. □

COROLLARY 6.4.4. *If $\{i\} \cup F(i) \cup \cdots \cup F^{h-1}(i)$ is a proper subset of N_n, then, joining $F^h(i)$ to this set, we add at least one more element. Thus, if*

$$\{i\} \cup F(i) \cup \cdots \cup F^h(i), \qquad h \le n-2,$$

contains exactly $h+1$ elements, then, consecutively joining $F(i), F^2(i), \ldots, F^h(i)$ to $\{i\}$, at each step we add exactly one more element.

COROLLARY 6.4.5. *For any $i \in N_n$,*

$$\{i\} \cup F(i) \cup \cdots \cup F^{n-1}(i) = N_n. \tag{6.4.15}$$

An indecomposable matrix $A = (a_{ij})$, $i, j = 1, \ldots, n$, is called deterministic if for each $i \in N_n$ there exists a unique j such that $a_{ij} > 0$. It is clear that a deterministic matrix is periodic with period $d = n$; that is, each subclass of the only essential class contains a single element.

PROPOSITION 6.4.6. *If an indecomposable matrix A is not deterministic, then for any $i \in N_n$ the set $F^h(i)$ contains at least two elements of N_n for some h, $1 \le h \le n$; that is, $|F^h(i)| \ge 2$.*

PROOF. Assume that each $F^h(i)$, $h = 0, 1, \ldots, n-1$, contains exactly one element. By Corollary 6.4.5, these elements are different, since

$$\{i\} \cup F(i) \cup \cdots \cup F^{n-1}(i) = N_n.$$

From this equality it follows that

$$F(i) \cup \cdots \cup F^{n-1}(i) \cup F^n(i) = N_n.$$

Thus, either $F^n(i) = \{i\}$, that is, the period d of the matrix is equal to n and the matrix is deterministic, which is impossible, or $F^n(i)$ contains at least two elements. □

As we have seen in Theorem 6.2.10, for the exponent of a primitive matrix A of order n the bound

$$\gamma \le n^2 - 2n + 2 \tag{6.4.16}$$

is true. This bound can be derived from the above properties of the associated operator. We can even derive sharper bounds. To do this, we need the following lemma.

LEMMA 6.4.7. *For any fixed $i \in N_n$, the set of nonnegative integers h such that*

$$\{i\} \subseteq F^h(i) \tag{6.4.17}$$

is a semigroup with respect to addition.

PROOF. The condition $\{i\} \subseteq F^h(i)$ means that $i \in F^h(i) = \{j : a_{ij}^{(h)} > 0\}$; that is, $a_{ii}^{(h)} > 0$. Assume that $\{i\} \subseteq F^t(i)$. This means that $a_{ii}^{(t)} > 0$. From the conditions $a_{ii}^{(h)} > 0$ and $a_{ii}^{(t)} > 0$ it follows that $a_{ii}^{(h+t)} > 0$, and therefore $\{i\} \subseteq F^{h+t}(i)$. □

LEMMA 6.4.8. *Let a matrix A of order n be indecomposable, and let there exist a nonnegative integer k such that the inclusion $\{i\} \subseteq F^l(i)$ holds for some $i \in N_n$ and for all $l \leq k$. Then*

$$F^{n+k-1}(i) = N_n. \tag{6.4.18}$$

PROOF. From the hypotheses of the lemma it follows that $\{i\} \subseteq F^{n_k-h-1}(i)$ for $0 \leq h \leq n-1$. Applying the mapping F^h to both sides of this inclusion, we arrive at $F^h(i) \subseteq F^{n+k-1}(i)$, $0 \leq h \leq n-1$. By Corollary 6.4.5,

$$\{i\} \cup F(i) \cup \cdots \cup F^{n-1}(i) = N_n.$$

Therefore, we have the inclusions

$$N_n = \{i\} \cup F(i) \cup \cdots \cup F^{n-1}(i) \subseteq F^{n+k-1}(i) \subseteq N_n,$$

which prove the lemma. □

THEOREM 6.4.9. *Let $A = (a_{ij})$, $i, j = 1, \dots, n$, be a primitive matrix, and let the matrix $A^h = (a_{ij}^{(h)})$ have at least $d > 0$ positive elements on the main diagonal for all $h \geq k$, where k is some nonnegative integer. Then for the exponent of A we have*

$$\gamma(A) \leq 2n - d + k - 1. \tag{6.4.19}$$

PROOF. Let $h \geq k$ and $a_{i_1 i_1}^{(h)} > 0, \dots, a_{i_d i_d}^{(h)} > 0$. By Lemma 6.4.3, the set $\{i\} \cup F(i) \cup \cdots \cup F^h(i)$, $0 \leq h \leq n-d$, contains at least $h+1$ distinct elements. From Corollary 6.4.4 it follows that $F^h(i)$ contains at least one additional element compared to the union $\{i\} \cup F(i) \cup \cdots \cup F^{h-1}(i)$. Hence at least one of the sets $F^h(i)$, $0 \leq h \leq n-d$, contains at least one of the elements $i_1, \dots, i_d$. For definiteness, let $\{i_\nu\} \subseteq F^h(i)$, $1 \leq \nu \leq d$. By Lemma 6.4.8,

$$F^{n+k-1}[F^h(i)] \supseteq F^{n+k-1}(i_\nu) = N_n \tag{6.4.20}$$

for $0 \leq k \leq h$. Using this relation, we find that

$$F^{2n-d+k-1}(i) = F^{n-d-h}\{F^{n+k-1}[F^h(i)]\} \supseteq F^{n-d-h}(N_n) = N_n.$$

Thus,

$$F^{2n-d+k-1}(i) = N_n \tag{6.4.21}$$

for any $i \in N_n$; that is,

$$\gamma(A) \leq 2n - d + k - 1.$$

□

COROLLARY 6.4.10. *If at least d elements on the main diagonal of a primitive matrix A are positive, then*

$$\gamma(A) \leq 2n - d - 1. \tag{6.4.22}$$

To obtain the assertion, it suffices to set $k = 0$ in Theorem 6.4.7.

COROLLARY 6.4.11. *If all elements on the main diagonal of a primitive matrix A are positive, then*

$$\gamma(A) \leq n - 1. \tag{6.4.23}$$

Corollary 6.4.11 follows from Corollary 6.4.10 with $d = n$.

THEOREM 6.4.12. *Let a matrix A be primitive, and for some $h \geq 1$ let the matrix $A+A^2+\cdots+A^h$ have at least $d>0$ positive elements on the main diagonal. Then the exponent of A satisfies*

$$\gamma(A) \leq n-d+h(n-1). \tag{6.4.24}$$

PROOF. Let $i_1, \ldots, i_d$ be the indices of the positive elements on the main diagonal of $A+A^2+\cdots+A^h$. Then for any i_ν, $1 \leq \nu \leq d$, there exists p, $1 \leq p \leq h$, such that $\{i_\nu\} \subseteq F^p(i_\nu)$. We apply Lemma 6.4.8 to A^p, setting $k=0$. Then $F_{A^p}^{n-1}(i_\nu)=N_n$. In view of Property 6.4.6, we obtain

$$F^{(n-1)p}(i_\nu)=F_{A^p}^{n-1}(i_\nu)=N_n. \tag{6.4.25}$$

Now we take an arbitrary element $i \in N_n$. By Lemma 6.4.3, $\{i\} \cup F(i) \cup \cdots \cup F^{n-d}(i)$ contains at least $n-d+1$ elements. Therefore, at least one of the sets $F^l(i)$, $0 \leq l \leq n-d$, contains at least one of the elements $i_1, \ldots, i_d$. Let $\{i_\nu\} \subseteq F^l(i)$, $1 \leq \nu \leq d$. Then

$$F^{(n-1)p}(F^l(i)) \supseteq F^{(n-1)p}(i_\nu)=N_n, \qquad 1 \leq p \leq h. \tag{6.4.26}$$

Thus,

$$\begin{aligned} F^{n-d+h(n-1)}(i) &= F^{n-d-l+(n-1)(h-p)}(F^{(n-1)p}(F^l(i))) \\ &\supseteq F^{n-d-l+(n-1)(h-p)}(N_n)=N_n, \end{aligned} \tag{6.4.27}$$

where $n-d-l+(h-p)(n-1) \geq 0$. Therefore,

$$\gamma(A) \leq n-d+h(n-1),$$

and the theorem is proved. □

A nonnegative matrix is called positively symmetric if the condition $a_{ij}>0$ implies that $a_{ji}>0$, $i,j=1,\ldots,n$.

LEMMA 6.4.13. *If $A=(a_{ij})$ is an indecomposable positively symmetric matrix, then $A^2=(a_{ij}^{(2)})$ is a primitive matrix, and all its elements on the main diagonal are positive.*

PROOF. If $A^2=(a_{ij}^{(2)})$, then

$$a_{ii}^{(2)}=\sum_{k=1}^{n} a_{ik}a_{ki}>0, \qquad 1 \leq i \leq n,$$

because by the indecomposability of A there exists k, $1 \leq k \leq n$, such that $a_{ik}>0$. Therefore, by the positive symmetricity of A we have $a_{ik}a_{ki}>0$. In view of the indecomposability of A^2, it follows that this matrix is primitive. □

THEOREM 6.4.14. *If a primitive matrix of order n is positively symmetric, then*

$$\gamma(A) \leq 2(n-1). \tag{6.4.28}$$

PROOF. By Lemma 6.4.13, all elements on the main diagonal of the matrix $A+A^2$ are positive. The required bound for the exponent now follows from Theorem 6.4.12 with $h=2$ and $d=n$. □

We introduce the row function $\delta(i)$ of the ith row of a primitive matrix $A = (a_{ij})$, $i, j = 1, \dots, n$, setting

$$\delta(i) = |\{j\colon a_{ij} > 0,\ i \neq j\}|, \qquad q \leq i \leq n. \tag{6.4.29}$$

Note that

$$\delta(i) = |F(i) \setminus \{i\}|. \tag{6.4.30}$$

By Corollary 6.4.4, this equality implies that $\delta(i) \geq 1$ for $n \geq 2$.

LEMMA 6.4.15. *If a matrix A of order n is indecomposable and $\delta(i)$ takes values δ_i, $1 \leq i \leq n$, then*

$$|\{i\} \cup F(i) \cup \dots \cup F^h(i)| \geq h + \delta_i \tag{6.4.31}$$

for $1 \leq h \leq n - \delta_i$.

PROOF. For $\delta_i = 1$, inequality (6.4.31) is proved in Lemma 6.4.3. For the case $\delta_i \geq 2$, we use induction on h. Since $\{i\} \cup F(i)| \geq 1 + \delta_i$, (6.4.31) holds for $h = 1$.

Now let h be an integer, $1 < h < n - \delta_i$. We set $L = \{i\} \cup F(i) \cup \dots \cup F^h(i)$. Assume that $|L| \geq h + \delta_i$. If $L = N_n$, then $n = |L| \geq h + 1 - \delta_i$. If $L \subset N_n$, then, in view of the inequality $h < n - 2$, using Corollary 6.4.4 we obtain

$$|L \cup F^{h+1}(i)| \geq |F(L)| \geq |L| + 1 \geq h + \delta_i + 1.$$

The lemma is proved. □

COROLLARY 6.4.16. *If $\{i\} \cup F(i) \cup \dots \cup F^{h-1}(i) \subset N_n$ and $\delta(i) = \delta_i$, $1 \leq i \leq n$, then*

$$|\{i\} \cup F(i) \cup \dots \cup F^{h-1}(i) \cup F^h(i)| \geq |\{i\} \cup F(i) \cup \dots \cup F^{h-1}(i)| + 1 \tag{6.4.32}$$

for $1 \leq h \leq n - \delta_i - 2$.

This inequality follows from Corollary 6.4.4.

COROLLARY 6.4.17. *For any $i \in N_n$, we have*

$$\{i\} \cup F(i) \cup \dots \cup F^{n-\delta_i}(i) = N_n, \tag{6.4.33}$$

where $\delta(i) = \delta_i$.

Using Lemma 6.4.15 with $h = n - \delta_i$, we obtain the inequality

$$|\{i\} \cup F(i) \cup \dots \cup F^{n-\delta_i}(i)| \geq n, \tag{6.4.34}$$

which proves the corollary.

LEMMA 6.4.18. *Let a matrix A of order n be indecomposable, and let there exist a nonnegative integer k such that the inclusion $\{i\} \subseteq F^h(i)$ holds for some $i \in N_n$ and for all $h \geq k$. Then*

$$F^{n+k-\delta_i}(i) = N_n, \tag{6.4.35}$$

where $\delta(i) = \delta_i$.

PROOF. For $0 \leq h \leq n - \delta_i$, from the hypotheses of the lemma it follows that $\{i\} \subseteq F^{n+k-h-\delta_i}(i)$. We apply the mapping F^h to both sides of this inclusion, and obtain

$$F^h(i) \subseteq F^{n+k-\delta_i}(i), \qquad 1 \leq h \leq n - \delta_i. \tag{6.4.36}$$

From this inclusion, using Corollary 6.4.17, we find that

$$N_n = \{i\} \cup F(i) \cup \cdots \cup F^h(i) \cup \cdots \cup F^{n-\delta_i}(i) \subseteq F^{n+k-\delta_i}(i). \tag{6.4.37}$$

Thus, the lemma is proved. □

THEOREM 6.4.19. *Let $A = (a_{ij})$ be a primitive matrix of order n, let k be a positive integer, and for $h \geq k$ and the matrix $A^h = (a_{ij}^{(h)})$ let there exist a set $J = \{i_1, \dots, i_d\} \subseteq N_n$, $1 \leq d \leq n$, such that*

$$a_{i_1 i_1}^{(h)} > 0,\ a_{i_2 i_2}^{(h)} > 0, \dots, a_{i_d i_d}^{(h)} > 0. \tag{6.4.38}$$

If

$$\Delta_1 = \min_{i\in J} \delta(i), \quad \delta = \min_{i \in N_n \setminus J} \delta(i), \quad \Delta_2 = \min(n-d, \delta - 1),$$

then

$$\gamma(A) \leq 2n - d + k - \Delta_1 - \Delta_2. \tag{6.4.39}$$

PROOF. If $i \in J$, then $F^{n+k-\delta(i)}(i) = N_n$ by Lemma 6.4.18. Let $i \in N_n \setminus J$ and $\delta(i) - 1 > n - d$. Then the inclusion $\{i_\nu\} \subseteq F(i)$ holds for some ν, $1 \leq \nu \leq d$, since $\delta(i) > n - d + 1$. Hence, in view of Lemma 6.4.18,

$$F^{n+k-\delta(i_\nu)}(i) \supseteq F^{n+k-\delta(i_\nu)-1}(i_\nu) = N_n,$$

since the values of δ do not exceed $n-1$.

Now assume that $i \in N_n \setminus J$ and $n - d \geq \delta(i) - 1$. Then by Lemma 6.4.15

$$|\{i\} \cup F(i) \cup \cdots \cup F^{n-d-\delta(i)+1}(i)| \geq n - d + 1. \tag{6.4.40}$$

Hence for any $i \in N_n$ there exists $i_\nu \in J$, $1 \leq \nu \leq d$, such that

$$\{i_\nu\} \subseteq F^p(i), \qquad 0 \leq p \leq n - d - \delta(i) + 1. \tag{6.4.41}$$

Therefore, according to Lemma 6.4.18, we have

$$F^{n+k-\delta(i_\nu)+p}(i) \supseteq F^{n+k-\delta(i_\nu)}(i_\nu) = N_n. \tag{6.4.42}$$

Hence

$$F^{2n-d+k-\delta(i)-\delta(i_\nu)+1}(i) = F^{n-d-\delta(i)-p+1}(F^{n+k-\delta(i_\nu)+p}(i)) = N_n \tag{6.4.43}$$

for all $i \in N_n \setminus J$, provided that $n - d \geq \delta(i) - 1$. Thus, we obtain the following bound for the exponent:

$$\begin{aligned} \gamma(A) &\leq \max[n - k - \Delta_1, 2n - d + k - \Delta_1 - (\delta - 1)] \\ &= 2n - d + k - \Delta_1 - \min(n - d, \delta - 1). \end{aligned} \tag{6.4.44}$$

The theorem is proved. □

6.4.3. The index of a nonnegative matrix. We consider powers of a nonnegative matrix $A = (a_{ij})$ of order n:

$$A,\ A^2, \dots, A^k, \dots. \tag{6.4.45}$$

We recall that the least k for which the matrix A^k has the maximum number of positive elements in comparison with the other powers of this matrix is called the index of the matrix A and is denoted by $\operatorname{ind} A$.

Let $A_1^k, \dots, A_n^k$ be the rows of the nonnegative matrix A^k, $k = 1, 2, \dots$. The least k for which the row A_i^k has the greatest number of positive elements is called the index of the ith row of A, and is denoted by $\operatorname{ind} A_i$, $1 \leq i \leq n$. Let $W(A^k)$ and

$W(A_i^k)$ be the number of positive elements of the whole matrix A and of the row A_i respectively. Then

$$\operatorname{ind} A = \min_{k>0}\{k\colon W(A^k) = \max_{m>0} W(A^m)\}, \tag{6.4.46}$$

$$\operatorname{ind} A_i = \min_{k>0}\{k\colon W(A_i^k) = \max_{m>0} W(A_i^m)\}, \qquad i = 1, \dots, n. \tag{6.4.47}$$

If a matrix A is primitive, then the maximum values of $W(A^k)$ and $W(A_i^k)$ are equal to n^2 and n, respectively, and in this case

$$\operatorname{ind} A = \gamma(A), \tag{6.4.48}$$

where $\gamma(A)$ is the exponent of A.

We consider the associated operator $F = F_A$. By definition,

$$W(A_i^k) = |F^k(i)|, \tag{6.4.49}$$

$$W(A^k) = \sum_{i=1}^{n} |F^k(i)|. \tag{6.4.50}$$

For $L \subseteq N_n$, we set

$$\operatorname{ind} F(L) = \min_{k>0}\{k\colon |F^k(L)| = \max_{m>0} |F^m(L)|\}. \tag{6.4.51}$$

In particular, we set

$$\operatorname{ind} F(i) = \operatorname{ind} F(\{i\}), \qquad 1 \le i \le n, \tag{6.4.52}$$

and define the index of the associated operator as

$$\operatorname{ind} F = \min_{k>0}\left\{k\colon \sum_{i=1}^{n} |F^k(i)| = \max_{m>0} \sum_{i=1}^{n} |F^m(i)|\right\}. \tag{6.4.53}$$

From (6.4.46) and (6.4.47), in view of (6.4.49) and (6.4.50), and from (6.4.51), (6.4.52), (6.4.53) it follows that

$$\operatorname{ind} A = \operatorname{ind} F, \tag{6.4.54}$$

$$\operatorname{ind} A_i = \operatorname{ind} F(i), \qquad 1 \le i \le n. \tag{6.4.55}$$

We say that a set $L \subseteq N_n$ possesses the absorption property with respect to the operator F if there exists $k > 0$ such that

$$L \subseteq F^k(L). \tag{6.4.56}$$

The least k satisfying condition (6.4.56) is called the absorption index, and is denoted by $d = d(L)$. Since the associated operator preserves inclusion of sets, condition (6.4.56) implies the chain of inclusions

$$L \subseteq F^d(L) \subseteq F^{2d}(L) \subseteq \cdots \subseteq F^{kd}(L) \subseteq F^{(k+1)d}(L) \subseteq \dots. \tag{6.4.57}$$

The associated operator $F = F_A$ is said to be nonsingular if the inequality

$$|L| \le |F(L)| \tag{6.4.58}$$

holds for any subset $L \subseteq N_n$. If F is nonsingular, then $|F^k(L)|$ with fixed L is a monotone nondecreasing function of k.

LEMMA 6.4.20. *If the operator F is nonsingular, then any set $L \subseteq N_n$ possesses the absorption property with respect to F.*

PROOF. Since F preserves unions of sets (Property 6.4.3), it suffices to demonstrate that the inclusion

$$i \in \bigcup_{k=1}^{n} F^k(i) \tag{6.4.59}$$

holds for any $i \in N_n$. Introducing

$$T = \bigcup_{k=1}^{n} F^k(i),$$

we find that $F(T \cup \{i\}) = T$. Hence

$$|T \cup \{i\}| \leq |T|, \tag{6.4.60}$$

since F is nonsingular. On the other hand, $T \subseteq T \cup \{i\}$; that is,

$$|T| \leq |T \cup \{i\}|. \tag{6.4.61}$$

By (6.4.60) and (6.4.61), $T = T \cup \{i\}$, and therefore, $i \in T$. □

For a nonsingular operator F, the inclusions (6.4.58) hold for all $L \subseteq N_n$ in view of Lemma 6.4.20, where d is the absorption index of F. Since there are only a finite number of subsets of N_n, there exists a k such that

$$F^{kd}(L) = F^{(k+1)d}(L). \tag{6.4.62}$$

The least k which satisfies (6.4.62) is called the saturation index, and is denoted by $b = b(L)$.

LEMMA 6.4.21. *If the operator F is nonsingular, then*

$$(b-1)d < \operatorname{ind} F(L) \leq bd \tag{6.4.63}$$

for all $L \subseteq N_n$, where $b = b(L)$, $d = d(L)$.

PROOF. From (6.4.62) and the definition of $b = b(L)$ it follows that

$$F^{bd}(L) = F^{(b+m)d}(L), \qquad m = 0, 1, \ldots. \tag{6.4.64}$$

Thus,

$$|F^{bd}(L)| = |F^{(b+1)d}(L)| = \cdots = |F^{(b+m)d}(L)| = \ldots.$$

Since $|F^k(L)|$ is a monotone nondecreasing function of k, we obtain

$$\begin{aligned} |F^k(L)| &\leq |F^{bd}(L)|, \qquad 1 \leq k < bd, \\ |F^k(L)| &= |F^{bd}(L)|, \qquad k \geq bd. \end{aligned} \tag{6.4.65}$$

Hence, in view of (6.4.51), it follows that

$$\operatorname{ind} F(L) \leq bd.$$

The inequality

$$(b-1)d < \operatorname{ind} F(L)$$

follows from the definition of $b = b(L)$ as the least k for which (6.4.62) holds. □

LEMMA 6.4.22. *If $X_1, X_2, \dots, X_m, \dots$ are finite nonempty sets, then either*

$$\left|\bigcup_{i=1}^{\nu+1} X_i\right| \geq |X_1| + \nu \tag{6.4.66}$$

for all $\nu \leq m$, or

$$X_{\nu+1} \subseteq \bigcup_{i=1}^{\nu} X_i \tag{6.4.67}$$

for some $\nu \leq m$.

PROOF. The first case means that

$$X_i \nsubseteq \bigcup_{\mu=1}^{i-1} X_\mu, \qquad 1 \leq i \leq \nu + 1.$$

The second case corresponds to the existence of i, $1 \leq i \leq \nu + 1$, such that

$$X_i \subseteq \bigcup_{\mu=1}^{i-1} X_\mu.$$

□

Let $L \subseteq N_n$, and let F be a nonsingular operator. We set

$$X_\nu = F^{(\nu-1)d}(L), \qquad \nu = 1, 2 \dots.$$

From (6.4.57) it follows that $X_1 \subseteq X_2 \subseteq \cdots \subseteq X_m \subseteq \dots$. We thus obtain

$$X_m = \bigcup_{i=1}^{m} X_i, \tag{6.4.68}$$

and therefore,

$$X_1 \nsubseteq \bigcup_{\mu=1}^{i-1} X_\mu, \qquad 1 \leq i \leq b + 1.$$

Now, setting $m = b + 1$, from (6.4.66) and (6.4.68) we obtain

$$\max_{k>0} |F^k(L)| = |F^{bd}(L)| \geq |L| + b(L). \tag{6.4.69}$$

We set $L = \{i\}$, $F(i) = F(\{i\})$, and consider the sequence

$$F(i),\ F^2(i), \dots, F^{d(i)}(i), \dots, \tag{6.4.70}$$

where $d(i)$ is the absorption index of $\{i\}$. From the definition of $d(i)$ and (6.4.56) it follows that

$$F^{(d(i)}(i) \nsubseteq \bigcup_{\mu=1}^{d(i)-1} F^\mu(i). \tag{6.4.71}$$

Applying Lemma 6.4.22 with $X_m = F^m(i)$ to the sequence (6.4.70) and taking (6.4.71) into account, from (6.4.66) we obtain

$$\left|\bigcup_{\mu=1}^{d(i)} F^\mu(i)\right| \geq |F(i)| + d(i) - 1. \tag{6.4.72}$$

From (6.4.69) and (6.4.72) it follows that

$$b(i) \le \max_{k>0} |F^k(i)| - 1, \tag{6.4.73}$$

$$d(i) \le \left| \bigcup_{\mu=1}^{d(i)} F^\mu(i) \right| - |F(i)| + 1. \tag{6.4.74}$$

THEOREM 6.4.23. *Let a matrix $A \ge 0$ of order n be given. If the associated operator $F = F_A$ is nonsingular, then*

$$\operatorname{ind} A_i = \operatorname{ind} F(i) \le n(n-1), \qquad 1 \le i \le n, \tag{6.4.75}$$

where $A_1, \ldots, A_n$ are the rows of A.

PROOF. Inequalities (6.4.63) yield

$$\operatorname{ind} F(i) \le b(i)d(i). \tag{6.4.76}$$

With the use of the obvious bounds

$$\max_{k>0} |F^k(i)| \le n, \quad \left| \bigcup_{\mu=1}^{d(i)} F^\mu(i) \right| \le n, \quad |F(i)| \ge 1,$$

from (6.4.73), (6.4.74), and (6.4.76) we obtain (6.4.75). □

In the following theorem, we give a criterion for nonsingularity of the associated operator $F = F_A$.

THEOREM 6.4.24. *Let a matrix $A = (a_{ij})$, $i, j = 1, \ldots, n$, be given. The associated operator $F = F_A$ is nonsingular if and only if* $\operatorname{per} A > 0$.

PROOF. Let the operator F be nonsingular. Then $|F(L)| \le |L|$ for any set $L \subseteq N_n$. Therefore, if $L = \{i_1, \ldots, i_k\}$, then in view of the relation

$$F(L) = \bigcup_{i \in L} F(i) \tag{6.4.77}$$

we have

$$|F(i_1) \cup F(i_2) \cup \cdots \cup F(i_k)| \ge k. \tag{6.4.78}$$

By Theorem 1.2.7, this means that the system of sets $F(1), \ldots, F(n)$ has some transversal $(r_1, \ldots, r_n)$. We define the permutation s of degree n by the equalities $s(i) = r_i$, $1 \le i \le n$. Since $F(i) = \{j \colon a_{ij} > 0\}$, we see that $a_{ir_i} > 0$, $1 \le i \le n$. Hence

$$\operatorname{per} A \ge a_{1r_1} \ldots a_{nr_n} > 0.$$

Now let $\operatorname{per} A > 0$. Then the system of sets $F(1), \ldots, F(n)$ has a transversal and condition (6.4.78) is satisfied (see Section 1.2). In view of (6.4.77), from (6.4.78) it follows that $|F(L)| \le |L|$; that is, F is nonsingular. □

6.5. Sequences of powers of a nonnegative matrix

The study of powers of a nonnegative matrix can usually be reduced to the study of powers of the corresponding supports. Therefore, we begin by considering the properties of sequences of powers of (0, 1)-matrices.

6.5.1. Sequences of powers of a $(0,1)$-matrix. We agree that, while multiplying $(0,1)$-matrices, the multiplication and addition of elements are over the Boolean algebra $B_2 = \{0,1\}$ (see Section 1.1.12). For $n \geq 2$, we consider the infinite sequence $\{A^k,\ k = 1,2,\dots\}$ of powers of a $(0,1)$-matrix $A = (a_{ij})$ of order n. Such a sequence is called convergent if there exists a positive integer t such that

$$A^{t+1} = A^t. \tag{6.5.1}$$

If $A^t = (a_{ij}^{(t)})$, then condition (6.5.1) means that $a_{ij}^{(t)} = a_{ij}^{(t+\nu)}$, $\nu = 0,1,\dots$, for all $1 \leq i,j \leq n$.

The sequence $\{A^k,\ k = 1,2,\dots\}$ is called periodic, or oscillating, if there exist positive integers $t, p > 1$ such that

$$A^{t+p} = A^t, \tag{6.5.2}$$

where p is the least number satisfying (6.5.2). The least number t in (6.5.2) is called the convergence index, and is denoted by $\gamma^{(p)}$, whereas p is the period of A. We agree that for a convergent sequence the period is equal to one. Equality (6.5.2) means that $a_{ij}^{(t)} = a_{ij}^{(t+p)}$ for all i, j, $1 \leq i,j \leq n$, and $t = \gamma^{(p)}, \gamma^{(p)} + 1, \dots$.

We note that the period p and the convergence index $\gamma^{(p)}$ always exist, because the number of $(0,1)$-matrices of order n is finite. For a convergent sequence $\{A^k,\ k = 1,2,\dots\}$, all elements of the matrix A^k have limits as $k \to \infty$. For a periodic sequence, some elements oscillate, whereas others may tend to limit values. Each matrix $\tilde{A}$ which occurs infinitely many times in the sequence $\{A^k,\ k = 1,2,\dots\}$ is called a limiting matrix of this sequence. If such a matrix $\tilde{A}$ is unique, it is called the limit of this sequence.

Since equalities (6.5.1) and (6.5.2) exhaust the possible cases, any sequence of powers of a $(0,1)$-matrix either has a limit or is periodic. The limit of a convergent sequences may be either the zero matrix $\mathbf{0}_{nn}$, or the matrix $\mathbf{J}_{nn}$ consisting of ones only, or an idempotent matrix $\tilde{A}$ such that $\mathbf{0}_{nn} < \tilde{A} < \mathbf{J}_{nn}$.

Let $\Gamma(A)$ be the directed graph whose adjacency matrix coincides with the $(0,1)$-matrix $A = (a_{ij})$ of order n. We assume that A contains a strong component H with $m \geq 1$ vertices. A strong component H is said to be q-universal if there exists a path of length q to each vertex of H from some vertex i.

LEMMA 6.5.1. *In a q-universal strong component H with m vertices, there exists a path of length q^*, $q^* \geq q + m$, from each vertex to any other vertex*

PROOF. We use the normal form (6.1.1) of the matrix A (see Section 6.1) and assume that the adjacency matrix A_{kk} of order $n_k = m$, $1 \leq k \leq t$, corresponds to the q-universal component H. In view of the rules for operations on block matrices, we obtain

$$\Pi A^{q+n_k} \Pi^T = \begin{pmatrix} A_{11}^{q+n_k} & & & & & 0 \\ B_{21} & A_{22}^{q+n_k} & & & & \\ & & \ddots & & & \\ B_{k1} & \dots & B_{k,k-1} & A_{kk}^{q+n_k} & & \\ & & & & \ddots & \\ B_{t1} & \dots & & & B_{t,t-1} & A_{tt}^{q+n_k} \end{pmatrix}.$$

Let $A_{kk}^{\nu} = (a_{ij}^{(\nu)})$, $i,j = 1,\dots,n_k$, $1 \leq k \leq t$, $\nu = 1,2,\dots$.

If in the q-universal component H there is a path to any vertex from a given vertex i, then the ith row of the matrices A_{kk}^{ν}, $\nu = q, q+1, \dots$, is positive. Let r and s be arbitrary vertices of the strong component H. In order to prove the lemma, it suffices to show that $a_{rs}^{(q+n_k)} = 1$.

If there is a chain of length μ, $1 \le \mu \le n_k$, from the vertex r to the vertex i, then $a_{ri}^{(\mu)} = 1$. Furthermore, $a_{is}^{(q+n_k-\mu)} = 1$, since the ith row of the matrix $A_{kk}^{q+n_k-\mu}$ is positive. Therefore,

$$a_{rs}^{(q+n_k)} = \sum_{u=1}^{n_k} a_{ri}^{(\mu)} a_{is}^{(q+n_k-\mu)} \ge a_{ri}^{(\mu)} a_{is}^{(q+n_k-\mu)} = 1.$$

The lemma is proved. □

The structure of a q-universal strong component is described in the following lemma.

LEMMA 6.5.2. *Let a q-universal strong component H contain m vertices, $m \ge 2$, and r simple circuits of lengths $l_1, \dots, l_r$. Then all circuits passing through a vertex i and belonging to H are of length $l = l_1x_1 + \cdots + l_rx_r$, where $x_1, \dots, x_r$ are nonnegative integers.*

In fact, the proof of this lemma coincides with that of Lemma 6.2.5, so we omit it.

THEOREM 6.5.3. *A strong component H with h vertices, $h \ge 2$, in the digraph $\Gamma(A)$ is q-universal if and only if the greatest common divisor of the lengths of all simple circuits in H is equal to one.*

PROOF. Let a strong component H contain simple circuits of lengths $l_1, \dots, l_r$, and let $\gcd(l_1, \dots, l_r) = 1$. By Lemma 6.5.2, there exists a circuit of length $l = l_1x_1 + \cdots + l_rx_r$, passing through a vertex i in H, where $x_1, \dots, x_r$ are nonnegative integers. From the condition $\gcd(l_1, \dots, l_r) = 1$ it follows that for the set $Y = N_r$ there exist a partition $Y = Y_1 \cup Y_2$, $Y_1 \cap Y_2 = \emptyset$, and nonnegative integers $y_1, \dots, y_r$ such that

$$\sum_{k \in Y_1} l_k y_k - \sum_{k \in Y_2} l_k y_k = 1.$$

We introduce the numbers

$$z_{\nu k} = \begin{cases} (h - \nu + 1)y_k, & k \in Y_1, \\ (\nu - 1)y_k, & k \in Y_2, \end{cases} \qquad \nu = 1, \dots, h, \quad k = 1, \dots, r.$$

It is clear that

$$\nu + \sum_{k=1}^{r} l_k z_{\nu k} = h \sum_{k \in Y_1} l_k y_k + 1, \qquad \nu = 1, \dots, h.$$

Hence for any vertex j such that there exists a path of length ν, $\nu = 1, \dots, h$, in H from the vertex i to j, there also exists a path from i to j of length

$$q = l + h \sum_{k \in Y_1} l_k y_k + 1.$$

Therefore, the strong component H is q-universal.

Let us prove the converse statement. If a strong component H is q-universal, then there exist circuits passing through a chosen vertex i of lengths

$$L_1 = \sum_{k=1}^{r} l_k x_k, \qquad L_2 = \sum_{k=1}^{r} l_k y_k,$$

where x_k, y_k, $1 \le k \le r$, are nonnegative integers.

By Lemma 6.5.1, we can take

$$L_1 = q^*, \quad L_2 = q^* + 1, \quad q^* \ge q + h.$$

Then

$$\sum_{k=1}^{r} l_k y_k - \sum_{k=1}^{r} l_k x_k = 1$$

This relation is possible only when $\gcd(l_1, \dots, l_r) = 1$. □

COROLLARY 6.5.4. *If a strong component H contains at least two simple circuits whose lengths are relatively prime, then this component is q-universal.*

Now we present necessary and sufficient conditions for convergence of the powers of a $(0,1)$-matrix to the corresponding limits, and conditions for oscillation. We consider $(0,1)$-matrices A of order $n \ge 2$.

THEOREM 6.5.5. *The sequence $\{A^k,\ k = 1, 2, \dots\}$ converges to the zero matrix $\mathbf{0}_{nn}$ if and only if the digraph $\Gamma(A)$ contains no strong components.*

PROOF. If $\Gamma(A)$ contains no strong components, then from an arbitrarily chosen vertex i we arrive in at most $n-1$ steps at a vertex j from which none of the vertices of $\Gamma(A)$ are attainable. Therefore, $A^k = \mathbf{0}_{nn}$ for all integers $k \ge n-1$.

Now let $A^q = \mathbf{0}_{nn}$ for some positive integer q. Hence for any i and j the digraph $\Gamma(A)$ contains no chains of length at least q from i to j. This means that $\Gamma(A)$ has no strong components. □

THEOREM 6.5.6. *The sequence $\{A^k,\ k = 1, 2, \dots\}$ converges to the matrix $\mathbf{J}_{nn}$ whose elements are all ones if and only if the digraph $\Gamma(A)$ is strongly connected and q-universal.*

PROOF. If the digraph $\Gamma(A)$ is strongly connected and q-universal, then, by Lemma 6.5.1, an arbitrary vertex is attainable from any other vertex in at most q^* steps, $q^* \ge q+n$. Hence $a^{q^*} = \mathbf{J}_{nn}$, and therefore $A^\nu = \mathbf{J}_{nn}$, $\nu = q^*, q^*+1, \dots$. On the other hand, if the matrix $\mathbf{J}_{nn}$ is the limit of the sequence $\{A^k,\ k = 1, 2, \dots\}$, then there exists a positive integer q such that $A^{q+\nu} = \mathbf{J}_{nn}$, $\nu = 0, 1, \dots$. This means that there exists a path from an arbitrary vertex to another arbitrary vertex of length q; that is, $\Gamma(A)$ is strongly connected and q-universal. □

THEOREM 6.5.7. *The sequence $\{A^k,\ k = 1, 2, \dots\}$ converges to an idempotent matrix $\tilde{A}$, $\mathbf{0}_{nn} < \tilde{A} < \mathbf{J}_{nn}$, if and only if the digraph $\Gamma(A)$ is not strongly connected but contains at least one strong component, and each strong component is q-universal.*

PROOF. Let $H_1, \dots, H_r$ be the strong components of the digraph $\Gamma(A)$ whose vertices form the sets $V_1, \dots, V_r$, and let matrices $A_1, \dots, A_r$ of orders $n_1, \dots, n_r$

be the adjacency matrices of the components. Since the strong components are q-universal, there exist $q_1^*, \ldots, q_r^*$ such that for all $q^* \geq \max(q_1^*, \ldots, q_r^*)$ the equalities

$$A_1^{q^*} = \mathbf{J}_{n_1 n_1}, \ldots, A_r^{q^*} = \mathbf{J}_{n_r n_r}$$

hold. Starting from an arbitrary vertex i of the digraph $\Gamma(A)$, in at most $n-1$ steps we arrive at either a vertex j belonging to a strong component, or at a vertex k from which none of the vertices of $\Gamma(A)$ are attainable. Hence $a_{ij}^{(\nu)} = 1$, $a_{ik}^{(\mu)} = 0$ for $\mu, \nu \geq q^* + n - 1$, and the limit of the sequence of the matrices is of the form mentioned in the theorem.

The necessity is proved by the reversing the above arguments. □

THEOREM 6.5.8. *The sequence* $\{A^k,\ k = 1, 2, \ldots\}$ *oscillates if and only if the digraph contains at least one strong component which is not* q*-universal. The period of the submatrix corresponding to this strong component is equal to the greatest common divisor of the lengths of all simple circuits contained in this component. The oscillation period of the whole matrix is equal to the least common multiple of the periods of those strong components which are not* q*-universal.*

PROOF. Let H be a strong component which is not q-universal.

By Lemma 6.5.2, all circuits in H passing through some vertex i are of length $l = l_1 x_1 + \cdots + l_r x_r$, where $l_1, \ldots, l_r$ are the lengths of the simple circuits of H and $x_1, \ldots, x_r$ are nonnegative integers. If $A^l = (a_{ij}^{(l)})$, $i, j = 1, \ldots, n$, then $a_{ii}^{(l)} = 1$. Assume that the sequence $\{A^k,\ k = 1, 2, \ldots\}$ has a limit. Then there exists a q such that $a_{ii}^{(q+\nu)} = 1$, $\nu = 0, 1, \ldots$. Therefore, there exist circuits of lengths

$$L_1 = \sum_{i=1}^{r} l_i x_i, \qquad L_2 = \sum_{i=1}^{r} l_i y_i,$$

where x_k, y_k, $1 \leq k \leq r$, are nonnegative integers such that $L_1 = q$, $L_2 = q + 1$. Then, as follows from the proof of Theorem 6.5.3, $\gcd(l_1, \ldots, l_r) = 1$, and therefore, H is a q-universal component. This contradiction means that the equality $a_{ii}^{(q)} = 1$ implies $a_{ii}^{(q+1)} = 0$ for all q.

Similarly we prove that for arbitrary distinct i and j the relation $a_{ij}^{(q)} = 1$ implies $a_{ij}^{(q+1)} = 0$. Thus, the sequence $\{A^k,\ k = 1, 2, \ldots\}$ oscillates. For the simple circuits of lengths $l_1, \ldots, l_r$ of the strong component H, we introduce the notation $p(H) = \gcd(l_1, \ldots, l_r)$. It is clear that the least oscillation period is equal to $p(H)$. Let p be the least common multiple of the oscillation periods $P(H)$ of all strong components H which are not q-universal, and let $\gamma = \gamma(p)$ be the convergence index of the matrix A. It is clear that $A^{\gamma+p} = A^{\gamma}$ for all $\nu \geq p$, and that p is the least number possessing this property. □

COROLLARY 6.5.9. *For a* $(0,1)$*-matrix* A *of order* $n \geq 2$ *the sequence* $\{A^k,\ k = 1, 2, \ldots\}$ *converges to a nonzero matrix if and only if the digraph* $\Gamma(A)$ *contains at least one strong component and for each strong component of* $\Gamma(A)$ *the greatest common divisor of the lengths of its simple circuits is equal to one.*

COROLLARY 6.5.10. *If the digraph* $\Gamma(A)$ *contains at least one strong component, then the sequence of powers of each submatrix of the* $(0,1)$*-matrix* A *corresponding to a strong component either oscillates or converges to the matrix whose elements are all ones.*

THEOREM 6.5.11. *If the digraph* $\Gamma(A)$ *is strongly connected and the sequence* $\{A^k,\ k = 1, 2, \dots\}$ *has period* p, *then there exists a unique idempotent limiting matrix* B_p *whose digraph* $\Gamma(B_p)$ *consists of* p *strongly connected components.*

PROOF. In the case $p = 1$, by Theorem 6.5.6 there exists the limiting matrix $\mathbf{J}_{nn}$ with all elements equal to one.

In the case $p > 1$, argueing as in the proof of Theorem 6.5.3, we see that, starting from an arbitrary vertex of $\Gamma(A)$, one can return to it in a number of steps divisible by p. Therefore, there exists a positive integer ν such that the matrix $A^{p\nu}$ contains an identity submatrix $\mathbf{I}_n$. From Theorem 6.5.3 it follows that the sequence $\{A^{pk},\ k = 1, 2 \dots\}$ is convergent.

If B_p is a limiting matrix for this sequence, then it is idempotent, and by Theorem 6.5.7 all strong components of $\Gamma(B_p)$ are q-universal. It remains to show that $\Gamma(B_p)$ consists of p strong components.

Let $B_1, \dots, B_p$ be the limiting matrices of the sequence $\{A^k,\ k = 1, 2, \dots\}$. Then $B_\mu = B_p A^\mu$, $\mu = 1, \dots, p$, and the elements $B_1, \dots, B_p$ are distinct and form a cyclic group, where B_p plays the part of the identity.

Arguing as in the proof of Theorem 6.5.3, we see that for any pair i, j of vertices of $\Gamma(A)$ the lengths of all paths from i to j belong to the same residue class modulo p. Therefore, a vertex i is joined with a vertex j of $\Gamma(B_\mu)$ if and only if the length l of the corresponding path satisfies the condition $l \equiv \mu \pmod p$, $1 \le \mu \le p$. In $\Gamma(B_p)$, each vertex i is joined with itself; therefore, if in $\Gamma(A)$ a vertex i transits to j in $l \equiv \mu \pmod p$ steps, then j transits to i in $l' \equiv p - \mu \pmod p$ steps. Hence if in $\Gamma(B_p)$ a vertex i transits to j, then j transits to i as well. Let k be a vertex in $\Gamma(A)$ differing from i and j, and let k transit to i in $\Gamma(B_\mu)$ and to j in $\Gamma(B_\nu)$, $1 \le \mu, \nu \le p$. Then i transits to j in $\Gamma(B_p)$ if and only if $p - \mu + \nu \equiv p \pmod p$; that is, $\mu = \nu$. Hence any two vertices i, j of the digraph $\Gamma(A)$ belong to the same strong component of $\Gamma(B_p)$ if and only if there exist paths from some vertex k to these vertices of lengths l and l', respectively, where $l, l' \equiv \nu \pmod p$, $1 \le \nu \le p$. But from a vertex k there is a path to any vertex of $\Gamma(A)$ of length $l \equiv \nu \pmod p$, $1 \le \nu \le p$. Hence $\Gamma(B_p)$ contains exactly p strong components. □

To study the powers of stochastic matrices, we need the following lemma.

LEMMA 6.5.12. *If in the digraph* $\Gamma(A)$ *of a* $(0, 1)$*-matrix* A *each vertex is joined with at least one vertex of* $\Gamma(A)$, *then there exists at least one strong component in* $\Gamma(A)$, *and each vertex of* $\Gamma(A)$ *is joined with at least one strong component.*

PROOF. Let i_1 be an arbitrary vertex of the digraph $\Gamma(A)$. The hypotheses of the lemma imply the existence of a chain $i_1 \to i_2 \to \cdots \to i_\mu \to \cdots \to i_\nu$. Since the number of vertices of $\Gamma(A)$ is finite, there are μ and ν such that $i_\mu = i_\nu$. This means that there exists an elementary circuit belonging to a strong component, and the vertex i_1 is joined with this circuit. Since i_1 is arbitrary, the lemma is proved. □

6.5.2. Simple homogeneous Markov chains. We apply the above results to study the limit behaviour of sequences of stochastic matrices. We consider a stochastic matrix

$$P = (p_{ij}), \quad p_{ij} \ge 0, \quad i, j = 1, \dots, n, \quad n \ge 2, \quad \sum_{j=1}^{n} p_{ij} = 1, \quad 1 \le i \le n.$$

Such a matrix, together with a vector of initial states

$$P^{(0)} = (p_1^{(0)}, \dots, p_n^{(0)}), \quad p_i^{(0)} \geq 0, \quad \sum_{i=1}^{n} p_i^{(0)} = 1,$$

defines a simple homogeneous Markov chain whose states are $1, \dots, n$ and in which p_{ij} is the transition probability to the state j in a single step provided that the chain is in the state i at the preceding instant of time. If $P^r = (p_{ij}^{(r)})$, then $p_{ij}^{(r)}$ is the probability of transition from the state i to the state j in r steps. If $p_i^{(k)}$ is the probability that the Markov chain at the kth step is in the state i and $P^{(k)} = (p_1^{(k)}, \dots, p_n^{(k)})$, then

$$P^{(k)} = P^{(0)} P^k, \qquad k = 1, 2, \dots.$$

Let $A = (a_{ij})$ be the support of the matrix $P = (p_{ij})$; that is,

$$a_{ij} = \begin{cases} 1, & p_{ij} > 0, \\ 0, & p_{ij} = 0. \end{cases}$$

The digraph $\Gamma(P)$ of the matrix P is the digraph with vertices $1, \dots, n$ and adjacency matrix A. It is clear that $\Gamma(P)$ coincides with the digraph $\Gamma(A)$.

THEOREM 6.5.13. *Let $P = (p_{ij})$ be a stochastic matrix of order $n \geq 2$ with digraph $\Gamma(P)$. The sequence $\{P^k,\ k = 1, 2, \dots\}$ converges if and only if each strong component of $\Gamma(P)$ is q-universal.*

PROOF. Let $H_1, \dots, H_g$ be the strong components of the digraph $\Gamma(P)$ which contain $a_1, \dots, a_g$ vertices respectively. For the matrix $P^q = (p_{ij}^{(q)})$, $q = 1, 2, \dots$, let $S_k^{(q)}(P)$ denote the set of elements $p_{ij}^{(q)}$ such that the states i and j belong to the strong component H_k, $1 \leq k \leq g$. Assume that the strong components $H_1, \dots, H_g$ are q-universal. Then by Theorem 6.5.7 there exists a positive integer q_k such that all elements of $S_k^{(q_k)}(P)$ are positive, $1 \leq k \leq g$. Hence there exist positive numbers

$$\pi_{1k}, \pi_{2k}, \dots, \pi_{q_k k}, \quad \sum_{\nu=1}^{q_k} \pi_{\nu k} = 1, \quad 1 \leq k \leq g,$$

such that

$$\lim_{t \to \infty} p_{ij}^{(t)} = \pi_{kj}$$

for all vertices i and j belonging to H_k, $1 \leq k \leq g$. From q-universality of the strong components $H_1, \dots, H_g$ it follows that $\lim_{t\to\infty} p_{ij}^{(t)} > 0$ if and only if i and j are joined to each other and j belongs to some strong component. Hence $\lim_{t\to\infty} p_{ij}^{(t)} > 0$ is proportional to π_{ik}, where $j \in H_k$, $1 \leq k \leq g$.

Thus, the sequence $\{P^k,\ k = 1, 2, \dots\}$ converges.

If the sequence $\{P^k,\ k = 1, 2, \dots\}$ converges, then for each strong component H_k, $1 \leq k \leq g$, the elements of $S_k^{(t)}(P)$ have limits as $t \to \infty$. If there exists a strong component H_d which is not q-universal, then, according to Corollary 6.5.10, the elements of $S_d^{(t)}(A)$ oscillate in the same way as the powers of the support A of the matrix P. Therefore, the elements of $S_d^{(t)}(P)$ cannot have nonzero limits, which contradicts the fact that H_d is a strong component. Therefore, every strong component of $\Gamma(P)$ is q-universal. □

THEOREM 6.5.14. *If P is a stochastic matrix of order $n \geq 2$ with digraph $\Gamma(P)$, then the sequence $\{P^k,\ k = 1, 2, \dots\}$ asymptotically oscillates if and only if at least one strong component of $\Gamma(P)$ is not q-universal. The oscillation period is equal to the least common multiple of the periods of the strong components of $\Gamma(P)$ which are not q-universal.*

PROOF. Since for the sequence of powers of a stochastic matrix the existence of a limit is an alternative to oscillation, the sufficiency follows from the reasoning used in the proof of the sufficiency in Theorem 6.5.13.

Consider the asymptotic oscillation of powers of the matrix P with support A. Let s denote the least common multiple of the periods of all strong components of $\Gamma(A)$ which are not q-universal. By Theorem 6.5.11 and Corollary 6.5.10, there exists a power $B_s = A^s$ such that $\Gamma(B_s)$ consists of strong components which are q-universal. Hence if

$$Q_s = \lim_{t \to \infty} P^{st},$$

then the set of limiting matrices is of the form $Q_s P, \dots, Q_s P^s$. □

If a stochastic matrix P of order $n \geq 2$ has limiting matrices $Q_1, \dots, Q_h$, then the corresponding supports $A_1, \dots, A_h$ are called asymptotic forms of the matrix P. It is clear that the asymptotic forms of P are determined by the limit behaviour of the sequence of powers of its support $\{A^k,\ k = 1, 2, \dots\}$ analyzed above.

6.6. Ergodicity of sequences of nonnegative matrices

6.6.1. Weak ergodicity. We consider a sequence of nonnegative matrices

$$\{A_k,\ k = 1, 2, \dots\}, \tag{6.6.1}$$

where $A_k = (a_{ij}(k))$, $i, j = 1, \dots, n$, $a_{ij}(k) \geq 0$, $k = 1, 2, \dots$.

We set

$$T_{pr} = A_{p+1} \cdots A_{p+r}. \tag{6.6.2}$$

We say that the sequence (6.6.1) satisfies the r-primitivity condition if

(a) there exists a positive integer r_0 such that the inequality

$$T_{pr} > 0 \tag{6.6.3}$$

holds for all $r \geq r_0$, $p \geq 0$, and

(b) the bounds

$$\min_{i,j}{}^{+} a_{ij}(k) \geq \lambda > 0, \tag{6.6.4}$$

$$\max_{i,j} a_{ij}(k) \leq \gamma < \infty \tag{6.6.5}$$

hold uniformly in $k = 1, 2, \dots$, where $\min^+$ means that the minimum is over all positive elements.

Note that if $A_k = A$, $k = 1, 2, \dots$, then the former condition coincides with the requirement that A is primitive, while the latter condition becomes trivial.

The objective of this section is to study the limit

$$T = \lim_{r \to \infty} T_{0r}.$$

We begin by proving a series of lemmas.

Let

$$P = (p_{ij}), \quad p_{ij} \geq 0, \quad i, j = 1, \dots, n, \quad \sum_{j=1}^{n} p_{ij} = 1,$$

be a stochastic matrix, $x = (x_1, \dots, x_n)^T$ a column vector, $y = (y_1, \dots, y_n)^T = Px$. We set

$$m_0 = \min_{1 \leq i \leq n} x_i, \qquad M_0 = \max_{1 \leq i \leq n} x_i,$$
$$m_1 = \min_{1 \leq i \leq n} y_i, \qquad M_1 = \max_{1 \leq i \leq n} y_i.$$

LEMMA 6.6.1. *If*

$$\varepsilon = \min_{i,j} p_{ij}$$

for a stochastic matrix $P = (p_{ij})$, $i, j = 1, \dots, n$, then

$$m_1 \geq m_0, \qquad M_1 \leq M_0, \tag{6.6.6}$$

$$M_1 - m_1 \leq (1 - 2\varepsilon)(M_0 - m_0). \tag{6.6.7}$$

PROOF. We consider the column vector x^* whose ith element is equal to m_0 and whose other elements are M_0. Then $x \leq x^*$, and each element of the vector Px^* is of the form $\alpha m_0 + (1-\alpha)M_0 = M_0 - \alpha(M_0 - m_0)$, $\alpha \geq \varepsilon$. Thus, each element of Px^* does not exceed $M_0 - \varepsilon(M_0 - m_0)$. Now the condition $x \leq x^*$ implies that $Px \leq Px^*$, and therefore

$$M_1 \leq M_0 - \varepsilon(M_0 - m_0). \tag{6.6.8}$$

The same arguments applied to the vector $-x$ show that each coordinate of $P(-x^*)$ does not exceed $-m_0 - \varepsilon(-m_0 + M_0)$. Hence

$$-m_1 \leq -m_0 - \varepsilon(-m_0 + M_0). \tag{6.6.9}$$

Inequalities (6.6.8) and (6.6.9) imply (6.6.6). Adding (6.6.8) to (6.6.9), we arrive at (6.6.7). □

Note that by the stochasticity of P the bound $\varepsilon \leq 1/2$ is always true.

We consider the matrix

$$T_{pr} = (t_{ij}(p, r)), \qquad i, j = 1, \dots, n, \tag{6.6.10}$$

defined by (6.6.2). We give a uniform bound for the elements of this matrix for sufficiently large r, provided that sequence (6.6.1) is r-primitive.

LEMMA 6.6.2. *If the sequence* (6.6.1) *is r-primitive, then the elements of the matrix* (6.6.10) *defined by* (6.6.2) *satisfy the inequalities*

$$0 < C_1 \leq t_{ij}(p, r_0) \leq C_2 < \infty, \tag{6.6.11}$$

where r_0 is defined by (6.6.3) *and the constants C_1 and C_2 do not depend on i, j, or p.*

PROOF. From (6.6.2) it follows that

$$t_{ij}(p, r_0) = \sum_{k_1, \dots, k_{r_0 - 1}} a_{ik_1}(p+1) a_{k_1 k_2}(p+2) \cdots a_{k_{r_0-1} j}(p + r_0), \tag{6.6.12}$$

where the summation is over all $k_1, \ldots, k_{r_0-1}$ such that $1 \le k_i \le n$, $1 \le i \le r_0 - 1$. Since $t_{jj}(p, r_0) > 0$, the right-hand side of (6.6.12) contains at least one positive summand, and therefore, in view of the conditions (6.6.4) and (6.6.5),

$$\lambda^{r_0} \le t_{ij}(p, r_0) \le n^{r_0-1}\gamma^{r_0}, \qquad 1 \le i, j \le n.$$

Hence we obtain (6.6.11). □

Now we formulate a theorem, usually called the theorem on weak ergodicity of a sequence of nonnegative matrices.

THEOREM 6.6.3. *If the sequence of nonnegative matrices* (6.6.1) *satisfies the* r*-primitivity conditions* (6.6.3), (6.6.4), (6.6.5), *then for the elements of the matrix* (6.6.10) *defined by* (6.6.2) *the limit*

$$\lim_{k\to\infty} \frac{t_{is}(p, k)}{t_{js}(p, k)} = v_{ij}(p) \tag{6.6.13}$$

exists and does not depend on s. *In other words, the rows of the matrix* T_{pk} *become proportional to one another as* $k \to \infty$.

PROOF. We begin with the obvious equality

$$t_{is}(p, k+1) = \sum_{r=1}^{n} t_{ir}(p, k) a_{rs}(k+1), \qquad 1 \le i, s \le n. \tag{6.6.14}$$

We consider the matrices $D_{pk} = (d_{sr}(k))$ whose elements are defined by the equalities

$$d_{sr}(p, k) = \frac{t_{jr}(p, k) a_{rs}(k+1)}{t_{js}(p, k+1)}, \qquad 1 \le s, r \le n, \quad k = 1, 2, \ldots.$$

With the use of (6.6.14) it is easy to check that the matrices D_{pk} are stochastic. Consider the column vector $x_k = (x_k^{(1)}, \ldots, x_k^{(n)})^T$ such that

$$x_k^{(s)} = \frac{t_{is}(p, k)}{t_{js}(p, k)}, \qquad 1 \le s \le n.$$

From (6.6.14) it follows that

$$x_{k+1}^{(s)} = \sum_{r=1}^{n} d_{sr}(p, k) x_k^{(r)}.$$

Therefore,

$$x_{k+1} = D_{pk} x_k, \qquad k = 1, 2, \ldots. \tag{6.6.15}$$

We introduce the notation

$$m_k = \min_{1\le s\le n} x_k^{(s)}, \qquad M_k = \max_{1\le s\le n} x_k^{(s)}.$$

Using Lemma 6.6.1, from (6.6.6) we obtain

$$m_k \le m_{k+1} \le M_{k+1} \le M_k. \tag{6.6.16}$$

Hence the sequence $m_1, m_2, \ldots, m_k, \ldots$ is monotone nondecreasing and bounded from above. Therefore, this sequence has a limit m^*. Similarly, the sequence $M_1, M_2, \ldots, M_k, \ldots$ is monotone nonincreasing and bounded from below. Therefore, it has a limit M^*.

It is clear that $m^* \leq M^*$. Let us prove that $m^* = M^*$. This is equivalent to the equality

$$\lim_{k\to\infty} (M_k - m_k) = 0 \tag{6.6.17}$$

for the nonincreasing sequence

$$M_1 - m_1, M_2 - m_2, \dots, M_k - m_k, \dots. \tag{6.6.18}$$

In order to prove (6.6.17), it suffices to point out a positive integer r_0 such that

$$\lim_{k\to\infty} (M_{kr_0} - m_{kr_0}) = 0, \tag{6.6.19}$$

that is, to demonstrate that a subsequence of (6.6.18) of the form

$$M_{r_0} - m_{r_0}, M_{2r_0} - m_{2r_0}, \dots, M_{kr_0} - m_{kr_0}, \dots \tag{6.6.20}$$

converges to zero as $k \to \infty$.

As the required r_0 we take the parameter r_0 from the definition of r-primitivity of the sequence (6.6.1), and consider the sequence of matrices $\{\bar{A}_k,\ k = 1, 2, \dots\}$, where

$$\bar{A}_k = A_{1+(k-1)r_0} A_{2+(k-1)r_0} \cdots A_{kr_0}. \tag{6.6.21}$$

From condition (6.6.3) it follows that $\bar{A}_k > 0$, $k = 1, 2, \dots$, and the sequence $\{\bar{A}_k,\ k = 1, 2, \dots\}$ satisfies the hypotheses of Lemma 6.6.2. If

$$\bar{A}_k = (\bar{a}_{ij}(k)), \quad i, j = 1, \dots, n, \quad k = 1, 2, \dots,$$

then there exist constants $\bar{C}_1$ and $\bar{C}_2$, which do not depend on i, j, or k, such that

$$0 < \bar{C}_1 \leq \bar{a}_{ij}(k) \leq \bar{C}_2 < \infty. \tag{6.6.22}$$

Consider the sequence of matrices $\{B_{pk},\ k = 1, 2, \dots\}$, where

$$B_{pk} = (b_{ij}(p, k)) = \bar{A}_{p+1} \cdots \bar{A}_{p+k}, \qquad i, j = 1, \dots, n. \tag{6.6.23}$$

From (6.6.23) we obtain

$$b_{is}(p, k+1) = \sum_{r=1}^{n} b_{ir}(p, k) \bar{a}_{rs}(p+k+1). \tag{6.6.24}$$

We consider the matrix $\bar{D}_{pk} = (\bar{d}_{sr})p, k))$, where

$$\bar{d}_{sr}(p, k) = \frac{b_{jr}(p, k)}{b_{js}(p, k+1)} \bar{a}_{rs}(p+k+1). \tag{6.6.25}$$

From (6.6.24) it follows that the matrix $\bar{D}_{pk}$ is stochastic. Furthermore, if the coordinates of the column vector $z_k = (z_k^{(1)}, \dots, z_k^{(n)})^T$ are equal to

$$z_k^{(s)} = \frac{b_{is}(p, k)}{b_{js}(p, k)}, \quad k = 1, 2, \dots, \quad 1 \leq i, j \leq n,$$

then

$$z_{k+1}^{(s)} = \sum_{r=1}^{n} \bar{d}_{sr}(p, k) z_k^{(r)}; \tag{6.6.26}$$

that is,

$$z_{k+1} = \bar{D}_{pk} z_k. \tag{6.6.27}$$

Let us find an estimate for the number

$$\bar{\varepsilon} = \min_{1\le s,r\le n} \bar{d}_{sr}(p,k) \tag{6.6.28}$$

which does not depend on p and k.

From (6.6.24) and (6.6.25) we find that

$$\bar{d}_{sr}(p,k) = \frac{\bar{a}_{rs}(p+k+1)b_{jr}(p,k)}{\sum_{q=1}^{n} b_{jq}(p,k)\bar{a}_{qs}(p+k+1)}, \qquad 1 \le s,r \le n. \tag{6.6.29}$$

Let us estimate $\max_{1\le j\le n}(b_{jq}(p,q)/b_{jr}(p,k))$. From (6.6.23) we find that

$$b_{jq}(p,k+1) = \sum_{l=1}^{n} \bar{a}_{jl}(p+1)(b_{lq}(p+1,k). \tag{6.6.30}$$

We consider the matrix $\tilde{D}_{pk} = (\tilde{d}_{jl}(p,k))$, where

$$\tilde{d}_{jl}(p,k) = \frac{b_{lr}(p+1,k)}{b_{jr}(p,k+1)}\bar{a}_{jl}(p+1). \tag{6.6.31}$$

If $w_k(p) = (w_k^{(1)}(p),\dots,w_k^{(n)}(p))^T$ is the column vector whose elements are

$$w_k^{(l)}(p) = \frac{b_{lq}(p,k)}{b_{lr}(p,k)}, \qquad 1 \le l \le n,$$

then from (6.6.30) we obtain

$$w_{k+1}^{(r)}(p) = \sum_{l=1}^{n} \tilde{d}_{jl}(p,k) w_k^{(l)}(p+1);$$

that is,

$$w_{k+1}(p) = \tilde{D}_{pk} w_k(p+1).$$

Now from Lemma 6.6.1 it follows that

$$\begin{aligned} \max_{1\le j\le n} \frac{b_{jq}(p,k+1)}{b_{jr}(p,k+1)} &\le \max_{1\le j\le n} \frac{b_{jq}(p+1,k)}{b_{jr}(p+1,k)} \\ &\le \cdots \le \max_{1\le j\le n} \frac{b_{jq}(p+k,1)}{b_{jr}(p+k,1)}. \end{aligned} \tag{6.6.32}$$

From the obvious equality $B_{p+k,1} = \bar{A}_{p+k+1}$ and inequalities (6.6.32) we find that

$$\max_{1\le j\le n} \frac{b_{jq}(p+k,1)}{b_{jr}(p+k,1)} \le \max_{1\le j\le n} \frac{\bar{a}_{jq}(p+k+1)}{\bar{a}_{jr}(p+k+1)}.$$

From (6.6.22) and (6.6.32) we find that the bound

$$\frac{b_{jq}(p,k+1)}{b_{jr}(p,k+1)} \le \frac{\bar{C}_2}{\bar{C}_1}$$

holds uniformly in all p, q, r, k, j. Using this bound, (6.6.22), and (6.6.29), we obtain

$$\bar{\varepsilon}_k \ge \varepsilon = \frac{1}{n}\left(\frac{\bar{C}_1}{\bar{C}_2}\right)^2 > 0. \tag{6.6.33}$$

We introduce the notation

$$z_k = (z_k^{(1)},\dots,z_k^{(n)})^T, \quad \bar{m}_k = \min_{1\le i\le n} z_k^{(i)}, \quad \bar{M}_k = \max_{1\le i\le n} z_k^{(i)}.$$

In view of (6.6.27), Lemma 6.6.1 yields

$$\bar{M}_{k+1} - \bar{m}_{k+1} \leq (1 - 2\varepsilon)(\bar{M}_k - \bar{m}_k). \tag{6.6.34}$$

Iterating (6.6.34), we obtain

$$\bar{M}_{k+1} - \bar{m}_{k+1} \leq (1 - 2\bar{\varepsilon})^k(\bar{M}_0 - \bar{m}_0). \tag{6.6.35}$$

Taking (6.6.21) and (6.6.23) into account, we obtain

$$\bar{m}_k = m_{kr_0}, \qquad \bar{M}_k = M_{kr_0}. \tag{6.6.36}$$

Therefore, (6.6.35) implies (6.6.19), which, in its turn, yields (6.6.17). Hence the limit (6.6.13) exists, which completes the proof of the theorem. □

COROLLARY 6.6.4. *If a sequence of nonnegative matrices* $\{A_k,\ k = 1, 2, \dots\}$ *satisfies the r-primitivity condition and*

$$T_{pk} = (t_{ij}(p,k)) = A_{p+1} \cdots A_{p+k}, \quad i,j = 1, \dots, n, \quad k = 1, 2, \dots,$$

then there exists a vector $\nu = (\nu_1, \dots, \nu_n)$ *with positive coordinates such that*

$$\lim_{k\to\infty} \frac{t_{is}(p,k)}{t_{js}(p,k)} = \frac{\nu_i}{\nu_j}, \qquad 1 \leq i,j \leq n, \tag{6.6.37}$$

for all i, j, p, s; *and* $\nu_1, \dots, \nu_n$ *do not depend on* s.

PROOF. It follows from (6.6.13) that for all q, $1 \leq q \leq n$,

$$v_{ij}(p) = \lim_{k\to\infty} \frac{t_{is}(p,k)}{t_{js}(p,k)} = \lim_{k\to\infty} \frac{t_{is}(p,k)/t_{qs}(p,k)}{t_{js}(p,k)/t_{qs}(p,k)} = \frac{v_{iq}(p)}{v_{jq}(p)}.$$

Hence

$$v_{ij}(p) \sum_{q=1}^{n} v_{jq}(p) = \sum_{q=1}^{n} v_{iq}(p).$$

Setting

$$\nu_i = \sum_{q=1}^{n} v_{iq}(p), \qquad 1 \leq i \leq n,$$

we arrive at (6.6.37). □

COROLLARY 6.6.5. *The rate of convergence of the ratio* $t_{is}(p,k)/t_{js}(p,k)$ *to* ν_i/ν_j *as* $k \to \infty$ *is estimated by the inequality*

$$\left| \frac{t_{is}(p,k)}{t_{js}(p,k)} - \frac{\nu_i}{\nu_j} \right| \leq C\rho^k, \tag{6.6.38}$$

where $0 < \rho < 1$, $0 \leq C < \infty$, *and* ρ *and* C *do not depend on* i *and* j.

PROOF. From (6.6.35), in view of (6.6.36) and (6.6.33), we obtain

$$|M_{kr_0} - m_{kr_0}| \leq C_{ij}(1 - 2\varepsilon)^k, \tag{6.6.39}$$

where

$$\varepsilon = \frac{1}{n}\left(\frac{\bar{C}_1}{\bar{C}_2}\right)^2$$

and

$$C_{ij} = M_0 - m_0$$

is a constant depending on i and j. Setting

$$\tilde{C} = \max_{1 \le i,j \le n} C_{ij}$$

for all $k \ge r_0$ and using the fact that the sequence $M_1 - m_1, \dots, M_k - m_k$ is positive and monotone decreasing, we obtain the uniform bound

$$M_k - m_k \le \tilde{C}(1 - 2\varepsilon)^{k/r_0 - 1}. \tag{6.6.40}$$

We set

$$\rho = (1 - 2\varepsilon)^{1/r_0}, \qquad \varepsilon = n\left(\frac{\lambda}{n\gamma}\right)^{2r_0}.$$

Then (6.6.40) yields

$$M_k - m_k \le C\rho^k$$

for all $k \ge r_0$, which implies (6.6.38). □

6.6.2. Strong ergodicity. Before we formulate and prove a theorem on the strong ergodicity of a sequence of nonnegative matrices, we give the following lemma.

LEMMA 6.6.6. *For a sequence of nonnegative matrices* $\{A_k,\ k = 1, 2, \dots\}$ *to satisfy the r-primitivity condition, it suffices that*

(1) *every matrix* A_k, $k = 1, 2, \dots,$ *of the sequence has no zero rows, and*
(2) *there exists a primitive matrix* A *such that, element-wise,*

$$\lim_{k \to \infty} A_k = A.$$

PROOF. Condition (2) for any $\varepsilon > 0$ implies the existence of r_1 such that the inequality

$$\|A_{p+1} \cdots A_{p+r} - A^r\| \le \varepsilon$$

holds for all $r \ge r_1$ and all $p \ge 0$, where $\|A\|$ is some norm of A. Since A is primitive, there exists r_2 such that $A^r > 0$ for all $r \ge r_2$. Then $A_{p+1} \cdots A_{p+r} > 0$ for all $r \ge r_0 = \max(r_1, r_2)$. Thus, condition (a) of r-primitivity is satisfied (see Section 6.6.1). From conditions (1) and (2) of the lemma it follows that condition (b) of r-primitivity is satisfied as well. □

Conditions (1) and (2) will be referred to as the strong r-primitivity conditions for a sequence of nonnegative matrices $\{A_k,\ k = 1, 2, \dots\}$.

Let $A = (a_{ij})$, $A_k = (a_{ij}(k))$, $i, j = 1, \dots, n$, $k = 1, 2, \dots$. If the sequence $\{A_k,\ k = 1, 2, \dots\}$ satisfies the strong r-primitivity conditions, then there exists k_0 such that for all $k \ge k_0$ the matrix A_k has positive elements at least in those positions which are occupied by positive elements in A. We consider the matrix $A_k^{(l)} = (a_{ij}^{(l)}(k))$, $i, j = 1, \dots, n$, where

$$a_{ij}^{(l)}(k) = \begin{cases} a_{ij}(k), & 1 \le k \le k_0, \\ \min(a_{ij}(k), a_{ij}), & k \ge k_0. \end{cases}$$

It is obvious that the sequence $\{A_k^{(l)},\ k = 1, 2, \dots\}$ satisfies the r-primitivity conditions, and therefore

$$A_k^{(l)} \le A \tag{6.6.41}$$

for all $k \geq k_0$. Since

$$A_k^{(l)} \leq A_k \tag{6.6.42}$$

for all $k \geq 1$, we find that

$$A_{p+1}^{(l)} \cdots A_{p+k}^{(l)} \leq T_{pk} = A_{p+1} \cdots A_{p+k}. \tag{6.6.43}$$

By the Perron–Frobenius Theorem [**87**, Chapter 13], if u is the right eigenvector of A corresponding to the maximum eigenvalue λ and the matrix R is defined by $R = uv^T > 0$, where the vector v satisfies the condition $v^T u = 1$ and is the left eigenvector of A^T, then

$$\lim_{k \to \infty} \left(\frac{1}{\lambda} A \right)^k = R. \tag{6.6.44}$$

Hence there exists a sequence $\delta_1, \delta_2, \ldots$, $0 \leq \delta_k \leq 1$, $k = 1, 2, \ldots$, such that

$$\lim_{k \to \infty} \delta_k = 0$$

and, furthermore,

$$(1 - \delta_k) R \leq \left(\frac{1}{\lambda} A \right)^k \leq (1 + \delta_k) R. \tag{6.6.45}$$

Since the sequences $\{A_k,\ k = 1, 2, \ldots\}$ and $\{A_k^{(l)},\ k = 1, 2, \ldots\}$ are strong and r-primitive, there exist sequences $\alpha_1, \alpha_2, \ldots$ and $\beta_1, \beta_2, \ldots$ such that $\alpha_k \geq 0$, $\beta_k \geq 0$, $k = 1, 2, \ldots$, $\alpha_k \to 0$, $\beta_k \to 0$ as $k \to \infty$, and

$$A - \alpha_k \lambda R \leq A_k^{(l)} \leq A + \alpha_k \lambda R, \tag{6.6.46}$$

$$A - \beta_k \lambda R \leq A_k \leq A + \beta_k \lambda R. \tag{6.6.47}$$

From these inequalities it follows that

$$A_k^{(l)} \geq A - \alpha_k \lambda R, \qquad k \geq k_0, \tag{6.6.48}$$

$$A_k \leq A + \beta_k \lambda R, \qquad k \geq 1. \tag{6.6.49}$$

LEMMA 6.6.7. *If for a sequence of matrices $\{C_k,\ k = 1, 2, \ldots\}$ there exists a sequence $\gamma_1, \gamma_2, \ldots$, $\gamma_k \geq 0$, $k \geq 1$, such that*

$$0 \leq C_k \leq \frac{1}{\lambda} A, \quad C_k \leq \gamma_k R, \quad k = 1, 2, \ldots, \tag{6.6.50}$$

where $R = uv^T$, u being the right eigenvector of A corresponding to the maximum eigenvalue λ, then

$$\left(\frac{1}{\lambda} A - C_1 \right) \cdots \left(\frac{1}{\lambda} A - C_k \right) \geq \left(\frac{1}{\lambda} A \right)^k - \sum_{q=1}^k \gamma_q R. \tag{6.6.51}$$

PROOF. We use induction on k. For $k = 1$, the inequality reduces to the second condition in (6.6.50).

Assume that (6.6.51) is true for some $k \geq 1$. Multiplying both sides of the inequality by

$$\left(\frac{1}{\lambda} A - C_{k+1} \right) \geq 0,$$

we obtain

$$\begin{aligned}\left(\frac{1}{\lambda}A - C_1\right)\cdots\left(\frac{1}{\lambda}A - C_{k+1}\right) \\ \geq \left(\frac{1}{\lambda}A\right)^{k+1} - \left(\frac{1}{\lambda}A\right)^{k} C_{k+1} - \sum_{q=1}^{k}\gamma_q\frac{1}{\lambda}RA.\end{aligned} \tag{6.6.52}$$

From the definition of R it follows that

$$AR = RA = \lambda R. \tag{6.6.53}$$

Using (6.6.50) and (6.6.53), we obtain

$$\left(\frac{1}{\lambda}A\right)^{k} C_{k+1} \leq \gamma_{k+1}\left(\frac{1}{\lambda}A\right)^{k} R = \gamma_{k+1}R. \tag{6.6.54}$$

Relations (6.6.53) and (6.6.54) imply that

$$\left(\frac{1}{\lambda}A - C_1\right)\cdots\left(\frac{1}{\lambda}A - C_{k+1}\right) \geq \left(\frac{1}{\lambda}A\right)^{k+1} - \sum_{q=1}^{k+1}\gamma_q R.$$

□

LEMMA 6.6.8. *The inequalities*

$$\begin{aligned}\left(1 - \delta_m - \sum_{q=p+k-m+1}^{p+k}\alpha_q\right)R \leq \frac{1}{\lambda^m}A^{(l)}_{p+k-m+1}\cdots A^{(l)}_{p+k} \\ \leq \left(\delta_m + \prod_{q=p+k-m-1}^{p+k}(1+\beta_q)\right)R.\end{aligned} \tag{6.6.55}$$

are true for all $k \geq k_0 + m - 1$ *and all* $p \geq 0$.

PROOF. Setting $C_k = \alpha_k R$, $\gamma_k = \alpha_k$ in (6.6.51), we find that for $k \geq k_0 + m - 1$

$$\begin{aligned}\left(\frac{1}{\lambda}A\right)^{m} - \sum_{q=p+k-m+1}^{p+k}\alpha_q R \leq \frac{1}{\lambda^m}A^{(l)}_{p+k-m+1}\cdots A^{(l)}_{p+k} \\ \leq \frac{1}{\lambda^m}A_{p+k-m+1}\cdots A_{p+k}.\end{aligned} \tag{6.6.56}$$

From (6.6.47) it follows that

$$\frac{1}{\lambda^m}A_{p+k-m+1}\cdots A_{p+k} \leq \left(\frac{1}{\lambda}A + \beta_{p+k-m+1}R\right)\cdots\left(\frac{1}{\lambda}A + \beta_{p+k}R\right). \tag{6.6.57}$$

By induction on k, let us show that

$$\begin{aligned}\left(\frac{1}{\lambda}A + \beta_{p+k-m+1}R\right)\cdots\left(\frac{1}{\lambda}A + \beta_{p+k}R\right) \\ \leq \left(\frac{1}{\lambda}A\right)^{m} + \left(\prod_{q=p+k-m+1}^{p+k}(1+\beta_q) - 1\right)R.\end{aligned} \tag{6.6.58}$$

Without loss of generality we set $p = 0$. Then we have to prove the inequality

$$\begin{aligned}(6.6.59)\quad &\left(\frac{1}{\lambda}A + \beta_{k-m+1}R\right)\cdots\left(\frac{1}{\lambda}A + \beta_k R\right)\\ &\qquad \le \left(\frac{1}{\lambda}A\right)^m + \left(\prod_{q=k-m+1}^{p+k}(1+\beta_q) - 1\right)R.\end{aligned}$$

We use induction on m. For $m = 1$ inequality (6.6.59) reduces to an identity. Using the induction assumption, from (6.6.59) we obtain

$$\begin{aligned}(6.6.60)\quad &\left(\frac{1}{\lambda}A + \beta_{k-m}R\right)\left(\frac{1}{\lambda}A + \beta_{k-m+1}R\right)\cdots\left(\frac{1}{\lambda}A + \beta_k R\right)\\ &\le \left(\frac{1}{\lambda}A\right)^{m+1} + \left(\prod_{q=k-m+1}^{p+k}(1+\beta_q) - 1\right)R + \beta_{k-m}\left(\prod_{q=k-m+1}^{p+k}(1+\beta_q)\right).\end{aligned}$$

We used above the obvious equalities

$$R^2 = R, \qquad \frac{1}{\lambda}AR = \frac{1}{\lambda}RA = R.$$

Inequality (6.6.60) yields (6.6.59), which, in turn, proves (6.6.58).

Now from (6.6.56), (6.6.57), and (6.6.60) we obtain

$$\begin{aligned}(6.6.61)\quad &\left(\frac{1}{\lambda}A\right)^m - \sum_{q=p+k-m+1}^{p+k}\alpha_q R \le \frac{1}{\lambda^m}A_{p+k-m+1}\cdots A_{p+k}\\ &\qquad \le \left(\frac{1}{\lambda}A\right)^m + \left(\prod_{q=p+k-m+1}^{p+k}(1+\beta_q) - 1\right)R.\end{aligned}$$

With the use of (6.6.45), from (6.6.61) we derive the required relation (6.6.55). □

Now we formulate and prove the theorem on the strong ergodicity of a sequence of nonnegative matrices.

THEOREM 6.6.9. *If a sequence of nonnegative matrices $\{A_k,\ k = 1, 2, \dots\}$ of order n satisfies the strong r-primitivity conditions and*

$$T_{pk} = (t_{ij}(p,k)) = A_{p+1}\cdots A_{p+k}, \quad p \ge 0, \quad k = 1, 2, \dots,$$

then the limit

$$(6.6.62)\qquad \lim_{k\to\infty}\frac{t_{ij}(p,k)}{\sum_{q=1}^{n} t_{iq}(p,q)} = v_j, \qquad i, j = 1, \dots, n,$$

exists, where

$$v^T = (v_1, \dots, v_n), \qquad \sum_{i=1}^{n} v_i = 1,$$

and v^T is the left eigenvector of A corresponding to the maximum eigenvalue λ.

PROOF. We consider a nonzero nonnegative column vector x and define the column vector w as

$$w^T = x^T A_{p+1}\cdots A_{p+k-m}.$$

Due to r-primitivity, $w \geq 0$ and $w \neq 0$. Using (6.6.43) and (6.6.55), we obtain

(6.6.63)
$$\lambda^m \left(1 - \delta_m - \sum_{q=p+k-m+1}^{p+k} \alpha_q\right) w^T R \leq x^T T_{pk} \leq \lambda^m \left(\delta_m + \prod_{q=p+k-m+1}^{p+k} (1+\beta_q)\right) R.$$

Multiplying this by the right eigenvector u corresponding to the maximum eigenvalue λ of A and dividing the parts of (6.6.63) by the corresponding parts of the inequality just obtained, we find that

$$\begin{aligned}\frac{\left(1 - \delta_m - \sum_{q=p+k-m+1}^{p+k} \alpha_q\right) w^T R}{\left(\delta_m - \prod_{q=p+k-m+1}^{p+k}(1+\beta_q)\right) w^T Ru} \\ \leq \frac{x^T T_{pk}}{x^T T_{pk} u} \leq \frac{\left(\delta_m + \prod_{q=p+k-m+1}^{p+k}(1+\beta_q)\right) w^T R}{\left(1 - \delta_m - \sum_{q=p+k-m+1}^{p+k} \alpha_q\right) w^T Ru}.\end{aligned} \tag{6.6.64}$$

It is clear that for all x and w we have

$$\frac{w^T R}{w^T Ru} = v^T,$$

where $v^T u = 1$ and v^T is the left eigenvector corresponding to the maximum eigenvalue λ of the matrix A. Therefore, for arbitrary positive integers p, m, and k such that $k - m \geq k_0 - 1$, from (6.6.64), passing to the limit as $k \to \infty$ and then as $m \to \infty$, we find that

$$\lim_{k\to\infty} \frac{x^T T_{pk}}{x^T T_{pk} u} = v^T \tag{6.6.65}$$

uniformly in $x \geq 0$ and $x \neq \mathbf{0}$. From the condition $v_1 + \cdots + v_n = 1$ and (6.6.65) it follows that

$$\lim_{k\to\infty} \frac{x^T T_{pk} e}{x^T T_{pk} u} = 1, \tag{6.6.66}$$

where $e = (1, \ldots, 1)^T$. From (6.6.65) and (6.6.66) we obtain

$$\lim_{k\to\infty} \frac{x^T T_{pk}}{x^T T_{pk} e} = v^T. \tag{6.6.67}$$

For a given i, $1 \leq i \leq n$, we define $x = (x_1, \ldots, x_n)^T$ by the relations

$$x_j = \begin{cases} 1, & j = i, \\ 0, & j \neq i, \end{cases} \qquad j = 1, \ldots, n.$$

Then from (6.6.67) it follows that for $i = 1, \ldots, n$

$$\lim_{k\to\infty} \frac{1}{\sum_{s=1}^n t_{is}(p,k)} (t_{i1}(p,k), \ldots, t_{ij}(p,k), \ldots, t_{in}(p,k)) = v^T,$$

which proves (6.6.62). □

6.6.3. Markov chains. A sequence of random variables $\xi_0, \xi_1, \dots$ taking values in the set $X = \{x_1, x_2, \dots, x_n\}$ is called a finite Markov chain with the set of states X if for any $x_{i_0}, x_{i_1}, \dots, x_{i_{k+1}} \in X$ the equalities

$$\mathbf{P}(\xi_{k+1} = x_{i_{k+1}} \mid \xi_k = x_{i_k}, \dots, \xi_0 = x_{i_0}) = \mathbf{P}(\xi_{k+1} = x_{i_{k+1}} \mid \xi_k = x_{i_k})$$

hold, provided that $\mathbf{P}(\xi_k = x_{i_k}, \dots, \xi_0 = x_{i_0}) > 0$.

The probabilities $P_{ij}(k+1) = \mathbf{P}(\xi_{k+1} = x_j \mid \xi_k = x_i)$, $k = 0, 1, \dots$, are called the transition probabilities from the state x_i to the state x_j at the $(k+1)$th step. It is obvious that

$$\mathbf{P}(\xi_0 = x_0, \dots, \xi_k = x_k) = \pi_{i0} P_{i_0 i_1}(1) \dots P_{i_{k-1} i_k}(k),$$

where $\pi_{i0} = \mathbf{P}(\xi_0 = x_i)$, $i = 1, \dots, n$, and the row vector $\pi_0^T = (\pi_{10}, \dots, \pi_{n0})$ is called the initial probability distribution of the Markov chain. Let $P_{ij}(0) = \pi_{j0}\delta_{ij}$, $i, j = 1, \dots, n$. The stochastic matrix

$$P_k = (P_{ij}(k)), \quad i, j = 1, \dots, n, \quad \sum_{j=1}^{n} P_{ij}(k) = 1, \quad k = 0, 1, \dots,$$

is called the transition matrix of the Markov chain at the kth step.

If $P_1 = \dots = P_k = \dots = P$, where P is a stochastic matrix, then the Markov chain is called homogeneous. If this is not the case, then the chain is called nonhomogeneous.

We consider the row vector

$$\Pi_k^T = (\mathbf{P}(\xi_k = x_1), \dots, \mathbf{P}(\xi_k = x_n)), \qquad k = 0, 1, \dots,$$

which determines the probability distribution of the random variable ξ_k.

It is obvious that

$$\Pi_k^T = \Pi_0^T P_1 P_2 \dots P_k, \qquad k = 0, 1, \dots,$$

where Π_0^T is the initial distribution of the Markov chain. Let

$$P_{pk} = (P_{ij}(p,k)) = P_{p+1} \dots P_{p+k}, \quad p \geq 0, \quad k = 0, 1, \dots.$$

It is clear that

$$\Pi_k^T = \Pi_0^T P_{0k}, \qquad k = 0, 1, \dots,$$

and

$$\Pi_k^T = \Pi_p^T P_{p,k-p}$$

for all $k > p$.

For a homogeneous Markov chain,

$$P_{pk} = P^k, \quad p \geq 0, \quad k = 1, 2, \dots,$$

We say that the Markov chain with transition matrices P_k, $k = 0, 1, \dots$, is weakly ergodic if the equality

$$\lim_{k \to \infty} (P_{is}(p,k) - P_{js}(p,k)) = 0$$

is true for all i, j, s, p.

THEOREM 6.6.10. *The Markov chain with transition matrices P_k, $k = 0, 1, \dots$, is weakly ergodic if the sequence of stochastic matrices $\{P_k,\ k = 0, 1, \dots\}$ is r-primitive.*

PROOF. In view of (6.6.13)

$$\frac{P_{is}(p,k)}{P_{js}(p,k)} = v_{ij}(p) + \varepsilon(i,j,s,p,k),$$

where $\varepsilon(i,j,s,p,k) \to 0$ as $k \to \infty$.

Since the matrix $P_{pk} = (P_{ij}(p,k))$, $i,j = 1,\dots,n$, is stochastic, the relation

$$v_{ij}(p) + \sum_{s=1}^{n} P_{js}(p,k)\varepsilon(i,j,s,p,k) = 1$$

holds, and $0 < P_{js}(p,k) \leq 1$. Passing to the limit as $k \to \infty$, we find that $v_{ij}(p) = 1$ for all i, j, p.

Thus, for the sequence of stochastic matrices $\{P_k,\ k = 0,1,\dots\}$, not only are the rows of the matrices P_{pk} as $k \to \infty$ proportional to one another, but they asymptotically coincide. Therefore,

$$\lim_{k\to\infty}\left(\frac{P_{is}(p,k)}{P_{js}(p,k)} - 1\right) = \lim_{k\to\infty}\frac{P_{is}(p,k) - P_{js}(p,k)}{P_{js}(p,k)} = 0.$$

Since the $P_{js}(p,k)$ are bounded for all i, j, s, p, we conclude that

$$\lim_{k\to\infty}(P_{is}(p,k) - P_{js}(p,k)) = 0.$$

The theorem is thus proved. □

The Markov chain with transition matrices P_k, $k = 0,1,\dots$, is called strongly ergodic if it is weakly ergodic and the limit

$$v_{is}(p) = \lim_{k\to\infty} P_{is}(p,k)$$

exists for all i, s, p.

THEOREM 6.6.11. *The Markov chain with transition matrices P_k, $k = 0,1,\dots$, is strongly ergodic if the sequence of stochastic matrices $\{P_k,\ k = 0,1,\dots\}$ is strongly r-primitive.*

PROOF. Since strong r-primitivity implies r-primitivity, the Markov chain under consideration is weakly ergodic. By Theorem 6.6.9, from the strong r-primitivity and (6.6.62) it follows that the limit

$$v_s = v_{is} = \lim_{k\to\infty} P_{is}(p,k), \qquad \sum_{s=1}^{n} v_s = 1,$$

exists for all i, s, p; that is, the Markov chain is strongly ergodic. □

Bibliography

1. H. L. Abbott, D. Hanson, and A. C. Liu, *An extremal problem in hypergraph theory*, J. Austral. Math. Soc. **31A** (1981), 129–135.
2. C. M. Ablow and J. L. Brenner, *Roots and canonical forms for circulant matrices*, Trans. Amer. Math. Soc. **107** (1963), 360–376.
3. R. Ahlswede and D. E. Daykin, *The number of values of combinatorial functions*, Bull. London Math. Soc. **11** (1979), 49–51.
4. W. Ahn and B. Choi, *Asymptotics about permanent of random* $(0,1)$*-matrices*, Math. Japon. **31** (1986), 167–174.
5. M. Albert and A. Frieze, *Occupancy problems and random algebras*, Discrete Math. **87** (1991), 1–8.
6. A. Aleksandroff, *Zur Theorie der gemischten Volumina von konvexen Körpern*, Rec. Math. Moscou (= Mat. Sbornik) (N.S.) **3** (1938), 227–251. (Russian; German summary)
7. N. Alon, *On the density of sets of vectors*, Discrete Math. **46** (1983), 199–202.
8. ———, *Transversal numbers of uniform hypergraphs*, Graphs and Comb. **6** (1990), 1–4.
9. R. P. Anstee, *A forbidden configuration theorem of Alon*, J. Comb. Theory **47A** (1988), 16–27.
10. R. P. Anstee and U. S. R. Murty, *Matrices with forbidden subconfiguration*, Discrete Math. **54** (1985), 113–116.
11. L. D. Baumert, *Cyclic difference sets*, Springer, Berlin, 1971.
12. C. Berge, *The theory of graphs and its applications*, Wiley, New York, 1962.
13. ———, *Graphs and hypergraphs*, Elsevier, Amsterdam, 1973.
14. ———, *Packing problem and hypergraph theory: A survey*, Ann. Discrete Math. **4** (1979), 3–37.
15. A. Berman and R. J. Plemmons, *Non-negative matrices in the mathematical sciences*, Academic Press, New York, 1979.
16. B. Bollobás, *Extremal graph theory*, Academic Press, London, 1978.
17. W. G. Bridges, *Some results on* λ*-designs*, J. Comb. Theory **8** (1970), 350–356.
18. R. A. Brualdi, *A note on multipliers of difference sets*, J. Res. Nat. **69B** (1965), 87–89.
19. ———, *Matrices of zeros and ones with fixed row and column sum vectors*, Linear Algebra Appl. **33** (1980), 159–231.
20. ———, *On Haber's minimal term rank formula*, Eur. J. Comb. **2** (1981), 17–20.
21. R. A. Brualdi and Lin Boliau, *A lattice generated by* $(0,1)$*-matrices*, Ars Combinatoria **31** (1991), 183–190.
22. R. A. Brualdi and P. M. Gibson, *Convex polyhedra of doubly stochastic matrices.* I, J. Comb. Theory **22A** (1977), 194–230.
23. ———, *Convex polyhedra of doubly stochastic matrices.* III, J. Comb. Theory **22A** (1977), 338–351.
24. ———, *Convex polyhedra of doubly stochastic matrices.* II, J. Comb. Theory **22B** (1977), 175–198.
25. R. A. Brualdi and H. Perfect, *Extension of partial diagonals of matrices*, Monatsh. Math. **75** (1971), 385–397.
26. R. A. Brualdi and Li Qiao, *Small diameter interchange graphs of classes of matrices of zeros and ones*, Linear Algebra Appl. **46** (1982), 177–194.
27. R. A. Brualdi and J. A. Ross, *Invariant sets for matrices of zeros and ones*, Proc. Amer. Math. Soc. **80** (1980), 706–710.
28. R. G. Busacker and T. L. Saaty, *Finite graphs and networks: An introduction with applications*, McGraw–Hill, New York, 1965.

29. D. M. Cvetkovic, M. Doob, and H. Sachs, *Spectra of graphs. theory and applications*, J. A. Barth Verlag, Leipzig, 1995.
30. D. E. Daykin and L. Lovasz, *The number of values of Boolean functions*, J. London Math. Soc. **12** (1976), 225–230.
31. N. G. de Bruijn and P. Erdös, *On a combinatorial problem*, Indagationes Math. **10** (1948), 421–423.
32. D. de Cæn and D. A. Gregory, *Primes in the semigroup of Boolean matrices*, Linear Algebra Appl. **37** (1981), 119–134.
33. A. L. Dulmage and N. S. Mendelsohn, *The exponents of incidence matrices*, Duke Math. J. **31** (1964), 575–584.
34. ———, *Graphs and matrices*, Graph Theory and Theoretical Physics (F. Harary, ed.), Academic Press, London, 1967, pp. 167–229.
35. G. P. Egorychev, *Solution of the van der Waerden problem for permanents*, Soviet Math. Dokl, **23** (1981), 619–622.
36. ———, *Proof of the van der Waerden conjecture for permanents*, Siberian Math. J. **22** (1982), 854–859.
37. R. Ermers and B. Polman, *On the eigenvalues of the structure matrix of matrices of zeros and ones*, Linear Algebra Appl. **95** (1987), 17–41.
38. D. I. Falikman, *Proof of the van der Waerden conjecture regarding the permanent of a doubly stochastic matrix*, Math. Notes **29** (1981), 475–479.
39. W. Feller, *Introduction to probability theory and its applications*, Vol. 1, Wiley, New York, 1968.
40. W. Fenchel, *Inégalités quadratiques entre les volumes mixtes des corps convexes*, C. R. Acad. Sci. Paris **203** (1936), 647–650.
41. T. Fenner and G. Loizou, *Combinatorial aspects of rectangular non-negative matrices*, Discrete Math. **20** (1977), 217–234.
42. D. R. Fulkerson and H. J. Ryser, *Widths and heights of* $(0,1)$*-matrices*, Canad. J. Math. **13** (1961), 239–255.
43. ———, *Multiplicities and minimal widths for* $(0,1)$*-matrices*, Canad. J. Math. **14** (1962), 498–508.
44. Z. Füredi, *On* r*-graphs and* r*-multihypergraphs with given maximum degree*, J. Austral. Math. Soc. **50A** (1991), 204–212.
45. F. R. Gantmakher, *The theory of matrices*, Chelsea, New York, 1959.
46. I. P. Goulden and D. M. Jackson, *Combinatorial enumeration*, Wiley, New York, 1983.
47. R. M. Haber, *Term rank of* $(0,1)$*-matrices*, Rend. Sem. Mat. Univ. Padova **30** (1960), 24–51.
48. M. Hall, *Combinatorial theory*, Blaisdell, Waltham, MA, 1967.
49. H. Hanani, *Balanced incomplete block designs and related designs*, Discrete Math. **11** (1975), 255–369.
50. F. Harary, *Graph theory*, Addison–Wesley, Reading, MA, 1969.
51. D. J. Hartfiel, *A simplified form for nearly reducible and nearly decomposable matrices*, Proc. Amer. Math. Soc. **24** (1970), 388–393.
52. D. J. Hartfiel and C. J. Maxson, *The chainable matrix, a special combinatorial matrix*, Discrete Math. **12** (1975), 245–256.
53. B. R. Heap and M. S. Lynn, *The structure of powers of non-negative matrices*, SIAM J. Appl. Math. **14** (1966), 610–633.
54. ———, *The structure of powers of non-negative matrices*, SIAM J. Appl. Math. **14** (1966), 762–777.
55. T. Hearne and C. Warner, *Minimal covers of finite sets*, Discrete Math. **5** (1973), 247–251.
56. J. D. Henderson and R. A. Dean, *The* 1*-width of* $(0,1)$*-matrices having column row sum* 3, J. Comb. Theory **16A** (1974), 355–370.
57. ———, *A general upper bound for* 1*-widths*, J. Comb. Theory **18A** (1975), 236–238.
58. A. L. Hoffman, *On the extremal case in a characterization of the arcs of a complete graph*, IBM J. Res. Develop. **4** (1960), 487–496.
59. T. C. Holladay and R. S. Varga, *On powers of non-negative matrices*, Proc. Amer. Math. Soc. **9** (1958), 631–634.
60. Shao Jiayu, *The connectivity of the interchange graph of class* $\mathfrak{A}(R,S)$ *of* $(0,1)$*-matrices*, Acta Math. Appl. Sinica **2** (1985), 304–308.

61. M. G. Karpovsky and V. D. Milman, *Coordinate density of sets of vectors*, Discrete Math. **24** (1978), 177–184.
62. N. N. Kuzyurin, *The asymptotic investigation of the problem of covering*, Probl. Kibern. **37** (1980), 19–56. (Russian)
63. ———, *On the problem of the α-depth of* $(0,1)$*-matrices*, Vopr. Kibern. **86** (1982), 131–139. (Russian)
64. C. W. H. Lam, *A generalization of cyclic difference set*, J. Comb. Theory **19A** (1975), 51–65.
65. ———, *On some solutions of* $\delta^l = d\mathbf{I} + \lambda\mathbf{J}$, J. Comb. Theory **23A** (1977), 140–147.
66. ———, *The search for a finite projective plane of order* 10, Amer. Math. Monthly **98** (1991), 305–318.
67. V. K. Leont′ev, *An upper bound for the α-height of* $(0,1)$*-matrices*, Math. Notes **15** (1974), 245–250.
68. M. Levin, *On non-negative matrices*, Pacific J. Math. **36** (1971), 753–759.
69. D. London, *Some notes on the van der Waerden conjecture*, Linear Algebra Appl. **4** (1971), 155–160.
70. S. L. Ma and W. C. Waterhouse, *The g-circulant solutions of* $A^m = \lambda\mathbf{J}$, Linear Algebra Appl. **85** (1987), 211–220.
71. W. Mantel, *Vraagtuck XYIII*, Wiskunde Opgaven met Oplossingen **10** (1907), 60–61.
72. M. Marcus and H. Minc, *A survey of matrix theory and matrix inequalities*, Allyn and Bacon, 1964.
73. M. Marcus and M. Newman, *On the minimum of the permanent of a doubly stochastic matrix*, Duke Math. J. **26** (1959), 61–72.
74. J. S. Marica and J. Schönheim, *Difference of sets and a problem of Graham*, Canad. Math. Bull. **12** (1969), 635–638.
75. H. Minc, *Permanents*, Addison–Wesley, Reading, MA, 1978.
76. ———, *Recurrence formulas for permanents of* $(0,1)$*-matrices*, Linear Algebra Appl. **71** (1985), 241–265.
77. ———, *Permanential compound permanents of* $(0,1)$*-circulants*, Linear Algebra Appl. **86** (1987), 11–42.
78. J. W. Moon and L. Moser, *Almost all* $(0,1)$*-matrices are primitive*, Stud. Sci. Math. Hungar. **1** (1966), 1–2.
79. J. W. Moon and N. J. Pullman, *On the powers of tournament matrices*, J. Comb. Theory **3** (1967), 1–9.
80. P. O'Neil, *Asymptotics in random* $(0,1)$*-matrices*, Proc. Amer. Math. Soc. **25** (1970), 290–295.
81. P. Perkins, *A theorem on regular matrices*, Pacific J. Math. **11** (1961), 1529–1533.
82. N. J. Pullman, *On the number of positive entries in the powers of a non-negative matrix*, Canad. Math. Bull. **7** (1964), 525–537.
83. A. Renyi, *On random generating elements of finite Boolean algebra*, Acta Sci. Math. (Szeged) **22** (1961), 75–81.
84. D. J. Richman and H. Schneider, *Primes in the semigroup of non-negative matrices*, Linear Multilinear Algebra **2** (1974), 135–140.
85. D. Rosenblatt, *On the graphs and asymptotic forms of finite Boolean relation matrices and stochastic matrices*, Naval Res. Logist. Quarterly **4** (1957), 151–167.
86. H. J. Ryser, *The term rank of a matrix*, Canad. J. Math. **10** (1958), 57–65.
87. ———, *Combinatorial mathematics*, Wiley, New York, 1963.
88. ———, *Matrices of zeros and ones in combinatorial mathematics*, Recent Advances in Matrix Theory, University of Wisconsin Press, 1964, pp. 103–124.
89. ———, *An extension of a theorem of de Bruijn and Erdös on combinatorial designs*, J. Algebra **10** (1968), 248–261.
90. ———, *Indeterminates and incidence matrices*, Linear Multilinear Algebra **1** (1973), 149–157.
91. ———, *Matrices and differences*, Discrete Math. **49** (1984), 169–173.
92. V. N. Sachkov, *Probability methods in combinatorial analysis*, Nauka, Moscow, 1978 (Russian); rev. English transl., Cambridge Univ. Press, 1997.
93. ———, *Introduction to combinatorial methods of discrete mathematics*, Nauka, Moscow, 1982 (Russian); rev. English transl., **[96]**.

94. ———, *The asymptotic behaviour of the number of t-minimal coverings of a set*, Discrete Math. Appl. **3** (1993), 265–273.
95. ———, *Random minimal coverings of sets*, Discrete Math. Appl. **3** (1993), 201–212.
96. ———, *Combinatorial methods in discrete mathematics*, Cambridge University Press, 1995.
97. N. Sauer, *On the density of families of sets*, J. Comb. Theory **13A** (1972), 145–147.
98. E. Seneta, *Non-negative matrices*, Wiley, New York, 1973.
99. V. S. Shevelev, *Modern enumeration theory of permutations with restricted positions*, Discrete Math. Appl **3** (1993), 229–263.
100. G. Sierksma and E. Sterken, *The structure matrix of $(0,1)$-matrices: Its rank, trace and eigenvalues. an application to econometric models*, Linear Algebra Appl. **83** (1986), 151–166.
101. R. Sinkhorn and P. Knopp, *Problems involving diagonal products in non-negative matrices*, Trans Amer. Math. Soc. **136** (1969), 67–75.
102. R. P. Stanley, *Enumerative combinatorics*, vol. 1, Wadsworth, Monterey, California, 1986.
103. V. E. Tarakanov, *Relations of maximal depth of classes of square $(0,1)$-matrices with various parameters*, Math. USSR Sbornik **6** (1968), 53–63.
104. ———, *Maximal depth of a class of $(0,1)$-matrices*, Math. USSR Sbornik **4** (1968), 3–12.
105. ———, *Maximum height of arbitrary classes of $(0,1)$-matrices and some of its applications*, Math. USSR Sbornik **21** (1973), 467–484.
106. ———, *Combinatorial problems on binary matrices*, Komb. Anal. **5** (1980), 4–15. (Russian)
107. ———, *Depth of $(0,1)$-matrices with the same row and same column sums*, Math. Notes **34** (1983), 718–725.
108. ———, *Combinatorial problems and $(0,1)$-matrices*, Nauka, Moscow, 1985. (Russian)
109. ———, *Covering numbers of regular multigraphs*, Math. Notes **46** (1989), 798–804.
110. ———, *On the edge-independence number and the edge-covering number for regular graphs*, Discrete Math. Appl **2** (1992), 1–10.
111. ———, *On the properties of interchange operations in classes of $(0,1)$-matrices*, Math. Notes **53** (1993), 443–448.
112. P. Turan, *Eine Extremalaufgabe aus der Graphentheorie*, Mat. Fiz. Lapok **48** (1941), 436–453.
113. Z. Tuza, *Minimum number of elements representating a set system of given rank*, J. Comb. Theory **52A** (1989), 84–89.
114. B. L. van der Waerden, *Aufgabe 45*, Jahresber. Deutsch. Math.-Vereinig. **35** (1926), 117.
115. J. H. van Lint, *The van der Waerden conjecture: Two proofs in one year*, Math. Intelligencer **4** (1982), 72–77.
116. D. W. Walkup, *Minimal interchanges of $(0,1)$-matrices and disjoint circuits in graphs*, Canad. J. Math. **17** (1965), 831–838.
117. K. Wang, *On the matrix equation $A^m = \lambda \mathbf{J}$*, J. Comb. Theory **29A** (1980), 134–141.
118. R. M. Wilson, *An existence theory for pairwise balanced designs. III: Proof of the existence conjecture*, J. Comb. Theory **18A** (1975), 171–179.
119. D. R. Woodall, *Square λ-linked designs*, Proc. London Math. Soc. **20** (1970), 669–687.
120. M. Salvi Zagaglia, *Combinatorial structure corresponding to reflective circulant $(0,1)$-matrices*, Ann. Discrete Math. **30** (1986), 363–372.

Index

Titles in This Series

213 **V. N. Sachkov and V. E. Tarakanov,** Combinatorics of nonnegative matrices, 2002

212 **A. V. Mel′nikov, S. N. volkov, and M. L. Nechaev,** Mathematics of financial obligations, 2002

211 **Takeo Ohsawa,** Analysis of several complex variables, 2002

210 **Toshitake Kohno,** Conformal field theory and topology, 2002

209 **Yasumasa Nishiura,** Far-from-equilibrium dynamics, 2002

208 **Yukio Matsumoto,** An introduction to Morse theory, 2002

207 **Ken'ichi Ohshika,** Discrete groups, 2002

206 **Yuji Shimizu and Kenji Ueno,** Advances in moduli theory, 2002

205 **Seiki Nishikawa,** Variational problems in geometry, 2001

204 **A. M. Vinogradov,** Cohomological analysis of partial differential equations and Secondary Calculus, 2001

203 **Te Sun Han and Kingo Kobayashi,** Mathematics of information and coding, 2002

202 **V. P. Maslov and G. A. Omel′yanov,** Geometric asymptotics for nonlinear PDE. I, 2001

201 **Shigeyuki Morita,** Geometry of differential forms, 2001

200 **V. V. Prasolov and V. M. Tikhomirov,** Geometry, 2001

199 **Shigeyuki Morita,** Geometry of characteristic classes, 2001

198 **V. A. Smirnov,** Simplicial and operad methods in algebraic topology, 2001

197 **Kenji Ueno,** Algebraic geometry 2: Sheaves and cohomology, 2001

196 **Yu. N. Lin′kov,** Asymptotic statistical methods for stochastic processes, 2001

195 **Minoru Wakimoto,** Infinite-dimensional Lie algebras, 2001

194 **Valery B. Nevzorov,** Records: Mathematical theory, 2001

193 **Toshio Nishino,** Function theory in several complex variables, 2001

192 **Yu. P. Solovyov and E. V. Troitsky,** C^*-algebras and elliptic operators in differential topology, 2001

191 **Shun-ichi Amari and Hiroshi Nagaoka,** Methods of information geometry, 2000

190 **Alexander N. Starkov,** Dynamical systems on homogeneous spaces, 2000

189 **Mitsuru Ikawa,** Hyperbolic partial differential equations and wave phenomena, 2000

188 **V. V. Buldygin and Yu. V. Kozachenko,** Metric characterization of random variables and random processes, 2000

187 **A. V. Fursikov,** Optimal control of distributed systems. Theory and applications, 2000

186 **Kazuya Kato, Nobushige Kurokawa, and Takeshi Saito,** Number theory 1: Fermat's dream, 2000

185 **Kenji Ueno,** Algebraic Geometry 1: From algebraic varieties to schemes, 1999

184 **A. V. Mel′nikov,** Financial markets, 1999

183 **Hajime Sato,** Algebraic topology: an intuitive approach, 1999

182 **I. S. Krasil′shchik and A. M. Vinogradov, Editors,** Symmetries and conservation laws for differential equations of mathematical physics, 1999

181 **Ya. G. Berkovich and E. M. Zhmud′,** Characters of finite groups. Part 2, 1999

180 **A. A. Milyutin and N. P. Osmolovskii,** Calculus of variations and optimal control, 1998

179 **V. E. Voskresenskiĭ,** Algebraic groups and their birational invariants, 1998

178 **Mitsuo Morimoto,** Analytic functionals on the sphere, 1998

For a complete list of titles in this series, visit the AMS Bookstore at **www.ams.org/bookstore/**.